Contemporary Food
Engineering Series
Da-Wen Sun, Series Editor

OPERATIONS
IN FOOD
REFRIGERATION

T0361944

Contemporary Food Engineering

Series Editor

Professor Da-Wen Sun, Director

Food Refrigeration & Computerized Food Technology
National University of Ireland, Dublin
(University College Dublin)
Dublin, Ireland
http://www.ucd.ie/sun/

OPERATIONS
IN FOOD
REFRIGERATION

Edited by **Rodolfo H. Mascheroni**

CRC Press
Taylor & Francis Group
Boca Raton   London   New York

CRC Press is an imprint of the
Taylor & Francis Group, an **informa** business

CRC Press
Taylor & Francis Group
6000 Broken Sound Parkway NW, Suite 300
Boca Raton, FL 33487-2742

First issued in paperback 2016

© 2012 by Taylor & Francis Group, LLC
CRC Press is an imprint of Taylor & Francis Group, an Informa business

No claim to original U.S. Government works

Version Date: 20120417

ISBN 13: 978-1-138-19892-0 (pbk)
ISBN 13: 978-1-4200-5548-1 (hbk)

Library of Congress Cataloging-in-Publication Data

Mascheroni, Rodolfo H.
    Operations in food refrigeration / Rodolfo H. Mascheroni.
        p. cm. -- (Contemporary food engineering)
    "A CRC title."
    Includes bibliographical references and index.
    ISBN 978-1-4200-5548-1 (alk. paper)
    1. Food--Preservation. 2. Cold storage. 3. Refrigeration and refrigerating machinery.
    I. Title.

TP372.2.M38 2012
664'.028--dc23   2012008615

**Visit the Taylor & Francis Web site at
http://www.taylorandfrancis.com**

**and the CRC Press Web site at
http://www.crcpress.com**

Contents

PART I Basic Concepts and General Calculation Procedures

PART II Operations Used in Refrigeration Technologies

Series Preface

CONTEMPORARY FOOD ENGINEERING

Food engineering is the multidisciplinary field of applied physical sciences combined with the knowledge of product properties. Food engineers provide the technological knowledge transfer essential to the cost-effective production and commercialization of food products and services. In particular, food engineers develop and design processes and equipment to convert raw agricultural materials and ingredients into safe, convenient, and nutritious consumer food products. However, food engineering topics are continuously undergoing changes to meet diverse consumer demands, and the subject is being rapidly developed to reflect market needs.

In the development of food engineering, one of the many challenges is to employ modern tools and knowledge, such as computational materials science and nanotechnology, to develop new products and processes. Simultaneously, improving food quality, safety, and security continues to be a critical issue in food engineering study. New packaging materials and techniques are being developed to provide more protection to foods, and novel preservation technologies are emerging to enhance food security and defense. Additionally, process control and automation regularly appear among the top priorities identified in food engineering. Advanced monitoring and control systems are developed to facilitate automation and flexible food manufacturing. Furthermore, energy saving and minimization of environmental problems continue to be important food engineering issues, and significant progress is being made in waste management, efficient utilization of energy, and reduction of effluents and emissions in food production.

The *Contemporary Food Engineering Series*, consisting of edited books, attempts to address some of the recent developments in food engineering. The series covers advances in classical unit operations in engineering applied to food manufacturing as well as such topics as progress in the transport and storage of liquid and solid foods; heating, chilling, and freezing of foods; mass transfer in foods; chemical and biochemical aspects of food engineering and the use of kinetic analysis; dehydration, thermal processing, nonthermal processing, extrusion, liquid food concentration, membrane processes, and applications of membranes in food processing; shelf-life and electronic indicators in inventory management; sustainable technologies in food processing; and packaging, cleaning, and sanitation. These books are aimed at professional food scientists, academics researching food engineering problems, and graduate-level students.

The editors of these books are leading engineers and scientists from many parts of the world. All the editors were asked to present their books to address the market's need and pinpoint cutting-edge technologies in food engineering.

All contributions are written by internationally renowned experts who have both academic and professional credentials. All authors have attempted to provide critical, comprehensive, and readily accessible information on the art and science of a relevant topic in each chapter, with reference lists for further information. Therefore, each book can serve as an essential reference source to students and researchers in universities and research institutions.

Da-Wen Sun
Series Editor

Preface

Food preservation technologies such as chilling, freezing, or freeze drying, involve a series of unit operations, mainly related to heat and mass transfer, which are of particular importance for the final quality of the food product. But these operations are not restrained to those transfers because pretreatments applied to food products may include sizing, sorting, peeling, cutting, blanching, partial dehydration, smoking, curing, preconcentration, etc., among several other possibilities.

The adequate design and implementation of each of these treatments influences not only the final quality of the processed food but also the productivity of the involved equipment and the operation costs.

This book, as part of a series on "Contemporary Food Engineering," is aimed at professional food technologists or engineers working in the chilled and frozen food industry, as well as at researchers in food engineering and as a reference text for pre- or post-degree studies. It attempts to provide readers with the fundamental issues involved in heat and mass transfer in food refrigeration, as well as those related to other operations applied to foods that are chilled or frozen. Besides the application of these principles to the design and operation of the involved equipment, quality issues related to those technologies are covered, with the aim of specifying the most adequate operation conditions for each type of processed food.

The book is divided in two sections.

Part I covers *basic concepts and general calculation procedures* as well as tools for simplified prediction methods (thermophysical properties, heat transfer coefficients, simplified prediction formulae, respiratory heats) and is divided into three chapters:

- Chapter 1: Cooling
- Chapter 2: Freezing and Thawing
- Chapter 3: Freeze Drying

Part II covers operations used in refrigeration technologies and is organized as follows:

Pretreatments are discussed in seven chapters:
- Chapter 4: Sizing, Peeling, Cutting, and Sorting of Fruits and Vegetables
- Chapter 5: Blanching of Fruits and Vegetable Products
- Chapter 6: Pretreatments for Meats I (Tenderization, Electrical Stimulation, and Portioning)
- Chapter 7: Pretreatments for Meats II (Curing, Smoking)
- Chapter 8: Pretreatments for Fish and Seafood in General
- Chapter 9: Processing of Poultry
- Chapter 10: Air and Osmotic Partial Dehydration, Infusion of Special Nutrients, and Concentration of Juices

Processing, which is broken down as follows:

a) Chilling and Freezing includes five chapters:
   - Chapter 11: By Air (Static and Continuous Equipment)
   - Chapter 12: By Cryogenic Gases and Liquids (Static and Continuous Equipment)
   - Chapter 13: By Contact with Refrigerated Surfaces (Plate Freezers, Surface Hardeners, and Scraped Surface Freezers)
   - Chapter 14: By Immersion in Water and Aqueous Solutions (Hydrocooling, Brines, Ice Slurries, and Refrigerated Sea Water)
   - Chapter 15: Special Precooling Techniques
   - Chapter 16: Special (Emerging) Freezing Techniques (Dehydrofreezing, Pressure-Shift Freezing, Ultrasonic-Assisted Freezing, and Hydrofluidization Freezing)

b) Thawing
   - Chapter 17: Thawing

c) Freeze Drying
   - Chapter 18: Freeze-Drying Equipment

Series Editor

Born in Southern China, Professor Da-Wen Sun is a world authority in food engineering research and education; he is a Member of the Royal Irish Academy (RIA), which is the highest academic honor in Ireland; he is also a member of Academia Europaea (The Academy of Europe). His main research activities include cooling, drying, and refrigeration processes and systems, quality and safety of food products, bioprocess simulation and optimization, and computer vision technology. Especially, his innovative studies on vacuum cooling of cooked meats, pizza quality inspection by computer vision, and edible films for shelf-life extension of fruit and vegetables have been widely reported in national and international media. Results of his work have been published in about 600 papers including over 250 peer-reviewed journal papers (h-index = 35). He has also edited 12 authoritative books. According to Thomson Scientific's *Essential Science Indicators*^{SM} updated as of 1 July 2010, based on data derived over a period of ten years plus four months (1 January 2000–30 April 2010) from ISI Web of Science, a total of 2554 scientists are among the top 1% of the most cited scientists in the category of Agriculture Sciences, and Professor Sun tops the list with his ranking of 31.

He received a first class BSc Honors and MSc in Mechanical Engineering, and a PhD in Chemical Engineering in China before working in various universities in Europe. He became the first Chinese national to be permanently employed in an Irish university when he was appointed College Lecturer at the National University of Ireland, Dublin (University College Dublin [UCD]), in 1995, and was then continuously promoted in the shortest possible time to Senior Lecturer, Associate Professor, and Full Professor. Dr. Sun is now a Professor of Food and Biosystems Engineering and the Director of the Food Refrigeration and Computerised Food Technology Research Group at the UCD.

As a leading educator in food engineering, Professor Sun has significantly contributed to the field of food engineering. He has trained many PhD students, who have made their own contributions to the industry and academia. He has also given lectures on advances in food engineering on a regular basis in academic institutions internationally and delivered keynote speeches at international conferences. As a recognized authority in food engineering, he has been conferred adjunct/visiting/consulting professorships from ten top universities in China, including Zhejiang University, Shanghai Jiaotong University, Harbin Institute of Technology, China Agricultural University, South China University of Technology, and Jiangnan University. In recognition of his significant contribution to Food Engineering worldwide and for his outstanding leadership in the field, the International Commission of Agricultural and Biosystems Engineering (CIGR) awarded him the "CIGR Merit Award" in 2000, and again in 2006, the Institution of Mechanical Engineers based in the U.K. named him "Food Engineer of the Year 2004." In 2008, he was awarded

"CIGR Recognition Award" in honor of his distinguished achievements as the top 1% of Agricultural Engineering scientists in the world. In 2007, he was presented with the only "AFST(I) Fellow Award" in that year by the Association of Food Scientists and Technologists (India), and in 2010, he was presented with the "CIGR Fellow Award"; the title of Fellow is the highest honor in CIGR and is conferred to individuals who have made sustained, outstanding contributions worldwide.

He is a Fellow of the Institution of Agricultural Engineers and a Fellow of Engineers Ireland (the Institution of Engineers of Ireland). He has also received numerous awards for teaching and research excellence, including the President's Research Fellowship, and has twice received the President's Research Award of the UCD. He is the Editor-in-Chief of *Food and Bioprocess Technology—an International Journal* (Springer; 2010 Impact Factor = 3.576, ranked at the 4th position among 126 ISI-listed food science and technology journals), Series Editor of "Contemporary Food Engineering" book series (CRC Press/Taylor & Francis), former Editor of *Journal of Food Engineering* (Elsevier), and Editorial Board Member for *Journal of Food Engineering* (Elsevier), *Journal of Food Process Engineering* (Blackwell), *Sensing and Instrumentation for Food Quality and Safety* (Springer), and *Czech Journal of Food Sciences*. He is also a Chartered Engineer.

On 28 May 2010, he was awarded membership of the RIA, which is the highest honor that can be attained by scholars and scientists working in Ireland; at the 51st CIGR General Assembly held during the CIGR World Congress in Quebec City, Canada, on 13–17 June 2010, he was elected Incoming President of CIGR and will become CIGR President in 2013–2014—the term of his CIGR presidency is six years, two years each for serving as Incoming President, President, and Past President. On 20 September 2011, he was elected to Academia Europaea (The Academy of Europe), which is functioning as European Academy of Humanities, Letters and Sciences and is one of the most prestigious academies in the world; election to the Academia Europaea represents the highest academic distinction.

Editor

Mascheroni, Rodolfo Horacio
Professional qualifications:

- Doctor (PhD) in Chemistry, Honorable Mention in Technological Chemistry from La Plata National University (UNLP), Argentina (1977)
- Full Professor, Faculty of Engineering, UNLP
- Main Researcher, National Research Council (CONICET), Argentina
- Member of ASHRAE, Associate Member of IIR, Member of ASABE
- Vice-Director of the Center for Research and Development on Food Chryotechnology of Argentina (staff: 150 researchers, fellows, and technicians)

Main interests in R&D and expertise:

- Heat and mass transfer and quality issues in food refrigeration, freezing, storage, transport, thawing, dehydration, and heat treatment
- Heat and mass transfer coefficients in food refrigeration
- Thermophysical properties of foods at low temperatures
- Freezing and storage conditions and frozen product development
- Thawing and heating of frozen raw materials and prepared foods
- Food dehydration processes

Contributors

Miriam E. Agnelli
Centro de Investigación y Desarrollo en
    Criotecnología de Alimentos
and
MODIAL, Facultad de Ingeniería,
    National University of La Plata
La Plata, Argentina

Antonello A. Barresi
Dipartimento di Scienza Applicata e
    Tecnologia
Politecnico di Torino
Torino, Italy

A. Javier Borderías
Instituto del Frío (CSIC)
Madrid, Spain

Alicia N. Califano
Centro de Investigación y Desarrollo en
    Criotecnología de Alimentos
La Plata, Argentina

Paloma Vírseda Chamorro
Department of Food Technology
Public University of Navarra
Pamplona, Spain

Giovanni Cortella
Dipartimento di Ingegneria Elettrica,
    Gestionale e Meccanica (DIEG)
Università degli studi di Udine
University of Udine
Udine, Italy

Silvia Estrada-Flores
Food Chain Intelligence
North Sydney, Australia

Judith Evans
Food Refrigeration and Process
    Engineering Research Center
University of Bristol
Bristol, United Kingdom

Davide Fissore
Dipartimento di Scienza Applicata e
    Tecnologia
Politecnico di Torino
Torino, Italy

Brian A. Fricke
Building Equipment Research
Energy and Transportation Science
    Division
Oak Ridge National Laboratory
Oak Ridge, Tennessee

Enrique Moltó García
Centro de Agroingeniería
Instituto Valenciano de Investigaciones
    Agrarias
Moncada, Spain

Natalia G. Graiver
Centro de Investigación y Desarrollo en
    Criotecnología de Alimentos
La Plata, Argentina

David L. Hopkins
Department of Primary Industries
Center for Red Meat and Sheep
    Development
Cowra, Australia

José Blasco Ivars
Centro de Agroingeniería
Instituto Valenciano de Investigaciones
    Agrarias
Moncada, Spain

Miriam N. Martino
Centro de Investigación y Desarrollo en
    Criotecnología de Alimentos
and
Facultad de Ciencias Exactas
Universidad Nacional de La Plata
La Plata, Argentina

Rodolfo H. Mascheroni
Centro de Investigación y Desarrollo en
    Criotecnología de Alimentos
and
MODIAL, Facultad de Ingeniería
National University of La Plata
La Plata, Argentina

Daniela F. Olivera
Centro de Investigación y Desarrollo en
    Criotecnología de Alimentos
and
MODIAL, Facultad de Ingeniería
National University of La Plata
La Plata, Argentina

Casey M. Owens
Department of Poultry Science
University of Arkansas
Fayetteville, Arkansas

Quang T. Pham
School of Chemical Sciences and
    Engineering
University of New South Wales
Sydney, Australia

Adriana N. Pinotti
Centro de Investigación y Desarrollo en
    Criotecnología de Alimentos
and
Departamento de Ingeniería Química,
    Facultad de Ingeniería
National University of La Plata
La Plata, Argentina

Laura A. Ramallo
Facultad de Ciencias Exactas, Químicas
    y Naturales
Universidad Nacional de Misiones
Posadas, Argentina

Viviana O. Salvadori
Centro de Investigación y Desarrollo en
    Criotecnología de Alimentos and
MODIAL, Facultad de Ingeniería
National University of La Plata
La Plata, Argentina

Salvatore Velardi
Dipartimento di Scienza Applicata e
    Tecnologia
Politecnico di Torino
Torino, Italy

Cristina Arroqui Vidaurreta
Department of Food Technology
Public University of Navarra
Pamplona, Spain

Sonia Z. Viña
Centro de Investigación y Desarrollo en
    Criotecnología de Alimentos
La Plata, Argentina

Noemí E. Zaritzky
Centro de Investigación y Desarrollo en
    Criotecnología de Alimentos
and
Departamento de Ingeniería Química,
    Facultad de Ingeniería
National University of La Plata
La Plata, Argentina

Part I

Basic Concepts and General Calculation Procedures

1 Cooling

Giovanni Cortella

CONTENTS

1.1 INTRODUCTION

Cooling plays an essential role in food processing and storage, and design methods are required for its best exploitation. When dealing with food chilling, a crucial aspect is the prediction of the cooling time, which is related to the geometry and the thermal properties of the product, to the heat transfer coefficient and to the temperature of the cooling medium. Nevertheless, of great importance for a correct design of the refrigeration system is also the evaluation of the heat load required to chill the product. Food cooling can be achieved by means of various equipment, where the cooling medium, usually a gas or occasionally a liquid, is kept at low temperature. It should be pointed out that in certain applications, for example, in blast chillers, the temperature of the cooling medium is affected by the presence of food, so that it increases in the occurrence of food loading. In such a case, the time–temperature profiles of the product and of the cooling medium are strictly interrelated, and their accurate prediction requires modeling the whole system in a dynamic state, which can be accomplished only by means of complex techniques. In most practical cases, it can be assumed that the cooling medium temperature holds steady, and the methods described in this chapter are based on this assumption.

The estimation of the chilling time can be accomplished by means of numerical, analytical, and approximate methods. Numerical methods are based on finite differences or finite-element techniques, and aim toward the solution of the thermal conduction equation on the whole solid domain. When fluid motion has also to be investigated, computational fluid dynamics methods are required.

Since numerical methods require a lengthy and specific description [1], this chapter will discuss the analytical and the approximate methods only, which can be accurate enough under usual operating conditions.

1.2 DESCRIPTION OF THE PROBLEM

1.2.1 HEAT CONDUCTION EQUATION

Cooling of a solid is a transient conduction problem, which is described by the equation for heat conduction:

$$\rho c \frac{\partial T}{\partial t} = \nabla \cdot (\lambda \nabla T) \tag{1.1}$$

or

$$\rho c \frac{\partial T}{\partial t} = \frac{\partial}{\partial x}\left(\lambda \frac{\partial T}{\partial x}\right) + \frac{\partial}{\partial y}\left(\lambda \frac{\partial T}{\partial y}\right) + \frac{\partial}{\partial z}\left(\lambda \frac{\partial T}{\partial z}\right). \tag{1.2}$$

If the thermal conductivity is independent of temperature and position, Equations 1.1 and 1.2 can be simplified

$$\frac{\partial T}{\partial t} = \alpha \nabla^2 T \tag{1.3}$$

or

$$\frac{\partial T}{\partial t} = \alpha \left(\frac{\partial^2 T}{\partial x^2} + \frac{\partial^2 T}{\partial y^2} + \frac{\partial^2 T}{\partial z^2} \right), \tag{1.4}$$

introducing the thermal diffusivity

$$\alpha = \frac{\lambda}{\rho c}. \tag{1.5}$$

1.2.2 BOUNDARY CONDITIONS

Equations 1.1 through 1.4 describe the conservation of energy. Appropriate initial and boundary conditions are required for their application in order to determine the time–temperature distribution in a medium. Because the equations are first order in time, only one condition (the initial condition) needs to be specified. On the contrary, the equations are second order in the spatial coordinates; therefore, two boundary conditions are required for each coordinate.

Three kinds of boundary conditions can be encountered in heat transfer problems. The boundary condition of the first kind is

$$T = T_s, \tag{1.6}$$

which is a prescribed surface temperature. This condition is encountered when food is in direct contact with a surface at prescribed temperature, for example, in plate freezers.

The boundary condition of the second kind is

$$q = -\lambda A \frac{\partial T}{\partial n} = q_s, \tag{1.7}$$

which is a prescribed heat flux at the surface. This condition has very few practical applications because it is very difficult to enforce heat flux on a surface. It is, however, used in the case of symmetry, where on the symmetry surface, the heat flux = 0, thus leading to

$$\frac{\partial T}{\partial n} = 0. \tag{1.8}$$

The boundary condition of the third kind is

$$q = -\lambda A \frac{\partial T}{\partial n} = q_{conv} = h\left(T - T_f\right), \tag{1.9}$$

which is a prescribed convective exchange at the surface. This is the most common boundary condition, and the heat transfer coefficient h can take into account both convective and radiative heat transfer, provided that the fluid and the radiation source are at the same temperature T_f. The boundary condition of the first kind and the symmetry conditions can be treated as particular cases of Equation 1.9, with $h \to \infty$ and $h = 0$, respectively.

1.2.3 BIOT NUMBER

When a solid is cooled by a fluid, both conductive and convective heat transfer processes take place inside and outside the body, respectively. The Biot number

$$\mathrm{Bi} = \frac{R_{\mathrm{int}}}{R_{\mathrm{ext}}} = \frac{L/\lambda A}{1/hA} = \frac{hL}{\lambda} \tag{1.10}$$

provides the ratio of internal to external resistance and, consequently, also the ratio of internal to external temperature drop:

$$\mathrm{Bi} = \frac{R_{\mathrm{int}}}{R_{\mathrm{ext}}} = \frac{\Delta T_{\mathrm{int}}}{\Delta T_{\mathrm{ext}}}. \tag{1.11}$$

According to Equations 1.10 and 1.11, when the Biot number Bi << 1 (say Bi < 0.1), the internal resistance can be neglected. The heat transfer process is controlled by the external convective resistance, and a uniform temperature distribution across the body can be assumed at any time during the transient process. This is also the case in liquid foods, where mixing takes place, which ensures a certain degree of temperature uniformity. On the contrary, when Bi > 0.1, the temperature distribution across the solid must be accounted for. In this case, shape becomes a crucial factor in the solution.

1.3 ANALYTICAL SOLUTIONS

The choice of the analytical solution depends on the Biot number, i.e., on the fact that the internal conductive resistance can be neglected.

1.3.1 INTERNAL CONDUCTIVE RESISTANCE NEGLIGIBLE

In this case, as already mentioned, temperature gradients within the product can be reasonably neglected at any time during the cooling process. In this case, we can no longer consider the problem from within the framework of the heat conduction equation (Equation 1.2) and use only Newton's cooling law, which derives from the expression of the heat balance between the rate of change of the internal energy and the convective heat loss from the surface:

$$\rho c V \frac{\mathrm{d}T}{\mathrm{d}t} = -hA(T - T_f). \tag{1.12}$$

Once integrated, the solution gives the time–temperature behavior:

$$T' = \frac{T - T_f}{T_i - T_f} = e^{-\frac{hA}{\rho cV}t},$$ (1.13)

which is the classic exponential decay profile for an RC system, where R is the external convective resistance, and C is the lumped thermal capacitance, thus giving the time constant

$$\tau = R_{\mathrm{ext}} C = \frac{1}{hA}\left(\rho cV\right).$$ (1.14)

T' is the dimensionless residual temperature difference, equal to 1 at the beginning of the cooling process and approaching 0 as the body temperature reaches asymptotically the fluid temperature. In Figure 1.1, the transient temperature values are reported for systems with a different time constant.

1.3.2 INTERNAL CONDUCTIVE RESISTANCE NOT NEGLIGIBLE

Equations 1.1 through 1.4 describe heat conduction in a three-dimensional solid body. Such equations have an exact solution for regular shapes with constant thermophysical properties, uniform initial conditions, and for the first or third kind of boundary condition. Regular shapes are the infinite slab, the infinite cylinder, and the sphere, where the temperature distribution and the heat flux are one-dimensional, and the infinite rectangular rod, the rectangular brick, and the finite cylinder, which are combinations of the previous shapes.

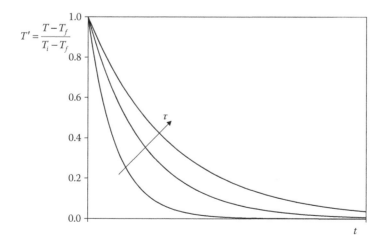

FIGURE 1.1   Cooling curve for Bi < 0.1, at different time constants.

1.3.2.1 Solution for Infinite Slab, Infinite Cylinder, and Sphere

With the above-mentioned assumptions, Equation 1.4 becomes

$$\frac{\partial T}{\partial t} = \alpha \frac{\partial^2 T}{\partial x^2}. \tag{1.15}$$

The Fourier number

$$\text{Fo} = \frac{t}{\tau_{\text{int}}} = \frac{t}{R_{\text{int}}C} = \frac{t}{(L/\lambda A)(\rho c L A)} = \frac{\lambda}{\rho c}\frac{t}{L^2} = \frac{\alpha t}{L^2} \tag{1.16}$$

is the ratio of time to the internal time constant and provides a dimensionless time, while

$$x' = \frac{x}{L} \tag{1.17}$$

provides a spatial dimensionless coordinate. The Biot number Bi, remains the ratio between the internal and the external resistance:

$$\text{Bi} = \frac{hL}{\lambda}. \tag{1.18}$$

In all these equations, L is the characteristic length defined in Table 1.1.
Equation 1.15 can thus be written in dimensionless terms:

$$\frac{\partial T'}{\partial Fo} = \frac{\partial^2 T'}{\partial x'^2}. \tag{1.19}$$

The boundary condition of the third kind, Equation 1.9, becomes

$$-\frac{\partial T}{\partial x'}\bigg|_{x'=1} = \text{Bi}T'. \tag{1.20}$$

TABLE 1.1

Characteristic Length L for Basic Shapes

Shape	L
Infinite slab of thickness $2H$ cooled by both sides	H
Infinite slab of thickness $2H$ cooled by one side only	$2H$
Infinite cylinder of radius r_{ext}	r_{ext}
Sphere of radius r_{ext}	r_{ext}

The solution to Equation 1.19 is available in the form of a series. The solution for the infinite slab is

$$T' = \sum_{n=1}^{\infty} 2 \frac{\sin \mu_n}{\mu_n + \sin \mu_n \cos \mu_n} e^{-\mu_n^2 Fo} \cos(\mu_n x') \tag{1.21}$$

where the values of μ_n are the roots of the equation

$$\mu_n \tan \mu_n = Bi. \tag{1.22}$$

Equation 1.21 predicts the product temperature in an infinite slab as a function of time (through the Fourier number, Fo), at a certain location (through x') and Biot number. A similar solution is obtained for the infinite cylinder:

$$T' = \sum_{n=1}^{\infty} \frac{2 J_1 \mu_n}{\mu_n \left[J_0^2(\mu_n) + J_1^2(\mu_n) \right]} e^{-\mu_n^2 Fo} J_0(\mu_n x'), \tag{1.23}$$

where the values of μ_n are the positive roots of the equation

$$\mu_n \frac{J_1(\mu_n)}{J_0(\mu_n)} = Bi \cdot \tag{1.24}$$

The quantities J_0 and J_1 are the Bessel functions of the zero order and of the first order, respectively.

The solution for the sphere is

$$T' = \sum_{n=1}^{\infty} 2 \frac{\sin \mu_n - \mu_n \cos \mu_n}{\mu_n - \sin \mu_n \cos \mu_n} e^{-\mu_n^2 Fo} \frac{\sin(\mu_n x')}{\mu_n x'}, \tag{1.25}$$

where the values of μ_n are the positive roots of the equation

$$1 - \mu_n \cot \mu_n = Bi. \tag{1.26}$$

Back-calculation, i.e., the prediction of the time at which a given temperature is reached in a certain position, requires iteration.

In Figures 1.2 through 1.7, the solutions are graphically represented. The temperature as a function of time is given at various Biot numbers, at the center of the body ($x' = 0$) and at its surface ($x' = 1$), respectively. The cooling rate increases with Bi, and the influence of the Biot number can be easily appraised. At the center, the response is obviously slower than at the surface, and when Bi $\to \infty$, the temperature at the surface reaches immediately the asymptotic value T_f, thus leading to $T' = 0$.

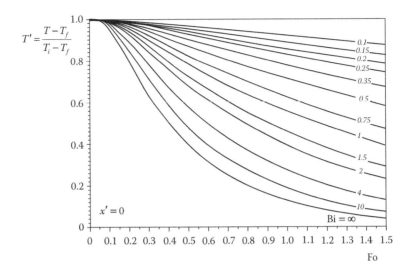

FIGURE 1.2 Cooling curves at the center of an infinite slab, at different Biot numbers.

The calculation of the exact solution requires advanced tools, and it is not accessible to all. Therefore, various charts similar to those of Figures 1.2 through 1.7 are available, for the infinite slab and cylinder and for the sphere, at various locations.

1.3.2.2 Solution for Regular Shapes

The infinite rectangular rod, the rectangular brick, and the finite cylinder are regular shapes, which can be obtained by intersection of the basic shapes investigated above.

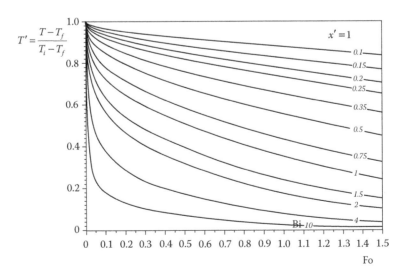

FIGURE 1.3 Cooling curves on the surface of an infinite slab, at different Biot numbers.

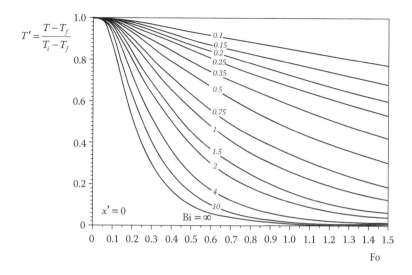

FIGURE 1.4 Cooling curves at the center of an infinite cylinder, at different Biot numbers.

For the regular shapes, there exists an analytical solution, which allows for the estimation of the temperature reached at any position at a given time. The dimensionless temperature can be obtained by multiplying the solutions for the corresponding locations in the bounding basic shapes. As an example, with reference to the finite upright cylinder of Figure 1.8, obtained by intersection of an infinite cylinder with radius r_{ext} and of an infinite slab with half thickness H, the solution for location A is

$$T'_A = T'_{\text{infinite cylinder, } x'=0} \cdot T'_{\text{infinite slab, } x'=1}, \tag{1.27}$$

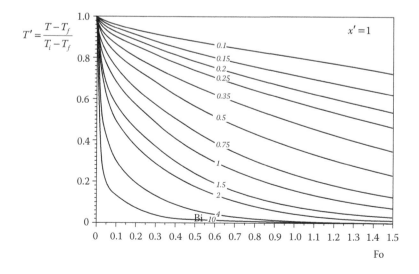

FIGURE 1.5 Cooling curves on the surface of an infinite cylinder, at different Biot numbers.

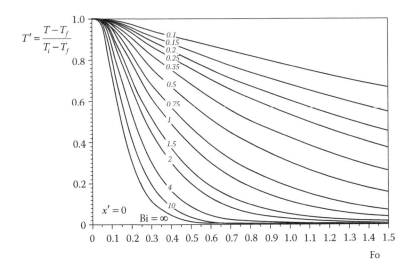

FIGURE 1.6 Cooling curves at the center of a sphere, at different Biot numbers.

while the solution for location B is

$$T'_B = T'_{\text{infinite cylinder, } x'=0} \cdot T'_{\text{infinite slab, } x'=0}. \tag{1.28}$$

It should be noted that the back-calculation for such regular shapes is not possible. Therefore, the calculation of the time at which a given temperature is obtained at a certain location requires iteration.

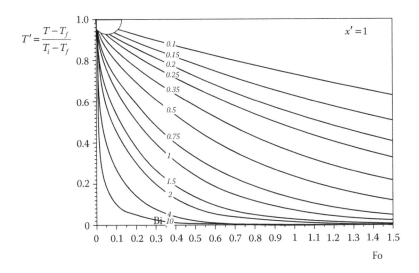

FIGURE 1.7 Cooling curves on the surface of a sphere, at different Biot numbers.

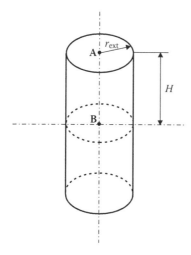

FIGURE 1.8 Finite cylinder obtained as intersection of basic shapes.

1.4 APPROXIMATE PREDICTION METHODS

Using the analytical solution is quite complicated, and back-calculations need iteration, while using charts prevents from a fully computerized calculation. Approximate
prediction methods are available to overcome such drawbacks, provided that the
Fourier number falls above 0.2. Such methods allow for an affordable computerized
calculation and can be applied to both regularly and irregularly shaped foods over a
wide range of Biot numbers.

1.4.1 Approximate Solution for Infinite Slab,
Infinite Cylinder, and Sphere

All the approximate methods are based on the evidence that the time–temperature
curve decreases in an exponential way, after an initial period of time when the temperature behaves quite differently depending on the location and the shape. This
means that the first term of the series is sufficient for a thorough description of the
phenomena after this initial period of time, that is, Fo > 0.2.

If only the first term of the series is considered, the solution for the infinite slab
becomes [2]

$$T' = \frac{2\sin\mu}{\mu + \sin\mu\cos\mu} e^{-\mu^2\text{Fo}}\cos(\mu x') = j e^{-\mu^2\text{Fo}} \tag{1.29}$$

at any location defined by x' and

$$T' = \frac{2\sin\mu}{\mu + \sin\mu\cos\mu} e^{-\mu^2\text{Fo}} = j_c e^{-\mu^2\text{Fo}} \tag{1.30}$$

at the center of the slab, where $x' = 0$.

From the comparison of Equations 1.29 and 1.30, it results that Equation 1.29 can be written as

$$T' = je^{-\mu^2 Fo} = \left[j_c \cos(\mu x') \right] e^{-\mu^2 Fo}, \qquad (1.31)$$

that is, the value of j at any location of the infinite slab can be derived from the value of j_c at its center as

$$j_{\text{slab}} = j_{c,\text{slab}} \cos(\mu x'). \qquad (1.32)$$

The approximate solution for the infinite cylinder at Fo > 0.2 is [2]

$$T' = \frac{2J_1\mu}{\mu\left[J_0^2(\mu) + J_1^2(\mu) \right]} e^{-\mu^2 Fo} J_0(\mu x') = je^{-\mu^2 Fo} \qquad (1.33)$$

at any location defined by x' and

$$T' = \frac{2J_1\mu}{\mu\left[J_0^2(\mu) + J_1^2(\mu) \right]} e^{-\mu^2 Fo} = j_c e^{-\mu^2 Fo} \qquad (1.34)$$

at the center (symmetry axis) of the cylinder, where $x' = 0$.
    It results thus:

$$j_{\text{cylinder}} = j_{c,\text{cylinder}} J_0(\mu x'). \qquad (1.35)$$

Finally, the approximate solution for the sphere at Fo > 0.2 is [2]

$$T' = 2\frac{\sin\mu - \mu\cos\mu}{\mu - \sin\mu\cos\mu} e^{-\mu^2 Fo} \frac{\sin(\mu x')}{\mu x'} = je^{-\mu^2 Fo} \qquad (1.36)$$

at any location defined by x' and

$$T' = 2\frac{\sin\mu - \mu\cos\mu}{\mu - \sin\mu\cos\mu} e^{-\mu^2 Fo} = j_c e^{-\mu^2 Fo} \qquad (1.37)$$

at the center of the sphere, where $x' = 0$.
    It results thus:

$$j_{\text{sphere}} = j_{c,\text{sphere}} \frac{\sin(\mu x')}{\mu x'}. \qquad (1.38)$$

The values of μ are obtained by the following regressions for the infinite slab:

$$\mu = 0.86213910 + 0.31577831\ \ln(Bi) + 0.00546354\ [\ln(Bi)]^2$$
$$- 0.01916579\ [\ln(Bi)]^3 - 0.00015960\ [\ln(Bi)]^4$$
$$+ 0.00103349\ [\ln(Bi)]^5 - 0.000112578\ [\ln(Bi)]^6; \qquad (1.39)$$

for the infinite cylinder:

$$\mu = 1.26002794 + 0.493872\ \ln(Bi) + 0.02107167\ [\ln(Bi)]^2$$
$$- 0.03130233\ [\ln(Bi)]^3 - 0.00124207\ [\ln(Bi)]^4$$
$$+ 0.00179771\ [\ln(Bi)]^5 - 0.00017900\ [\ln(Bi)]^6; \qquad (1.40)$$

and for the sphere:

$$\mu = 1.57777412 + 0.65046168\ \ln(Bi) + 0.04243499\ [\ln(Bi)]^2$$
$$- 0.04137967\ [\ln(Bi)]^3 - 0.00291009\ [\ln(Bi)]^4$$
$$+ 0.00244787\ [\ln(Bi)]^5 - 0.00021821\ [\ln(Bi)]^6. \qquad (1.41)$$

The values of j_c can be obtained from μ given their definition in Equations 1.30, 1.34, and 1.37.

Figure 1.9 reports the values of μ and Figure 1.10 the values of j_c as a function of Bi for the infinite slab, the infinite cylinder, and the sphere.

Figure 1.11 reports the values of j calculated at the surface (j_s) by means of Equations 1.32, 1.35, and 1.38, as a function of Bi for the infinite slab, the infinite cylinder, and the sphere.

An alternative description of the cooling curve is often used in food technology. When the cooling curves for the center of the objects—Equations 1.30, 1.34, and

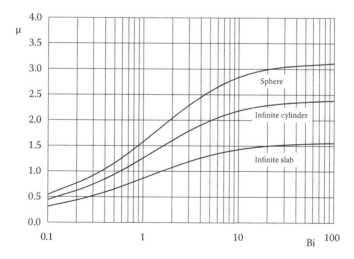

FIGURE 1.9 Plot of μ against the Biot number, for an infinite slab, an infinite cylinder, and a sphere.

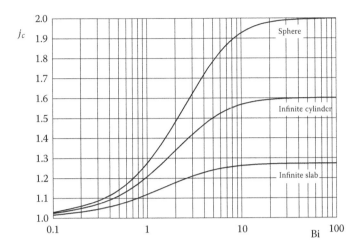

FIGURE 1.10 Plot of j_c against the Biot number, for an infinite slab, an infinite cylinder, and a sphere.

1.37—are plotted in a semilogarithmic chart, the exponential decrease results in a linear behavior (Figure 1.12).

For the infinite slab, the cooling curve, Equation 1.30, at the center is thus often written as [3, 4]

$$T' = \frac{2\sin\mu}{\mu + \sin\mu\cos\mu} e^{-\ln 10\frac{t}{f}} = j_c e^{-2.303\frac{t}{f}}, \tag{1.42}$$

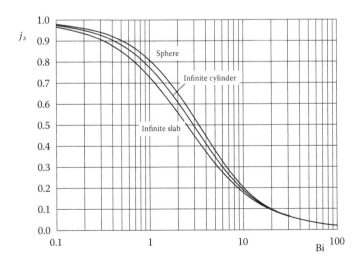

FIGURE 1.11 Plot of j_s against the Biot number, for an infinite slab, an infinite cylinder, and a sphere.

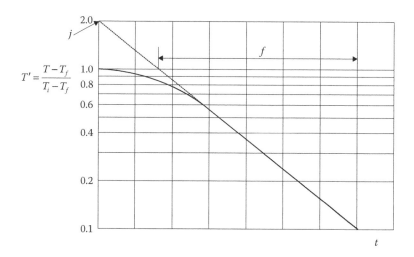

FIGURE 1.12 Cooling curve for Bi > 0.1 on a semilogarithmic chart.

where j_c is the intercept of the exponential decay curve with the y axis, and the f factor represents the time required for T' to decrease by a factor of 10, in the assumption of the exponential decay. Therefore, $(2.303/f = \ln(10)/f)$ is the slope of the linear portion of the curve. From Equations 1.30 and 1.42, it results thus

$$\frac{f\alpha}{L^2} = \frac{\ln(10)}{\mu_2}. \tag{1.43}$$

Similarly, the cooling curve at the center of the infinite cylinder, Equation 1.34, becomes

$$T' = \frac{2J_1\mu}{\mu\left[J_0^2(\mu) + J_1^2(\mu)\right]} e^{-\ln 10\frac{t}{f}} = j_c e^{-2.303\frac{t}{f}}, \tag{1.44}$$

and the cooling curve at the center of the sphere (Equation 1.37) becomes

$$T' = 2\frac{\sin\mu - \mu\cos\mu}{\mu - \sin\mu\cos\mu} e^{-\ln 10\frac{t}{f}} = j_c e^{-2.303\frac{t}{f}}. \tag{1.45}$$

The values of $f\alpha/L^2$ against Bi are reported in Figure 1.13 for the infinite slab, the infinite cylinder, and the sphere.

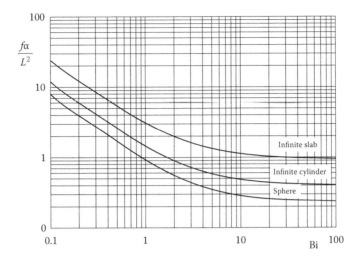

FIGURE 1.13 Plot of $f\alpha/L^2$ against the Biot number, for an infinite slab, an infinite cylinder, and a sphere.

1.4.2 SOLUTION FOR REGULAR SHAPES

The procedure to get the approximate solution for regular shapes that can be obtained by intersection of the basic shapes (i.e., the infinite rectangular rod, the rectangular brick, and the finite cylinder) is the same as the one illustrated for the analytical solution in Section 1.3.2.2. The dimensionless temperature T' is obtained by multiplying the solutions for the corresponding locations in the bounding basic shapes.

1.4.3 SOLUTION FOR IRREGULAR SHAPES

The solutions described above apply only to simple regular geometries where the one-dimensional assumption for the heat flux is realistic or to regular shapes where the superposition principle can be applied. However, foods are usually irregularly shaped, and various approaches have been used to solve this problem.

Smith et al. [5], in the case where $Bi \to \infty$, developed a geometry index G,

$$G = \frac{\mu^2}{\pi^2},\tag{1.46}$$

which can be obtained for an irregular shape using

$$G = 0.25 + \frac{0.375}{C_1^2} + \frac{0.375}{C_2^2},\tag{1.47}$$

where C_1 and C_2 are

$$C_1 = \frac{A_1}{\pi L^2} \qquad (1.48)$$

and

$$C_2 = \frac{A_2}{\pi L^2}, \qquad (1.49)$$

respectively, L being the shortest distance between the center of the product and the surface, A_1 the minimum cross-sectional area containing L, and A_2 the cross-sectional area orthogonal to A_1.

Once G has been found, a nomograph is required to evaluate the f factor [5, 6], but regressions to overcome this need have been published by Hayakawa and Villalobos [7].

Similarly, Fikiin [8] presented a method with a shape factor independent of the Biot number, quoting an accuracy within approximately 9% compared with experimental tests.

Another method was developed by Cleland and Earle [9] for the evaluation of the cooling time at the thermal center, based on the "equivalent heat transfer dimensionality." A shape factor E is defined, whose value is 1, 2, or 3 for the infinite slab, the infinite cylinder, or the sphere, respectively, and is dependent on the Biot number for irregular shapes, and which was derived from empirical equations at the early stage of their work. The authors, in collaboration with Lin et al. [10, 11], then developed a new method based on the first-term approximation to the analytical solution for convective cooling of a sphere. By means of this method, it is possible to evaluate the equivalent heat transfer dimensionality E for both complex regular shapes (e.g., ellipses and ellipsoids) and irregular shapes.

1.5 EFFECT OF DIFFERENT OPERATING CONDITIONS

The above-discussed methods are based on various assumptions both on operating and on boundary conditions. In certain cases, such assumptions are not realistic and deviations can arise, but such methods are still usable with appropriate modifications without making use of the most effective but expensive numerical methods.

1.5.1 EFFECT OF RADIATIVE HEAT TRANSFER

When the temperature difference between the food and the adjoining surfaces becomes significant, the fractions of radiative and convective heat transfer can be comparable. In this case, the heat transfer coefficient between the product and the fluid is corrected and named *effective* heat transfer coefficient:

$$h_{\text{eff}} = h_{\text{conv}} + h_{\text{rad}}, \qquad (1.50)$$

where h_{conv} is the convective coefficient and h_{rad} is estimated as

$$h_{rad} = \frac{q''_{rad}}{|T_s - T_f|},$$ (1.51)

q''_{rad} being the radiative heat flux per unit area.

In the particular case of a product of emissivity ε completely surrounded by a black body, or by a cavity of much greater surface at the same temperature as the fluid, the specific radiative heat flux can be expressed as

$$q_{rad} = \sigma\varepsilon \left|T_s^4 - T_{cavity}^4\right| = \sigma\varepsilon \left|T_s^4 - T_f^4\right|,$$ (1.52)

where σ is the Stefan–Boltzmann constant and all temperatures are in Kelvin. Thus, Equation 1.51 becomes

$$h_{rad} = \frac{\sigma\varepsilon \left|T_s^4 - T_f^4\right|}{|T_s - T_f|} = \sigma\varepsilon \left(T_s^2 + T_f^2\right)\left(T_s + T_f\right),$$ (1.53)

where all temperatures are in Kelvin.

1.5.2 EFFECT OF VARIABLE FLUID TEMPERATURE

In many important applications, like blast chillers, the fluid temperature is subject to a progressive and tangible decrease. In this instance, the use of an arithmetic mean value gives a conservative prediction. Otherwise, the process must be discretized in time, or a numeric solution must be adopted.

1.5.3 EFFECT OF WATER LOSS FROM THE PRODUCT

Cooling processes of unpackaged products frequently involve weight loss, which originates from evaporation of water on the surface of products, thus leading to major additional cooling power due to latent heat removal. For this reason, the prediction of cooling time cannot disregard the effect of evaporation.

Chuntranuluck et al. [12–14] observed that linear plots are no longer obtained when $\ln(T')$ is plotted against Fo. In fact, T approaches an equilibrium temperature T_{eq}, which is lower than T_f. When using T_{eq} instead of T_f in the definition of T', plots become sufficiently linearized to use the approximate solution Equation 1.30. The authors derived the values of j_c and μ^2 from comparison with the results of numerical simulations and gave correlations of these values to those with convection only as functions of Bi, ambient relative humidity $H_{r,}$ and product water activity a_w [12]:

$$\frac{j_{cevap}}{j_{cconv}} = 1 - \frac{0.0153a_w^{2.4}}{\mathrm{Bi}^{0.4}} + 0.0335 \left(\frac{\mathrm{Bi}^{\frac{4}{3}} + 1.85}{\frac{\mathrm{Bi}^{\frac{4}{3}}}{E} + \frac{1.85}{n}} \right) e^{-(\mathrm{Bi}-2.5)^2} + 0.0725 H_r \, e^{-(\mathrm{Bi}-0.7)^2}$$

$$+ T_f \left(0.00338 H_r + 0.00413 e^{-(\mathrm{Bi}-0.9)^2} \right) - T_i \left(0.00447 e^{-1.33\mathrm{Bi}} + 0.000599 \right);$$

$$(1.54)$$

$$\frac{\mu_{evap}^2}{\mu_{conv}^2} = 1 + \frac{\mathrm{Bi}}{15\left(\mathrm{Bi}^{1.5} + 1.5\right)} + \frac{T_a\left(H_r + 0.34\right) + \left(5H_r + 0.12T_i + 9.87\right)a_w^{0.8}}{19\left(\mathrm{Bi}^{1.2} + 1.2\right)}. \quad (1.55)$$

In Equation 1.54, E and n are shape factors. Values for E are 0.75, 1.76, and 3, and values for n are 1, 2, and 3 for the infinite slab, the infinite cylinder, and the sphere, respectively.

Values of j_{av} for the evaluation of the average temperature are also available.

NOMENCLATURE

A	Surface area (m²)
a_w	Water activity (–)
Bi	Biot number (Equation 1.10) [–]
c	Specific heat (J kg⁻¹ K⁻¹)
E	Equivalent heat transfer dimensionality (–)
f	Time for a T' decrease by a factor of 10 (Equation 1.42) (s)
Fo	Fourier number (Equation 1.16) (–)
h	Convective heat transfer coefficient (W m⁻² K⁻¹)
H_r	Relative humidity (–)
j	Lag factor (–)
J	Bessel function
L	Characteristic length (m)
n	Normal to the surface
q	Heat flux (W)
q''	Heat flux per unit area (W m⁻²)
R	Thermal resistance (K W⁻¹)
t	Time (s)
T	Temperature (°C)
V	Volume (m³)
x	Cartesian coordinate (m)
y	Cartesian coordinate (m)
z	Cartesian coordinate (m)

GREEK SYMBOLS

α   Thermal diffusivity (m^2 s^{-1})
λ   Thermal conductivity (W m^{-1} K^{-1})
μ   Root of (Equation 1.22)
ρ   Density (kg m^{-3})
τ   Time constant (s)

SUPERSCRIPTS

′   dimensionless

SUBSCRIPTS

c   Center
conv   Convection
eff   Effective
eq   Equilibrium
f   Fluid
i   Initial
int   Internal
n Index of summation
ext   External
rad   Radiation
s   Surface

REFERENCES

[1] Sun, D.W., ed. 2007. *Computational fluid dynamics in food processing*. CRC Press, Taylor & Francis Company, Boca Raton, FL.

[2] Incropera, F.P., De Witt, D.P., Bergman, T.L., Lavine, A.S. 2006. *Fundamentals of heat and mass transfer*. John Wiley & Sons, New York.

[3] Cleland, A.C. 1990. *Food refrigeration processes: Analysis, design and simulation*. Elsevier Applied Science, London.

[4] Cooling and freezing times of foods. 2006. In: *2006 ASHRAE Handbook—Refrigeration*. American Society of Heating, Refrigerating and Air-Conditioning Engineers (ASHRAE), Atlanta, GA.

[5] Smith, R.E., Nelson, G.L., Henrickson, R.L. 1968. Applications of geometry analysis of anomalous shapes to problems in transient heat transfer. *Transactions of the ASAE* 11(2):296–302.

[6] Smith, R.E. 1966. Analysis of transient heat transfer from anomalous shape with heterogeneous properties. PhD Thesis, Oklahoma State University, Stillwater, OK.

[7] Hayakawa, K., Villalobos, G. 1989. Formulas for estimating Smith et al. parameters to determine the mass average temperature of irregularly shaped bodies. *Journal of Food Process Engineering* 11(4):237–256.

[8] Fikiin, A.G. 1983. Application des methodes analytiques pour déterminer la durée du refroidissement des produits alimentaires (et d'autre corps solides). *Proceedings 16th International Congress of Refrigeration* 2:411–414.

[9] Cleland, A.C., Earle, R.L. 1982. A simple method for prediction of heating and cooling rates in solids of various shapes. *International Journal of Refrigeration* 5(2):98–106.

[10] Lin, Z., Cleland, A.C., Cleland, D.J., Serralach, G.F. 1996. A simple method for prediction of chilling times for objects of two-dimensional irregular shape. *International Journal of Refrigeration* 19(2):95–106.

[11] Lin, Z., Cleland, A.C., Cleland, D.J., Serralach G.F. 1996. A simple method for prediction of chilling times: Extensions to three-dimensional irregular shape. *International Journal of Refrigeration* 19(2):107–114.

[12] Chuntranuluck, S., Wells, C.M., Cleland, A.C. 1998. Prediction of chilling times of foods in situations where evaporative cooling is significant—Part 1: Method development. *Journal of Food Engineering* 37:111–125.

[13] Chuntranuluck, S., Wells, C.M., Cleland, A.C. 1998. Prediction of chilling times of foods in situations where evaporative cooling is significant—Part 2: Experimental testing. *Journal of Food Engineering* 37:127–141.

[14] Chuntranuluck, S., Wells, C.M., Cleland, A.C. 1998. Prediction of chilling times of foods in situations where evaporative cooling is significant—Part 3: Applications. *Journal of Food Engineering* 37:143–157.

2 Freezing and Thawing

Viviana O. Salvadori

CONTENTS

2.1 INTRODUCTION

Freezing is an important operation in food preservation, for it involves millions of tons of food per year (Pierce 2002).

During freezing, ice formation begins at a temperature T_{if}, characteristic to each type of foodstuff, and continues over a wide temperature range. The water phase change brings about an important and continuous variation of the characteristic physical properties of the phenomenon (ρ, Cp, k). It is important to take into account that this phase change does not occur at a constant temperature but instead takes place as a dynamic transformation over a range of temperatures, which affects the whole freezing process.

According to its definition (International Institute of Refrigeration 1972), effective freezing time is the time required to lower the temperature of the product from its initial temperature to a given temperature at its thermal center (generally −10°C or −18°C). When a foodstuff is industrially frozen, the process has to be controlled closely, since food reaching the storage stage partially frozen means an increase in energy consumption. Yet, it is economically unfavorable to refrigerate the product to a lower temperature than the storage one.

Thawing is another important food operation related to industrial processed foods. Thus, thawing times must be calculated properly to control this stage in a food processing plant.

Therefore, adequate methods for predicting freezing and thawing times should be available to design and evaluate plants for processed foods. In this sense, one of the main interests of design engineers and equipment users is to be able to count on simple and accurate prediction methods for the simulation of the process they are dealing with, mainly for the calculation of process times as a function of material characteristics and operating conditions.

2.2 THEORY

Detailed modeling of heat transfer in freezing or thawing of foods leads to strongly nonlinear differential balances due to the rapid variation of the thermal properties with temperature in the freezing range.

Generally, the balance to be solved is

$$\frac{\partial T}{\partial t} = \nabla(\alpha(\nabla T))$$

$$\frac{\partial T}{\partial \vec{n}} = Bi\left(T_s - T_a\right),$$

(2.1)

where the Laplacian ∇ can have one, two, or three components, depending on whether the heat transfer is one-dimensional (slab, infinite cylinder, or sphere) or multidimensional (finite cylinder, block, or irregular shapes).

Equation 2.1 is a simple mathematical model that involves several assumptions:

- Subcooling is neglected; freezing begins when the surface reaches T_{if}.
- Ice content follows an equilibrium curve; solute concentration effects are neglected.
- Thermal properties do not depend on freezing conditions.
- Chemical reactions (protein insolubilization and enzymatic reactions) and ice recrystallization during freezing are neglected.

All these assumptions introduce small errors, which up to now are very difficult to quantify.

2.2.1 THERMOPHYSICAL PROPERTIES

As stated in the previous paragraph, knowledge of the different physical properties is required in order to properly evaluate freezing and thawing times.

In literature, we can find several comprehensive works on the thermophysical properties of foods (Urbicain and Lozano 1997; Fricke and Becker 2001; ASHRAE 2006), presenting experimental values for a given foodstuff at a given condition as well as prediction equations as a function of temperature and, eventually, water content or other parameters.

Our aim in this chapter is to present those models that are easier to apply, are valid for a wide range of foods and independent variables, and have been used by different authors, thus showing their reliability.

In particular, partially frozen foods are made up of ice, water, and dry tissue (protein, carbohydrate, lipid, and so on). Therefore, the physical properties can be calculated as a function of the properties of individual components.

2.2.1.1 T_{if} Initial Freezing Temperature

This temperature separates food being frozen into two zones: in the unfrozen one, water is still in liquid state; in the frozen one, water has been transformed to ice.

Foods with high water content can be assumed as having ideal solutions behavior. According to Raoult law, the freezing point depression of an ideal binary solution is

$$\Delta T_{if} = T_w(K) - T_{if}(K) = \frac{RT_w^2 m}{\lambda}$$

$$T_{if}(°C) = -1.86\, m = \frac{RT_w^2}{\lambda M_w} \ln y_w. \tag{2.2}$$

Other empirical equations for specific foods (meat, fruit and vegetables, and fruit juices, respectively) have been developed (Sanz et al. 1989; Chang and Tao 1981):

$$T_{if} = \frac{1 - x_w}{0.06908 - 0.4393 x_w}$$

$$T_{if} = 20.289 - 60.365 x_w + 40.874 x_w^2 \tag{2.3}$$

$$T_{if} = -152.68 + 327.354 x_w - 176.49 x_w^2.$$

2.2.1.2 Ice Content

Riedel (1957) studied, through calorimetric experiments, food enthalpy variations associated with water content freezing, assuming enthalpy to be zero at $T_{ref} = -40°C$. Riedel's charts for different foods show ice content as a function of temperature.

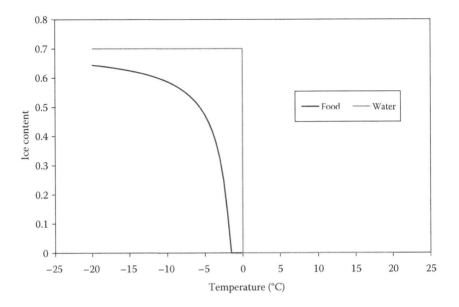

FIGURE 2.1   Dependence of ice content with temperature, initial water content = 0.7.

Mascheroni (1977) obtained the ice content of meat products using a regression of Riedel's curves:

$$x_i = 1.1866x_w - 0.1866 + 2.7013\frac{(1-x_w)}{T}.$$ (2.4)

Miles et al. (1990) have proposed a general equation for foods:

$$x_i = (x_w - x_{bw})\left(1 - \frac{T_{if}}{T}\right).$$ (2.5)

Both ice content increase and liquid water content decrease are continuous functions of temperature (see Figure 2.1) and promote significant variation of all the thermophysical properties involved in freezing time prediction.

2.2.1.3   Thermal Conductivity, Density, Specific Heat, and Thermal Diffusivity

For other thermophysical properties such as thermal conductivity, density, specific heat, and thermal diffusivity, different theoretical and semiempirical models were proposed by different authors. Urbicain and Lozano (1997) present a complete review of these models. As is shown in Figure 2.2, these properties depend strongly on temperature.

Between these options, Choi and Okos' model (1986) is the most used at this time. The authors present a collection of useful prediction equations (Equation 2.6) for the

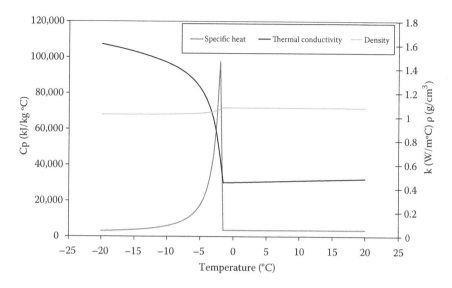

FIGURE 2.2 Dependence of physical properties with temperature.

principal thermal properties of foods, based on the mass fraction and the thermal properties of major pure components of food products (protein, lipid, carbohydrate, fiber, ash, and water), in the temperature range of −40°C to 150°C:

$$k = \sum x_j k_j$$

$$\rho = \frac{1}{\sum x_j^V / \rho_j}$$

$$C_p = \sum x_j C_{pj}$$

$$\alpha = \sum x_j \alpha_j.$$

(2.6)

Linear or quadratic functions of the temperature for the properties of each component are used.

The thermal conductivity and the thermal diffusivity also depend on the spatial structure of the food. (Food has layers of different components, and heat flux may travel parallel or normal to these layers.)

Considering prediction equations for specific products, Sanz et al. (1987, 1989) present a complete list of equations valid for meat products. The cited review of Urbicain and Lozano details other specific models for apples, potatoes, fruit juices, and so on.

Below the food's freezing point, the sensible heat from temperature change and the latent heat from ice formation must be considered. Because latent heat is not released at a constant temperature, but rather over a range of temperatures, an apparent specific heat, originally proposed by Bonacina et al. (1973), must be used to account for both the sensible and latent heat effects:

$$C_{p,ap}(T) = \frac{dH(T)}{dT} = C_p(T) + \lambda \delta\left(T - T_{if}\right), \quad C_p(T) = \begin{cases} C_1(T), & T < T_{if} \\ C_2(T), & T > T_{if} \end{cases}, \quad (2.7)$$

where δ is the Dirac function used to represent the heat jump observed near the initial freezing point.

A simplified expression for $C_{p,ap}$ was presented by Mascheroni (1977) for meat products, considering that enthalpy depends only on temperature and water content, x_i:

$$C_{p,ap} = C_p - \lambda Y_0 \frac{dx_i}{dT}. \quad (2.8)$$

2.2.2 BOUNDARY CONDITIONS

The most frequent initial condition assumes uniform initial temperature, greater than the initial freezing point, in the entire food product.

According to Equation 1.1, convective heat transfer is assumed at the food surface. Therefore, knowledge of the surface heat transfer coefficient is needed to solve the energy balance.

Newton's law of cooling defines the surface heat transfer coefficient, h, as a function of the surface temperature of the food, the surrounding fluid temperature, and the surface area of the food through which the heat transfer occurs. For most applications, h can be determined experimentally. In addition, several researchers have reported useful correlations, which give the Nusselt number, Nu, as a function of the Reynolds number, Re, and the Prandtl number, Pr. It is important to bear in mind that the Reynolds number is calculated in accordance with the freezing process, and the Prandtl number is determined while considering the physical properties of the surrounding medium. In freezing and thawing equipment, we can find very small h values such as 10 W/(m² °C) (cold store rooms), intermediate values of 30–50 W/(m² °C) (air blast tunnels), and high h values of 300 W/(m² °C) (plate, impingement, and cryogenic freezers).

The cited resources on thermal properties (Urbicain and Lozano 1997; Fricke and Becker 2001; ASHRAE 2006) also present excellent options for calculating heat transfer coefficients.

Figure 2.3 shows typical (numerical) freezing curves, considering different values of the boundary conditions.

2.2.3 EXPERIMENTAL DATA

Through many years, several authors have measured experimental freezing times using different food products or test substances and diverse operating conditions.

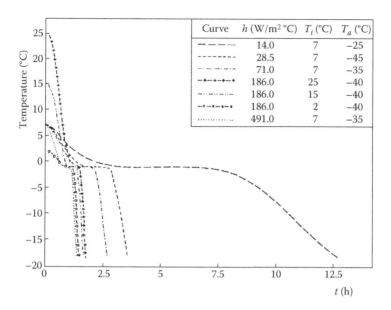

FIGURE 2.3 Influence of boundary conditions (h and T_a) and initial temperature in the freezing curve.

Figure 2.4 shows four experimental freezing curves measured on foods of different initial water content: hamburgers, cooked tagliatelle, osmotic dehydrated fruits, and beef muscle; or different freezing equipment: impingement, air blast tunnel, or cryogenic cabinet. As can be observed, the shape of the curve is similar, although the initial freezing point is different, depending on Y_o. Table 2.1 summarizes the source

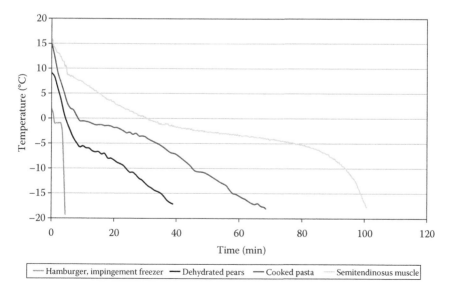

FIGURE 2.4 Typical freezing curves.

TABLE 2.1
Published Experimental Data Sets of Freezing Times

Shape	Material	Number of Data	Reference
Slab	Tylose	43	Cleland and Earle (1977)
	Tylose	32	Pham and Willix (1990)
	Lamb	4	Bazán and Mascheroni (1984)
	Beef[a]	10	Mascheroni and Calvelo (1982)
	Minced beef	6	Cleland and Earle (1977)
	Pig liver	44	Creed and James (1983)
	Mashed potatoes	6	Cleland and Earle (1977)
	Shark	6	Hense and Kieckbusch (1991)
	Beef[b]	4	Mascheroni and Calvelo (1982)
	Beef[b] (unwrapped)	9	Hung and Thompson (1983)
	Semiliquid foods	37	Michniewicz (1995)
	Tylose (unwrapped)	22	Hung and Thompson (1983)
	Mashed potato (unwrapped)	8	Hung and Thompson (1983)
	Carp (unwrapped)	7	Hung and Thompson (1983)
	Hamburgers	11	Salvadori and Mascheroni (2002)
Infinite cylinder	Tylose	30	Cleland and Earle (1979a)
	Tylose	13	Cleland et al. (1987)
	Shrimp	6	Wang and Kolbe (1987)
Sphere	Tylose	30	Cleland and Earle (1979a)
	Tylose	4	Cleland et al. (1987)
	Apple	36	Ilicali and Saglam (1987)
	Meat balls (unwrapped)	48	Tocci and Mascheroni (1995)
	Meat balls (unwrapped)	9	Campañone et al. (2002)
Blocks	Tylose (three sides)	72	Cleland and Earle (1979b)
	Beef (three sides)	17	De Michelis and Calvelo (1983)
	Minced beef (three sides)	18	Ilicali (1989a)
	Mashed potatoes (three sides)	20	Ilicali (1989a)
	Beef (two sides)	3	De Michelis and Calvelo (1983)
	Minced beef (three sides)	4	Cleland et al. (1987)
	Strawberry pulp	6	Salvadori et al. (1996)
Finite cylinder	Scallop	49	Chung and Merritt (1991)
	Beef	5	De Michelis and Calvelo (1983)
	Strawberry pulp	3	Salvadori et al. (1996)

[a] Heat transfer perpendicular to fibers.
[b] Heat transfer parallel to fibers.

TABLE 2.2
Published Experimental Data Sets of Thawing Times

Shape	Material	Number of Data	Reference
Slab	Tylose	35	Cleland et al. (1986)
	Agar gel	11	Flores and Mascheroni (1983)
	Beef	4	James et al. (1977)
	Beef	11	James and Bailey (1980)
	Lamb	12	Flores et al. (1993)
	Minced beef	6	Cleland et al. (1986)
	Minced beef	6	Ilicali (1989b)
	Minced beef	3	Flores et al. (1993)
	Cod	1	Vanichseni (1971)
	Sardine	4	Crepey and Becel (1978)
	Mackerel	5	Flechtenmacher (1983)
	Mashed potatoes	6	Ilicali (1989b)
	Tylose	12	Devres (1991)
Infinite cylinder	Tylose	34	Cleland et al. (1986)
	Tylose	10	Cleland et al. (1987c)
	Minced beef	5	Ilicali (1989b)
	Potato puree	6	Ilicali (1989b)
Sphere	Tylose	35	Cleland et al. (1986)
	Tylose	2	Cleland et al. (1987c)
	Minced beef	6	Ilicali (1989b)
	Potato puree	6	Ilicali (1989b)
Blocks	Tylose (three sides)	68	Cleland et al. (1987c)
	Minced beef (three sides)	4	Cleland et al. (1987c)

of each data set found in the literature. (Only data sets with complete information about food characteristic and freezing conditions were included in this table.) These data sets were employed by different authors to test their own prediction methods. As can be seen in Table 2.1, only foodstuffs with regular geometries, principally one-dimensional geometries, were used. The freezers used to perform the experiments included air blast tunnel, plate, immersion, and also impingement equipment.

In a similar way, Table 2.2 reports experimental data sets of thawing times.

2.3 NUMERICAL METHODS

Numerical methods are used to solve the thermal balance equations presented in Equation 2.1 in a more accurate way.

In literature, we can find a great quantity of work (more than 200 references) that utilizes different numeric methods, the most common being finite differences and finite elements. Finite differences are easier to program than finite elements, if regular geometries are involved. However, if the food geometry is irregular, finite differences are more complex to implement, making finite elements the preferred method.

The use of numeric methods is not easy, although there are several commercial software packages available that simplify their application, many of which require specific training.

First, a thorough knowledge of the variations of thermophysical properties is needed. Evaluations of the complete processing time, t_f or t_t, and knowledge of the thermal history at different points of the food are then possible. A wide range of operative conditions, including variable ones, can be considered.

When numerical methods were first being applied, it took several hours to process the results for a specific condition. What was once an important disadvantage in applying numerical methods has now become practically obsolete with the advances in computer technology.

Cleland et al. (1982) have shown that, even if a much-elaborated numerical method is used without introducing simplifying assumptions, a limit to the accuracy of the prediction exists, arising from uncertainties in the values of the thermophysical properties.

Besides, Salvadori and Mascheroni (1991) have observed that results with accuracy similar to that of the numerical method can be obtained by means of a suitable simplified method.

Details on different numerical methods are outside the scope of this chapter. The following reviews are recommended to readers who are interested in learning more on the topic: Cleland (1990) and Delgado and Sun (2001). Recently, Pham (2006) has presented a unified overview of the common numerical methods applied for modeling heat and mass transfer during the freezing, thawing, and frozen storage of foods. Mass transfer is also considered, involving different approaches for dense and porous foods. Supercooling, nucleation and transmembrane diffusion effects during freezing, recrystallization during frozen storage, and high-pressure freezing and thawing are considered. In concordance with the author's opinion, we can establish that the classic "pure thermal" freezing problem is solved; on the contrary, more research is needed to model other nonthermal phenomena associated with the freezing process in order to be able to predict changes in food quality at the micro level.

2.4   APPROXIMATE METHODS

Approximate prediction equations are a good alternative for predicting freezing and thawing times of foods with unidirectional heat conduction if the values of thermal properties, food characteristics, and process conditions are known. Through the last two decades, there has been a continuous growth in the interest in improving and simplifying prediction methods for food freezing and thawing times. Cleland (1990), Hung (1990), Becker and Fricke (1999a,b), and Delgado and Sun (2001), among others, present extensive reviews on both types of prediction methods. The major restriction of these methods is that their accuracy depends mainly on the simplifications introduced and is guaranteed only for the range of food sizes and operative conditions for which they were developed.

In general, simplified methods have the following characteristics:

- They simplify the differential balance (Equation 2.1) or present equations or graphical methods obtained experimentally or by means of numerical methods.
- They predict only freezing time, not the temperature evolution inside the food.
- The average value of physical properties is used.
- Their application is valid only for the range of operative conditions in which they were obtained. Symmetric boundary conditions are used in most cases. (The same value is used throughout the surface of the foodstuff for the external temperature and the heat transfer coefficient.) In general, they cannot be used in variable operative conditions.
- They are strongly dependent on the shape of the product, and many of them have been developed for only one particular food.
- The calculation of the process time is usually simple and can be performed even with a standard calculator.

2.4.1 PLANK'S METHOD

The first approximate method was proposed in 1913 and modified in 1941 (Plank 1963). It solves Equation 2.1 with the following assumptions:

- Constant phase change temperature.
- Uniform initial temperature, equal to the phase change temperature.
- Enthalpy involves only latent heat, without sensible heat before or after the phase change.
- Frozen ρC_p of the frozen phase tends to zero, and a pseudostationary state is obtained with a lineal temperature profile.
- Constant physical properties in each phase (unfrozen and frozen).
- Third-kind boundary condition (Newton's cooling law).

Equation 2.9 results from analytically solving Equation 2.1 with the mentioned assumptions:

$$t_f = \frac{\rho_f \Delta H}{\left(T_{if} - T_a\right)} \left(\frac{PD}{h} + \frac{RD^2}{k_f} \right). \tag{2.9}$$

P and R are constant values that depend on geometry; P is equal to 0.5, 0.25, and 0.166 for slab, infinite cylinder, and sphere, respectively, and R is equal to $P/4$.

As Plank's assumptions do not verify the most common freezing conditions, freezing times (predicted by Equation 2.9) are subestimated by 40%. However, this equation has a good theoretical basis, and the use of the shape factors P and R corresponding to the three simple regular shapes under the same operative conditions maintains a relationship that is verified experimentally: t_f (slab)/t_f (infinite cylinder)/t_f (sphere) = 6:3:2.

Therefore, several authors have proposed diverse modifications of Equation 2.9 that improve its precision, introduce factors that rectify the original formula, reevaluate the constants P and R, or add two terms (or one term and a correcting factor) to account for the cooling and tempering periods.

2.4.2 CLELAND AND EARLE'S METHOD

Cleland and Earle have done a thorough study on freezing and thawing times, measuring experimental data for a wide range of operative conditions, using tylose as a test substance. They also developed a useful finite-difference method using an implicit scheme. From their experimental and numerical data, Cleland and Earle (1984) modified Plank's equation, proposing new expressions of the constants P and R. In the following equation, valid for slabs, ΔH^* is the difference in enthalpy between the temperature at the beginning of the phase change and the reference temperature ($-10°C$), and D is the thickness or diameter of the food product:

$$t_f = \frac{H^*}{\left(T_{if} - T_a\right)}\left(\frac{PD}{h} + \frac{RD^2}{k_f}\right)\left(1 - \frac{1.65Ste}{k_f}\ln\left(\frac{T_c - T_a}{T_{ref} - T_a}\right)\right)$$

$$P = 0.5\left(1.026 + 0.5808Pk + Ste\left(0.2296Pk + 0.105\right)\right) \qquad (2.10)$$

$$R = 0.125\left(1.202 + Ste\left(3.41Pk + 0.7336\right)\right).$$

This method allows the evaluation of freezing times for different final center temperatures.

2.4.3 SALVADORI AND MASCHERONI'S METHOD

Another simplified methodology was developed by mathematical regression of the thermal histories obtained by means of a numerical simulation program, using a finite-difference scheme.

Emphasis was directed mainly to the analysis of the temperature at the thermal center, this being defined as the point of highest temperature during freezing or the point of lowest temperature during thawing. When regularly shaped foods are studied, the thermal center and the geometric center are coincident. The calculation program, coded in Fortran 90, was run numerous times in order to cover a wide range of operative conditions ($2°C \leq T_i \leq 25°C$; $-20°C \leq T_a \leq -45°C$; $1 \leq Bi \leq 50$).

The thermal behavior of the center in any one of the processing conditions was represented graphically versus a dimensionless variable X (defined to take into account both the influence of time and the other variables representative of each situation: L, α_o, h, T_o, and T_a). The definition of X is based upon previous studies or in literature equations developed for predicting t_f (or t_t). Empirical values of the parameters a, b, c, m, and n were determined using a nonlinear regression software (Salvadori and Mascheroni 1991) (Table 2.3). As a result, simple prediction equations for both freezing and thawing times were obtained:

TABLE 2.3

Empirical Constants of Equation 2.11

Geometry	a	b	c	m	N
Slab	−1.272	65.489	0.184	1.070	0.096
Infinite cylinder	−0.750	32.198	0.179	1.032	0.037
Sphere	−0.439	24.804	0.167	1.078	0.073

$$X = \frac{Fo(-1 - T_a)^m}{(1/Bi + c)(1 + T_i)^n}$$

$$X = aT_c + b$$

$$t_f = \frac{L^2}{\alpha_o}(aT_c + b)(1/Bi + c)(1 + T_o)^n(-1 - T_a)^m \qquad (2.11)$$

$$t_t = \frac{L^2}{\alpha_o}(aT_c + b)(1/Bi + c)(1 + T_o)^n(-1 - T_a)^m.$$

Although this method was developed from the numerical solution of the energy balance representing the freezing or thawing of beef, it had been proved to be applicable to foods with high water content and continuous structure. Furthermore, it was also shown that the method can be applied with the same accuracy to working conditions other than those within the range used to obtain the empirical parameters. This method is especially suitable for practical industrial calculations, since there is no need for complex numerical calculation programs. Nor does it require a thorough knowledge of the variations of thermophysical properties as a function of the temperature, since only properties of fresh food, readily available in the literature, are used.

Later, the same authors (Salvadori and Mascheroni 1997) developed a calculus program, coded in Visual Basic 4.5. This application has an attractive and very simple graphical interface. The software included a database of physical properties of different foods and allows the calculation of chilling, freezing, and thawing times.

2.4.4 MULTIDIMENSIONAL GEOMETRIES

To deal with multidimensional shapes, two courses of action have normally been followed:

- The development of specific methods for each shape
- The extension of a prediction method for simple shapes using shape factors that account for heat transfer in multiple spatial directions

As proposed by Cleland and his coworkers in successive papers (Cleland et al. 1987a,b; Hossain et al. 1992a,b,c), one of the ways to determine the freezing time of a multidimensional product is to know both the freezing time of a slab under the same operative conditions and a geometric shape factor. The authors initially named such a factor *equivalent heat transfer dimensionality*, more recently referred to as E.

Therefore, the freezing time of a food of any shape can be calculated by the following equation:

$$t_f = \frac{t_{f,\text{slab}}}{E} \tag{2.12}$$

The variable E compares the total contribution of heat transfer through each face of a multidimensional object to that of a slab with a thickness equal to the smallest of the object's dimensions, exposing both to the same conditions. For a slab, $E = 1$; for an infinite cylinder, $E = 2$; and for a sphere, $E = 3$. In objects of other shapes, the value of E lies between the extreme values of 1 and 3 and depends primarily on the overall shape and, to a lesser extent, on the heat transfer coefficients.

2.4.4.1 Calculation of Shape Factors

Through several years of development, different formulas to calculate the E factor were proposed, the following being the most general ones:

(1) An empirical equation, fitted with a set of 270 numerical freezing times, generated with a fully validated finite-difference method:

$$E = G1 + G2\ E1 + G3\ E2. \tag{2.13}$$

The factors $G1$, $G2$, and $G3$ depend solely on the geometry and are given in Table 2.4.

In turn, $E1$ and $E2$ are functions of the Biot number and of the dimension ratios β_1 and β_2:

$$E_1 = \frac{X\left(2.32/\beta_1^{1.77}\right)}{\beta_1} + \left[1 - X(2.32/\beta_1^{1.77})\right]\frac{0.73}{\beta_2^{2.5}}$$

$$E_2 = \frac{X\left(2.32/\beta_2^{1.77}\right)}{\beta_2} + \left[1 - X(2.32/\beta_2^{1.77})\right]\frac{0.50}{\beta_2^{3.69}} \tag{2.14}$$

$$X(x) = \frac{x}{\left(Bi^{1.34} + x\right)}.$$

TABLE 2.4

Geometric Constants for Calculation of E (Equation 2.13)

Geometry	G1	G2	G3
Bidimensional package	1	1	0
Tridimensional package	1	1	1
Finite cylinder	2	0	1

(2) Hossain et al. (1992a,b,c) developed analytical expressions of the shape factor E (E_{an} for regular shapes), deduced under the assumption of phase change at constant temperature.

Although the analytical expressions thus obtained for both freezing times can lead to results with high error (owing to the inherent assumption in their derivation), the number of equivalent dimensions E can be used in conjunction with more accurate prediction equations for $t_{f,slab}$ so as to calculate the freezing time t_f of multidimensional shapes. In addition to the equations for the calculation of E_{an}, the authors designed graphs of E_{an} versus Bi for the different geometries.

2.4.5 NEURAL NETWORKS

Artificial neural networks (ANNs) allow the modeling of complex real systems in a relatively simple manner and result in an efficient way of dealing with the system's intrinsic nonlinearities. In this way, ANNs may be used to estimate or predict process behavior without the need for a mathematical model or a prediction equation associated with the physical problem or any assumption about the nature of the underlying mechanisms (Ramesh et al. 1996). ANNs are based on the progress of neurobiology that has allowed researchers to build mathematical models of neurons to emulate neural behavior. Therefore, neural networks are recognized as helpful tools for dynamic modeling and have been extensively studied since the publication of the perceptron identification method by Rosenblat in 1958 (Basheer and Hajmeer 2000).

As regards food freezing and thawing, Mittal and Zhang (2000) developed a feedforward neural network to predict the freezing time of food products with simple regular shapes. The ANN was trained using numerous freezing time data generated with Pham's (1986) approximate model and was validated using reported experimental data.

Goñi et al. (2008) developed another ANN. The database was constituted only by reported experimental data on very different foods, composition, and freezing or thawing procedure, which enriches the implicit network knowledge. These ANNs were able to estimate with high accuracy the freezing and thawing times of foods of any shape, size, and composition. The capabilities of the neural network methodology of finding a set of parameters to represent in a unique model both processes, freezing and thawing, must be emphasized. This fact is—to date—impossible to obtain with other known approximate prediction methods.

2.4.6 FREEZING WITH SUBLIMATION

When unwrapped foods are frozen and/or stored in the frozen state or with nonadhering packaging, weight losses take place due to sublimation of the surface ice, as can be seen in Figure 2.5. This weight loss is not negligible. Ice sublimation produces a dehydrated surface layer that changes the appearance, color, texture, and taste. Furthermore, this weight loss becomes an important quality and economic factor.

Several authors have surveyed ice sublimation in frozen foods: lamb (Pham and Willix 1984), tylose and beef (Sukhwal and Aguirre Puente 1983), potatoes

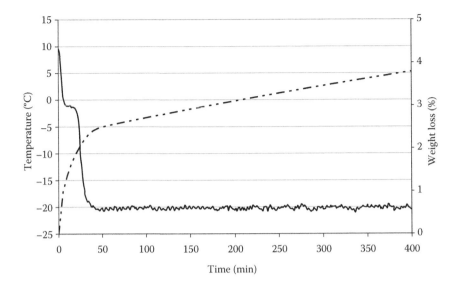

FIGURE 2.5 Temperature (solid line) and weight loss (dashed line) during freezing and frozen storage. Air temperature: −29.8°C. Sample: cylinder of semitendinosus muscle (2.7 cm diameter, 10 cm long).

(Lambrinos and Aguirre Puente 1983), beef and pork (Méndez Bustabad 1999), and mashed potatoes and meat balls (Campañone et al. 2002).

These losses have two different origins, but their effects are cumulative (Campañone et al. 2001).

During freezing, the food surface has a higher temperature than the circulating air, and thus, the surface water vapor pressure, P_{va}, is also higher than that of air. The lower the freezing speed, the slower the decrease in surface temperature, and this results in greater differences in vapor pressure between food and air. In addition, the duration of the freezing process will be longer, with both factors increasing the weight loss.

During storage, room temperature fluctuations are transferred, delayed in time, to the frozen products. Thus, there will be alternating periods in which their surface temperature is higher than the room temperature, with subsequent ice sublimation. The cumulative weight lost throughout long storage periods may cause an important quality loss. This second effect is usually much more important than the one caused during freezing.

Campañone et al. (2001) have proposed and solved a theoretical model that considers simultaneous heat and mass transfer during freezing and frozen storage. The model is solved using a finite-difference method, employing a variable grid to follow the movement of the sublimation front.

One of the most significant contributions of this model is the estimation of the depth of the dehydrated layer (and the associated weight losses) along the freezing and/or storage processes.

From the numerical results, empirical prediction equations for freezing times, dehydrated layer thickness, and weight losses were developed, accounting for the influence of the different process conditions (Campañone et al. 2005a,b): air temperature, air rate (affecting both the heat transfer coefficient and mass transfer coefficient), relative humidity (which determines the driving force for evaporation and sublimation), and food thickness. As an example, the following equation for prediction of weight loss during freezing is presented, with f, g, h, i, j, k, l, and m being empirical constants that depend on food shape:

$$WL(\%) = fL^g \, \mathrm{Re}^h (1 + iT_o)(1/Bi + j)(-1 - T_a)^k (1 - RH)^l Y_o^m. \tag{2.15}$$

NOMENCLATURE

Bi	Biot number, hL/k_o
Bi'	Biot number, hD/k_f
Cp	Specific heat, J/(kg °C)
D	Diameter, m
E	Equivalent heat transfer dimensionality
Fo	Fourier number, $\alpha t/L^2$
h	Heat transfer coefficient, W/(m² °C)
H	Enthalpy, J/kg
k	Thermal conductivity, W/(m °C)
L	Half-thickness or radius, m
m	Molality (solute moles/solvent kg)
Pk	Plank number, $\rho_o Cp_o (T_o - T_{if})/\Delta H$
Pr	Prandtl number, $\mu Cp/k$
R	Universal gas constant, 8.31 J/(mol K)
Re	Reynolds number, $\rho v \, D/\mu$
RH	Relative humidity, %
Ste	Stefan number, $\rho_f Cp_f (T_{if} - T_a)/\Delta H$
t	Time, s (or min)
T	Temperature, °C
T_K	Temperature, K
T_w	Freezing temperature of pure water (K)
w	Ice content, (kg of ice)/(kg of food)
WL	Weight loss (kg of water)/(kg of food)
x	Weight fraction
y	Mole fraction
Y_o	Initial water content, (kg of water)/(kg of food)

GREEK SYMBOLS

α	Thermal diffusivity, m²/s
ρ	Density, kg/m³
ε	Porosity
λ	Latent heat of ice solidification, J/kg

SUBSCRIPTS

a   Air
c   Thermal center
f   Freezing
if   Initial freezing
o   Initial
s   Surface
t   Thawing
w   Liquid water

ACKNOWLEDGMENTS

This study was supported by Consejo Nacional de Investigaciones Científicas y Técnicas, Agencia Nacional de Promoción Científica y Tecnológica, and Universidad Nacional de La Plata, Argentina.

REFERENCES

ASHRAE. 2006. *Refrigeration Handbook*, Chapter R09 Thermal properties of foods. Atlanta, GA, USA: American Society of Heating, Refrigerating and Air-Conditioning Engineers.

Basheer, I.A. and Hajmeer, M. 2000. Artificial neural networks: Fundamentals, computing, design, and application. *Journal of Microbiology Method* 43:3–31.

Bazán, H.C. and Mascheroni, R.H. 1984. Heat transfer with simultaneous change of phase in freezing boned mutton. *Latin American Journal of Heat Mass Transfer* 8:55–76.

Becker, B.R. and Fricke, B.A. 1999a. Evaluation of semi-analytical/empirical freezing time estimation methods. Part I: Regularly shaped food items. *International Journal of Heating, Ventilating, Air-Conditioning and Refrigeration Research* 5:149–167.

Becker, B.R. and Fricke, B.A. 1999b. Evaluation of semi-analytical/empirical freezing time estimation methods. Part II: Irregularly shaped food items. *International Journal of Heating, Ventilating, Air-Conditioning and Refrigeration Research* 55: 169–185.

Bonacina, C., Comini, G., Fasano, A. and Primicerio, M. 1973. Numerical solution of phase-change problems. *International Journal of Heat and Mass Transfer* 16:1825–1832.

Campañone, L.A., Salvadori, V.O. and Mascheroni, R.H. 2001. Weight loss during freezing and storage of unpackaged foods. *Journal of Food Engineering* 47:69–79.

Campañone, L.A., Salvadori, V.O. and Mascheroni, R.H. 2002. Monitoring of weight losses in meat products during freezing and frozen storage. *Food Science Technology International* 8:229–239.

Campañone, L.A., Salvadori, V.O. and Mascheroni, R.H. 2005a. Food freezing with simultaneous surface dehydration. Approximate prediction of weight loss during freezing and storage. *International Journal of Heat and Mass Transfer* 48:1195–1204.

Campañone, L.A., Salvadori, V.O. and Mascheroni, R.H. 2005b. Food freezing with simultaneous surface dehydration. Approximate prediction of freezing time. *International Journal of Heat and Mass Transfer* 48:1205–1213.

Chang, H.D. and Tao, L.C. 1981. Correlations of enthalpies of food systems. *Journal of Food Science* 46:1493–1497.

Choi, Y. and Okos, M.R. 1986. Effects of temperature and composition on the thermal properties of foods. In *Food Engineering and Process Applications,* M. Le Maguer and P. Jelen (eds.), Vol. I, Chapter 9, pp. 93–102. London: Elsevier Applied Science.

Chung, S.L. and Merritt, J.H. 1991. Freezing time modelling of small finite cylindrical shaped foodstuff. *Journal of Food Science* 56:1072–1075.

Cleland, A.C. 1990. *Food Refrigeration Processes. Analysis, Design and Simulation*, pp. 95–136. London: Elsevier.

Cleland, A.C., Earle, R.L. and Cleland, D.J. 1982. The effect of freezing rate on the accuracy of numerical freezing calculations. *International Journal of Refrigeration* 5: 294–301.

Cleland, A.C. and Earle, R.L. 1977. A comparison of analytical and numerical methods of predicting the freezing times of foods. *Journal of Food Science* 42:1390–1395.

Cleland, A.C. and Earle, R.L. 1979a. A comparison of methods for predicting the freezing times of cylindrical and spherical foodstuffs. *Journal of Food Science* 44:958–963.

Cleland, A.C. and Earle, R.L. 1979b. Prediction of freezing times for foods in rectangular packages. *Journal of Food Science* 44:964–970.

Cleland, A.C. and Earle, R.L. 1984. Freezing time prediction for different final product temperature. *Journal of Food Science* 49:1230–1232.

Cleland, D.J., Cleland, A.C., Earle, R.L. and Byrne, S.J. 1986. Prediction of thawing times for foods of simple shape. *International Journal of Refrigeration* 9:220–228.

Cleland, D.J., Cleland, A.C. and Earle, R.L. 1987a. Prediction of freezing and thawing times for multidimensional shapes by simple methods: Part I—Regular shapes. *International Journal of Refrigeration* 10:156–164.

Cleland, D.J., Cleland, A.C. and Earle, R.L. 1987b. Prediction of freezing and thawing times for multidimensional shapes by simple methods: Part II—Irregular shapes. *International Journal of Refrigeration* 10:234–240.

Cleland, D.J., Cleland, A.C., Earle, R.L. and Byrne, S.J. 1987c. Experimental data for freezing and thawing of multi-dimensional objects. *International Journal of Refrigeration* 10:22–31.

Creed, P.G. and James, S.J. 1983. The freezing times of liver in vertical plate freezers. *Proceedings of the 15th International Congress of Refrigeration, IV*, pp. 145–151. Paris, France: International Institute of Refrigeration.

Crepey, J.R. and Becel, P. 1978. Etude experimentale de certaines methodes de decongelation apliquées au thon et a la sardine. *IIR Comissions C2, D1 and D2*, pp. 201–210, Budapest, Hungary.

De Michelis, A. and Calvelo, A. 1983. Freezing time prediction for bricks and cylindrical-shaped foods. *Journal of Food Science* 48:909–913.

Delgado, A.E. and Sun, D.-W. 2001. Heat and mass transfer models for predicting freezing processes review. *Journal of Food Engineering* 47:157–174.

Devres, Y.O. 1991. An analytical model for thawing of frozen foodstuffs. Research Memorandum 130. South Bank Polytechnic, United Kingdom.

Flechtenmacher, W. 1983. Measurement of heat transfer in thawing of fish fillet block. Part 2: Thawing of mackerel fillets in running water. *International Zeits Lebensmitteltechnologie & Verfahren* 34:362–368.

Flores, E.S. and Mascheroni, R.H. 1983. Theoretical and experimental study of thawing of frozen food blocks by aspersion with water. *Latin American Journal of Heat Mass Transfer* 7:263–279.

Flores, E.S., Bazán, H., De Michelis, A. and Mascheroni, R.H. 1993. Thawing time of blocks of boneless or minced meats. Measure and prediction for different types of equipments. *Latin American Applied Research* 23:79–87.

Fricke, B.A. and Becker, B.R. 2001. Evaluation of thermophysical property models for foods. *International Journal of Heating, Ventilating, Air-Conditioning and Refrigeration Research* 7:311–330.

Goñi, S.M., Oddone, S., Segura, J.A., Mascheroni, R.H. and Salvadori, V.O. 2008. Prediction of foods freezing and thawing times: Artificial neural networks and genetic algorithm approach. *Journal of Food Engineering* 84:164–178.

Hense, H. and Kieckbusch, T.G. 1991. Congelamento de cacao: I-Resultados experimentais. *Proceedings of the IV Latin American Congress of Heat and Mass Transfer*, pp. 112.

Hossain, Md. M., Cleland, D.J. and Cleland, A.C. 1992a. Prediction of freezing and thawing times for foods of regular multi-dimensional shape by using an analytically derived geometric factor. *International Journal of Refrigeration* 15:227–234.

Hossain, Md. M., Cleland, D.J. and Cleland, A.C. 1992b. Prediction of freezing and thawing times for foods of two-dimensional irregular shape by using a semi-analytical geometric factor. *International Journal of Refrigeration* 15:235–240.

Hossain, Md. M., Cleland, D.J. and Cleland, A.C. 1992c. Prediction of freezing and thawing times for foods of three-dimensional irregular shape by using a semi-analytical geometric factor. *International Journal of Refrigeration* 15:241–246.

Hung, Y.C. 1990. Prediction of cooling and freezing times. *Food Technology* 44:137–153.

Hung, Y.C. and Thompson, D.R. 1983. Freezing time prediction for slab shape foodstuffs by an improved analytical method. *Journal of Food Science* 48:555–560.

Ilicali, C.A. 1989a. A simplified analytical model for freezing time calculation in brick-shaped foods. *Journal of Food Process Engineering* 11:177–191.

Ilicali, C.A. 1989b. A simplified analytical model for thawing time calculation in foods. *Journal of Food Science* 54:1031–1036, 1039.

Ilicali, C.A. and Saglam, N. 1987. A simplified analytical model for freezing time calculation in foods. *Journal of Food Process Engineering* 9:299–314.

International Institute of Refrigeration. 1972. *Recommendations for the Processing and Handling of Frozen Foods,* 2nd ed., pp. 16. Paris: IIR.

James, S.J. and Bailey, C. 1980. Air and vacuum thawing of unwrapped boneless meat blocks. *Refrigeration and Air Conditioning* 83:75–79.

James, S.J., Creed, P.G. and Roberts, T.A. 1977. Air thawing of beef quarters. *Journal of the Science of Food and Agriculture* 28:1109–1119.

Lambrinos, G.P. and Aguirre Puente, J. 1983. Deshydratation des milieux dispersés congelés. Influence des conditions d'entreposage sur les pertes de masse. *Proceedings of the XVI International Congress of Refrigeration* 2:567–573.

Mascheroni, R.H. 1977. Transferencia de calor con simultáneo cambio de fase en productos cárneos. Tesis doctoral, Universidad Nacional de La Plata.

Mascheroni, R.H. and Calvelo, A. 1982. A simplified model for freezing time calculations in foods. *Journal of Food Science* 47:1201–1207.

Méndez Bustabad, O. (1999). Weight loss during freezing storage of frozen meat. *Journal of Food Engineering* 41:1–11.

Michniewicz, M. 1995. Freezing of packed semi-liquid foodstuffs. *Proceedings of the 19th International Congress of Refrigeration, IIIa,* pp. 668–675, The Hague, Netherlands.

Miles, C.A., van Beek, G. and Veerkamp, C.H. 1990. Calculation of thermophysical properties of foods. In *Physical Properties of Foods,* pp. 269–312, London: Applied Science.

Mittal, G.S. and Zhang, J. 2000. Prediction of freezing time for food products using a neural network. *Food Research International* 33:557–562.

Pham, Q.T. 1986. Simplified equation for predicting the freezing time of foodstuffs. *Journal of Food Technology* 21:209–219.

Pham, Q.T. 2006. Modelling heat and mass transfer in frozen foods: a review. *International Journal of Refrigeration* 29:876–888.

Pham, Q.T. and Willix, J. 1984. A model for food desiccation in frozen storage. *Journal of Food Science* 49:1275–1281.

Pham, Q.T. and Willix, J. 1990. Effect of Biot number and freezing rate on accuracy of some food freezing time prediction methods. *Journal of Food Science* 55:1429–1434.

Pierce, J.J. 2002. EU frozen food consumption inches up, but mature markets face challenges. *Quick Frozen Foods International* 44:140–147.

Plank, R. 1963. El empleo del frío en la industria de la alimentación. Editorial Reverté.

Ramesh, M.N., Kumar, M.A. and Rao, P.N.S. 1996. Application of artificial neural networks to investigate the drying of cooked rice. *Journal of Food Process Engineering* 19:321–329.

Riedel, L. 1957. Calorimetric investigation of the meat freezing process. *Kaltetechnik* 9:38–40.

Salvadori, V.O. and Mascheroni, R.H. 1991. Prediction of freezing and thawing times by means of a simplified analytical method. *Journal of Food Engineering* 13:67–78.

Salvadori, V.O. and Mascheroni, R.H. 1997. Tiempos: Un software de cálculo de tiempos de refrigeración y congelación de alimentos. In *Herramientas de Cálculo para la Ingeniería de Alimentos* Vol. III, pp. 25–31. Valencia, Spain: Universidad Politécnica de Valencia.

Salvadori, V.O. and Mascheroni, R.H. 2002. Analysis of impingement freezers performance. *Journal of Food Engineering* 54:133–140.

Salvadori, V.O., Mascheroni, R.H. and De Michelis, A. 1996. Freezing of strawberry pulp in large containers: experimental determination and prediction of freezing. *International Journal of Refrigeration* 19:87–94.

Sanz, P.D., Domínguez, M. and Mascheroni, R.H. 1987. Thermophysical properties of meat products. General bibliography and experimental data. *Transactions of the ASAE* 30:283–289, 296.

Sanz, P.D., Dominguez, M. and Mascheroni, R.H. 1989. Equations for the prediction of thermophysical properties of meat products. *Latin American Applied Research* 19:155–163.

Sukhwal, R.N. and Aguirre Puente, J. 1983. Sublimation des milieux dispersés. Considerations theoriques et experimentation. *Revue General Thermique* 262:663–673.

Tocci, A.M. and Mascheroni, R.H. 1995. Freezing times of meat balls in belt freezers: Experimental determination and prediction by different methods. *International Journal of Refrigeration* 17:445–452.

Urbicain, M.J. and Lozano, J.E. 1997. Thermal and rheological properties of foods. In *Handbook of Food Engineering Practice*, K.J. Valentas, E. Rotstein, and R.P. Singh (eds.), pp. 425–486. New York: CRC Press.

Vanichseni, S. 1971. Thawing of frozen lamb shoulders. *MIRINZ Report No 233*, Hamilton, New Zealand.

Wang, D.Q. and Kolbe, E. 1987. Measurement and prediction of freezing times of vacuum canned Pacific shrimp. *International Journal of Refrigeration* 10:18–21.

3 Freeze Drying: Basic Concepts and General Calculation Procedures

Davide Fissore and Salvatore Velardi

CONTENTS

3.1 INTRODUCTION

Water is an important component of all biological systems: in fresh foods, the percentage of water can vary from 60% to more than 90%. During processing, storage, and distribution of foodstuffs, water can be responsible for changes in the physical, chemical, and/or biological characteristics: moisture content has thus to be reduced in order to avoid deterioration of aroma compounds or degradation of nutritional substances. Air drying is one of the oldest methods used to preserve foods: the product is exposed to a continuously flowing stream of hot air causing the evaporation of the moisture. The dehydrated products can have an extended life of a year, but the quality can be drastically lowered. Other processes can be used in order to improve the quality of the dried product, for example, vacuum freeze drying (FD). In FD, water is removed from a frozen product by sublimation: the driving force is given by the vapor pressure difference between the ice front and the surrounding environment.

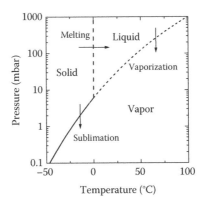

FIGURE 3.1 Phase diagram of water.

Figure 3.1 shows the equilibrium water vapor pressure over liquid water or ice as a function of the temperature: at 0°C and 6.1 mbar water vapor, water and ice coexist in equilibrium (triple point), while at temperatures below 0°C and vapor pressures below 6.1 mbar, liquid is never present and sublimation of ice to water vapor (and from vapor to ice) is the only phase change that can occur.

The low working temperature makes FD particularly suitable for those substances that could be damaged by the high temperature of the traditional drying treatments (Ratti 2001); moreover, FD maintains the original product characteristics such as shape, appearance, taste, color, flavor, texture, and biological activity. This is shown qualitatively in Figure 3.2, where FD and conventional drying of a wet product are compared: during FD, the ice crystals change state from solid to gas, without melting, and thus, vapor is transported within the product, without causing displacement of soluble substances such as sugars and salts. On the contrary, in conventional drying, the water evaporates from the product surface, the liquid water inside the product is transported to the surface by capillary action, and soluble substances are relocated to the product surface. FD of food is also applied for products where weight is an important factor (e.g., in space missions and in backpacking foods), as the freeze-dried products have a very low specific weight and can thus be transported and stored easily and economically. Finally, as the ice sublimates, it leaves voids in the dried residual material, making it easy to be rehydrated. For this reason, it is said to be "lyophilic," from two Greek words meaning "solvent-loving": the freeze-dried product is said to be lyophilized, and the FD process is also named lyophilization. This characteristic has also been exploited in the preparation of some dishes: when a freeze-dried product is eaten, the instantaneous rehydration results in a "flavor explosion" in the mouth.

D'Arsonval and Bordas (1906) demonstrated for the first time that it was possible to dry a frozen product using moderate vacuum. The technique was already well known to the Incas, who used to store meat that had been firstly frozen and then dried under the sun in the rarefied atmosphere of the Andes. Only in the twentieth century was the vacuum process applied to the drying of biological materials, showing that this process was particularly appropriate for the preparation and

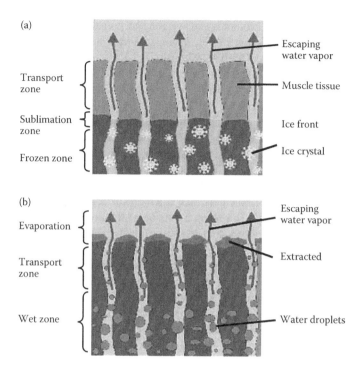

FIGURE 3.2 Comparison between the freeze-drying (a) and the conventional drying (b) of a wet structure. (Courtesy of GEA-Niro A/S, Søborg, Denmark.)

the preservation of sensitive materials. FD was extensively used in the years before World War II for human serum and plasma for injection (Flosdorf 1954). By the mid-1950s, many industries were already using FD to preserve pharmaceutical and biological products, and research and development in FD of foods began to be promoted for military purposes: freeze-dried foods were used in the emergency kit for military personnel. However, the application of FD to the food industry remained extremely limited because the process was considered an expensive operation. Today, FD is a prosperous field, and many freeze-dried foodstuffs are manufactured. Mellor (1978) reported on FD of peas, beans, beef, and spinach; Hammami and Rene (1997) studied the production of high-quality freeze-dried strawberry pieces; Krokida et al. (1998) reported the effect of FD conditions on agricultural products, and structural properties such as particle density, bulk density, and porosity of apple, banana, carrot, and potato were investigated after FD. Oetjen and Haseley (2004) have recently reviewed the application of FD to vegetables, fruits, juices, coffee, eggs, and rice.

3.2 FREEZE DRYING CYCLE

The material undergoing FD has to be processed immediately after its production in order to preserve its qualities, namely, the organoleptical, biological, nutritional, chemical, and physical characteristics. It should also be distributed in homogeneous batches of the same shape, volume, and weight and present a large surface and a small thickness in order to allow an easy drying: the treatments are a function of the characteristics of the product (Mellor 1978).

Meat and *fish* can be described as semisolid foods, comprising molecular tissue, characterized by labile proteins interdispersed with adipose tissue and about 1% of glycogen, and containing about 75% of water. Connective tissue separates long bundles of cylindrical fibers made up of thin threads (fibrils) and a complex matrix of liquid channels (sarcolemma); thus, the orientation of the muscle is important with respect to heat and vapor transport. Moreover, the heterogeneous nature of meat does not permit drying the bones, the adipose structure, and the muscle all together. Because adipose tissue contains only 10%–50% of water compared with 70%–75% of lean meat, some adipose tissue may melt if a high surface temperature is reached in the early stages of FD, thus blocking a portion of the porous dry meat by liquid fat, which inhibits vapor transfer and leads to thawing. Meat should therefore be trimmed of excess fat before FD. Lean meat is prepared for FD by being molded into blocks, frozen, and cut into thin slices 10–15 mm thick: the cut surface should be at right angles to the direction of the muscle fibers so that the ice interface moves remaining parallel to the direction of the heat flow.

In *fruits* and *vegetables*, the typical cell consists of a droplet of solution surrounded by an osmotic membrane and enveloped by a cellulose wall, and with free water on the outside; between adjacent cells, there is a pectic compound, the middle lamella, which cements the cells together. The effect of the structure in the plant material on the shape of the ice interface is not as much important as in the case of meat and fish. Apples and pears are peeled and cored, and stone fruits are split and the stone removed, and then they are sliced or diced and dipped in 0.1% sodium sulfite solution to prevent enzymatic browning prior to freezing. Strawberries, raspberries, and currants are washed, and in some cases, the skins are slit to improve the drying rate. The main preparative operations for vegetables are washing, peeling, trimming, cutting, final washing of cut material, and blanching.

Liquid foods have a nonspecific form and contain much water, removal of which leaves rehydratable powders. Such foods may be divided into two main categories: natural products (e.g., milk and egg white and yolk) and extracts (e.g., fruit juice, tea, and coffee infusions). The composition and concentration of liquid foods can vary widely. Multieffect evaporation (or freeze concentration, in case of heat-sensitive products) is required to concentrate this kind of food prior to freezing and drying.

The FD process is performed through three successive steps:

– *Freezing*: the product is frozen at low temperature.
– *Primary drying*: the ice is sublimated under low pressure; heat is supplied as the sublimation process is endothermic.
– *Secondary drying*: residual moisture (due to the water that did not previously crystallize and that is strongly bound by adsorption phenomena to the partially dried cake) is progressively brought down to a low level, thus allowing for long-term preservation at room temperature; this step requires vacuum and a moderate temperature ($+20°C$ to $+60°C$).

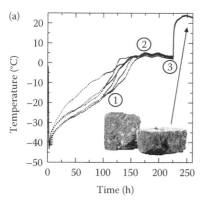
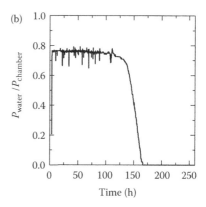

FIGURE 3.3 Freeze-drying of spinach samples (in a Lyobeta25 freeze-dryer by Telstar, Terrassa, Spain): cycle run using 32 samples having an almost cubic shape, with a side of 44 mm and a total weight of 1.935 kg (total pressure = 10 Pa). Time evolution of the product temperature (a) and of the water composition in the chamber given as the ratio between the partial pressure, and of the water vapor percentage in the chamber (b). The thermocouples are placed in various samples at 22 mm from the bottom.

Figure 3.3 shows an example of the time evolution of the product temperature and of the water vapor percentage in the chamber during an FD cycle of spinach samples. The blanched product, prepared in samples of almost cubic shape, was frozen before FD and dried in a small-scale plant at constant temperature and pressure. The operation was monitored using some thermocouples placed in the center of various samples, thus evidencing the great heterogeneity of the batch. The measured temperature increases more steeply after the passage of the dry/frozen interface. It must be evidenced that in the case shown, the primary drying ended after about 150 h, as indicated by the change of the composition in the chamber. When primary drying is finished, the shelf temperature is increased to carry out the secondary drying; however, also during primary drying, a part of the water is carried out by desorption from the dried product. Beside product temperature, the chamber pressure is also generally monitored using a capacitance manometer (Baratron). Moreover, it can be possible to evaluate the partial pressure of water by elaborating the different signals obtained from the capacitance manometer and from a thermal conductivity gauge (Pirani); other methods for measuring the water vapor concentration are discussed in Chapter 18.

The FD process is followed by appropriate operations aiming to ensure the final quality of the product. The porous structure of the lyophilized product is generally very hygroscopic; thus, if the vacuum is broken with ambient air, the dried material immediately sucks water vapor and oxygen, hence impairing the long-term storage. Vacuum has to be broken using a dry inert gas (e.g., carbon dioxide, nitrogen, argon) if direct sealing of the material under vacuum cannot be achieved. Once stocked, the product must be rehydrated before usage: due to the highly porous structure, rehydration is very fast (foodstuffs may require a few minutes to be reconstituted) and usually complete.

3.3 FREEZING

Every product to be frozen is a complex mixture of water and other substances. When cooled down to low temperatures, it undergoes various transformations and hardens due to the crystallization of water into ice. This step is of paramount importance in determining the evolution of the rest of the FD cycle. It must be emphasized that, whatever the conditions of freezing, there always remains a certain amount of the original water that never freezes: this is called "bound" water, as it is fixed on the internal structure of the material by adsorption phenomena. The freezing step involves the "free water," that is, the water that is susceptible to dissolve suspended inorganic and organic compounds and to undergo phase transition.

In case of liquid products such as fruit juices, coffee, or skim milk, during the freezing step most of the water separates into ice crystals throughout a matrix of glassy and/or crystalline solute. In this step, two main problems come into play: first, the temperature at which nucleation of ice occurs is stochastic and dependent on a high number of parameters, thus inducing heterogeneity in the batch and affecting the primary drying stage (Searles et al. 2001). Second, the concentrations of all dissolved materials increase dramatically as water freezes into ice, potentially imparting massive pH changes that can degrade the product. The precipitation of the solute as an amorphous mass, or as crystals, or as a combination of the two physic states can be observed.

In the case of more elaborate systems, like meat and vegetables, highly complex and intricate mixtures of crystals, hydrates, and glass can be obtained; in this case, the ice crystals can damage the tissues.

The state transition between the liquid phase and the solid phase occurring during the freezing step can be explained as a consequence of two different phenomena: the nucleation and the crystal growth. There are two mechanisms of nucleation: in homogeneous nucleation, the ice seeds are constituted by aggregates of molecules growing with casual direction of expansion; in heterogeneous nucleation, the formation of the ice crystals can occur around particles in suspension in the solution. Nucleation can also be promoted by cold surfaces and thermocouples when they are present. When the nuclei have been formed and are stable, they begin to grow: other molecules can join them, thus forming the crystal.

The most influencing parameter on the ice crystal morphology is the freezing rate. If a cooling surface is used to freeze the product, the freezing rate depends on the heat flux exchanged between the product and the cooling surface, and thus on the difference of temperature between the product and the surface and on the resistances that are opposed to the heat transfer. If the freezing rate is high, the nucleation rate is higher than the growth rate, thus resulting in a high number of crystals of small size. In this case, sublimation is difficult and the primary drying is not favored, while the secondary drying is faster because the dried product has got a high specific surface. If the freezing rate is low, the phenomenon of growth is more important than nucleation, and a lower number of crystals of high dimension are obtained: this structure favors primary drying, but it can damage the cell structure; secondary drying is slower and more difficult. In general, freezing as rapidly as possible is recommended, for example, 0.5°C/min to 1°C/min (Oetjen and Haseley 2004).

For pure water, the heat that has to be withdrawn for freezing (Q_{fr}) can be calculated using Equation 3.1 if the initial and final temperatures are known:

$$Q_{fr} = c_{p,w}\left(T^0 - T_{fr}\right) + Q_e + C_{p,ice}\left(T_{fr} - T_f\right). \tag{3.1}$$

In other cases, the solid content has to be taken into account:

$$Q_{fr} = \left(c_{p,w}x_w + c_{p,s}x_s\right)\left(T^0 - T_{fr}\right) + x_w Q_e + \left(c_{p,ice}x_w + c_{p,s}x_s\right)\left(T_{fr} - T_f\right). \tag{3.2}$$

x_w is the part of ice that freezes until temperature T_f is reached; $c_{p,s}$ is the specific heat of the solid ($\cong$ 1470 J/kg K for meat, $\cong$ 1340 J/kg K for vegetables). If not all water is frozen at that temperature, an additional term has to be introduced, taking into account the cooling of the unfrozen water. The unfrozen water content can vary between 6%–7% of total water, as in eggs and haddocks, and 11%–13%, as in lean beef, yolk, and yeast (Riedel 1972), but higher values can also be found in some products (e.g., 27% for glycerine, 29.1% for glucose, 35.9% for sucrose, 40.8% for lactose, and 49% for fructose).

The product to be frozen can be placed over trays (or in vials, in case of some liquid products) placed on a cooled surface or in a flow of cold air; liquid products can be frozen more quickly on a rotating cylinder. The transport of energy from the freezing zone of the product to the cooling medium can be described in a simplified way, assuming that the product is an infinite plate, cooled from one side only, and that the energy flows only perpendicularly to its infinite expansion. According to these assumptions, the freezing time is given by (Oetjen and Haseley 2004):

$$t_{fr} = \frac{\Delta J \rho_g}{\Delta T}\left(\frac{d^2}{2\lambda_g} + \frac{d}{K_{su}}\right). \tag{3.3}$$

Equation 3.3 shows that the time required to freeze the product is the sum of two terms: the first is a function of thermal conductivity, and the second is a function of heat transfer. This equation can be used to study the influence of various operating variables, namely, the thickness of the product and the temperature difference between the freezing point and the cooling medium, on the freezing time. As an example, Oetjen and Haseley (2004) took into consideration the freezing of a slice of lean beef at −20°C using a cooling medium at −43°C (K_{su} = 458 J/m² s K, ρ_g = 1.38 J/m s K): if the thickness is 2 cm, the time required is 12 min, whereas if the thickness is reduced to 0.2 cm, the time required is 28 s. When cold air is used to freeze the product, K_{su} is strongly variable with the gas velocity, the surface conditions of the product, and the geometry of the installation: about 20 J/m² s K can be obtained, but half of this value (or less) is not unusual. For the slice of lean beef previously considered, this means that the freezing time can be more than 90 min when cold air is used.

Further details about freezing of foodstuffs can be found in Chapter 2.

3.4   EVAPORATIVE FREEZING

Evaporative freezing is an alternative method for rapid-freezing solid products (Mellor 1978; Elia and Barresi 1998). This technique can be damaging for the product appearance and structure; positive exceptions are given by materials with a high moisture content at the surface: in fact, some ebullition may occur during the pressure reduction period, and large cracks may form. The main applications of FD coupled with evaporative freezing are provided by the food industry. High freezing velocity and low maintenance and equipment costs are the major advantages of this process; moreover, a significant fraction of the water content undergoes evaporation by the end of the freezing stage: a value of one seventh of the total water is reported by Rey (1975), although even higher values can result, depending on the material nature.

As an example, Figure 3.4 compares FD cycles with evaporative freezing and with conventional freezing. A capacitive balance (Rovero et al. 2001) located in the freezing chamber is used to monitor in-line the weight loss of a batch of apple sticks. As expected, the freezing rate is higher in case of evaporative freezing than in case of standard freezing carried out using cooling plates (Figure 3.4a) since the heat is transferred during the preliminary freezing step by convection, radiation, and conduction; these transfer phenomena are somehow limited by the physical conditions of the process, whereas evaporative freezing depends on the intrinsic latent heat demand promoted by the fast vacuum evaporation. Also, the lowest temperatures achieved with the two techniques are different as a consequence of the different heat transfer mechanisms involved in the freezing step. As far as the weight loss is concerned (Figure 3.4b), the evaporative freezing is characterized by an initial high loss rate followed by a slowly decaying rate. With conventional freezing, there was no water removal until the vacuum pump was started, and then, after a short transient, the release of water continued at a fairly constant rate for about 15 h. The shape of the two curves indicates a different number of consecutive steps: a first one of rapid weight loss corresponding to the initial freezing and vaporization of water, an intermediate one, and finally a progressively slowing down period of water sublimation, as the process approaches the end. On the

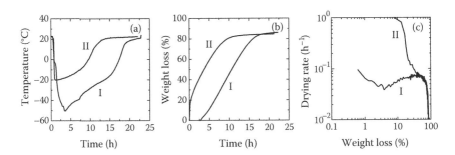

FIGURE 3.4   Comparison between freeze-drying cycles for apple sticks (10 × 10 × 70 mm) using conventional (curve I) and evaporative (curve II) freezing: product temperatures (a), weight loss (b), and drying rate (c). Chamber pressure: 6 Pa. (Data from Ghio, S., Barresi, A. A., and Rovero, G. 2000. *Transactions IchemE Part C, Food Bioproducts Processing* 78: 187–192.)

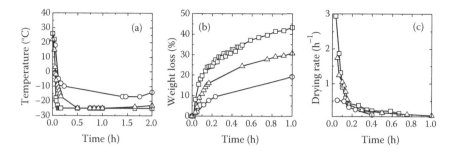

FIGURE 3.5 Comparison of the first part of freeze-drying cycles for some fruits and vegetables (o: plums, Δ: prickly pears, □: cépes, *Boletus edulis*) using evaporative freezing: product temperature (a), weight loss (b), and drying rate (c). Chamber pressure: 6 Pa; plumps and prickly pears were divided into two pieces and not peeled before drying, while small cepes were frozen in their original shape. (Data from Ghio, S., Baressi, A. A., and Rovero, G. 2000. *Transactions IchemE Part C, Food Bioproducts Processing* 78: 187–192.)

contrary, the conventional process was carried out for over 80% of the weight loss at an almost constant drying rate, followed by a final step in which the water release was progressively slowing down. If the drying rate as a function of the residual water is considered (Figure 3.4c), it is possible to see that, in case of evaporative freezing, the drying rate is higher until about 60% of the water has been removed. Despite the high freezing rate, the final size of the pores is influenced by water vapor generation in the bulk of the product. It must be evidenced that evaporative freezing is difficult to be controlled because the shape and the size of the ice crystals are affected by the rate of cooling, which depends on the velocity at which the pressure is reduced and is affected by the structure of the product.

Applications of this technique to various vegetables, such as tomatoes and mushrooms, and to various kinds of fruits, such as apricots, plums, pears, and bananas, have been discussed by Ghio et al. (2000). In Figure 3.5, temperature variation and weight loss during the evaporative freezing for some products characterized by different texture, water, and sugar content are shown. Water evaporation mechanisms controlled the freezing rate: Mushrooms, characterized by a fibrous and tubular texture, lost water in an extremely rapid way under vacuum and reached the lowest temperature in about 6 min; this low temperature is maintained for a while thanks to the high water content. Prickly pears also had a relatively fast freezing and a high drying rate at the beginning as a consequence of their grainy texture; after the first hour, the drying rate reduced because water diffusion was affected by sugar migration on the outer fruit portion. The swelling and the eruption of constituent liquors for plums built up a limiting action as far as the drying kinetics was concerned: the cell expanded foam, as a further resistance to mass transfer, prevented the diffusion of water vapor into the gas bulk.

3.5 PRIMARY DRYING

Primary drying begins when the pressure in the freeze-dryer chamber is reduced down to a value that allows the sublimation of the frozen water: a porous cake of dried material is obtained, and vapor originated at the sublimation interface flows through

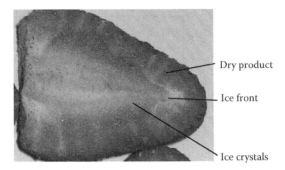

Dry product

Ice front

Ice crystals

FIGURE 3.6 Cross section of a strawberry midway through the sublimation process. (Courtesy of GEA-Niro A/S, Søborg, Denmark.)

the pores of the dried material into the chamber. Figure 3.6 illustrates one piece of a strawberry midway through the sublimation process: it is possible to see the frozen core, surrounded by a dried layer from which the frozen water has already been sublimated. Vapor is continuously removed by condensation into the cold coils/plates of a refrigerated trap, called ice condenser. At same extent during this phase, water vapor can also be originated by desorption of the (unfrozen) water in the dried layer.

The driving force for the sublimation process is provided by the partial pressure difference of water at the sublimating ice surface and at the condenser. Because vapor pressure is a function of the temperature, the driving force can also be expressed in terms of the temperatures at the sublimating surface and at the cold trap. Since it is not the total pressure in the freeze-dryer that determines the ice temperature but rather the partial pressure of water vapor, it follows that the use of a low total pressure is not strictly necessary, even if drying is usually operated under vacuum.

Sublimation is an endothermic process; thus, the temperature of the frozen product tends to decrease as the ice sublimates into the vapor phase. If no heat is supplied to the product, the vapor pressure at the sublimating interface would reach the same value of the water partial pressure in the chamber; when equilibrium is attained, no physical stimulus would exist anymore to have sublimation from the product. Thus, heat has to be continuously supplied to the product in order to balance the energy absorbed during sublimation. Heat can be transferred by several means, that is, by radiation from hot surfaces, by conduction from heated plates or gases, by gas convection, and by dielectric losses in a high-frequency field. Let us briefly discuss these methods.

An infinite plate at a temperature T_1 and with a radiant efficiency ε_1 transmits to a frozen product at a temperature T_2 and with a radiant efficiency ε_2 an amount of energy given by

$$Q = \sigma\left(T_1^4 - T_2^4\right)\left(\frac{1}{\varepsilon_1} - \frac{1}{\varepsilon_2} - 1\right)^{-1} \tag{3.4}$$

where $\sigma = 0.569 \times 10^{-7}$ J/s m^{2}K^4. The real amount of energy that is effectively transmitted by radiation is lower than the value given by Equation 3.4 as part of the energy hits the wall of the chamber. For most products, the value of ε is close to 1; for polished steel, $\varepsilon = 0.12$. Thus, at a shelf temperature of 100°C, about 40–80 J/s m² are transmitted, depending on product temperature, thus resulting in a sublimation rate of about 0.2 kg/h m² if the product temperature is −20°C.

Energy can be transferred also by conduction from the direct contact between the product container and the shelf. If the shelf and the tray are planar, at 0.2 mbar, the heat transfer coefficient is about 25 J/s m² K, whereas at 0.45 mbar, a value of 35 J/s m² K is more realistic: a temperature difference of 23°C between the sublimation interface and the shelf is thus required to sublimate 1 kg/h m².

Microwave heating provides an energy input that is essentially unaffected by the dried layer of the material that is undergoing FD and that is absorbed mainly in the frozen region; FD time can be reduced up to 60%–75%, and the product has similar quality with (or even higher quality than) that obtained in a conventional FD process using radiation and/or conduction (Sunderland 1982; Rosemberg and Bögl 1987). Nevertheless, microwave FD is not widely used in industry due to the many technical problems that can be encountered (corona discharges, melting and overheating of the frozen kernel, nonuniform heating) and to the cost that is higher than that of a conventional equipment.

Typically, primary drying is the longest stage of FD, and it has a large impact on process economics. The key to optimum drying stays in the balance between the temperature that maximizes the vapor pressure of the product and the temperature that maintains the integrity of the product. It has been reported that the quality of the foodstuffs is seriously affected when the temperature exceeds the glass transition temperature (T_g) of the food product (Peleg 1996). T_g can be defined as the temperature at which an amorphous system changes from the glassy to the rubbery state (Roos and Karel 1991; Karmas et al. 1992): in the glassy state, the molecules are frozen in place, and generally, the product is hard and rigid, whereas in the rubbery state, molecules are allowed to move, and the product is much more soft and flexible. T_g is also related to shrinkage, as a significant change in volume can be noticed when the temperature of the process is higher than the T_g of the material at that moisture content: above T_g, the viscosity of the material drops to a level that facilitates deformation (Roos and Himberg 1994; Levi and Karel 1995). T_g can be determined experimentally by following the variation of some physics, thermodynamic, or dielectric properties as a function of temperature (Slade and Levine 1991; Roos 1992). Finally, for very high temperature, the ice could melt, causing product deterioration and, in some cases, structural collapse. Thus, the temperature of the frozen material has to be controlled accurately in the course of the drying phase.

The primary drying time is directly related to the ice sublimation rate and is determined by numerous factors including chamber pressure, shelf temperature, heat transfer coefficient, volume, and product resistance. The following equation allows calculating the time required to complete the primary drying (Oetjen and Haseley 2004):

$$t_{PD} = \frac{\rho_g x_w \Delta H_s x_{w,ice} d}{\left(T_{shelf} - T_{max}\right)} \left(\frac{1}{K_v} + \frac{d}{2k_{fp}} + \frac{d}{2\Delta H_s \varphi} \right)^{-1}. \tag{3.5}$$

Equation 3.5 is based on the following simplifications:

- The layer is endless, and the energy is transmitted only from the shelf to one side of the layer.
- The vapor is transported from the ice front through the porous dried layer.
- The frozen layer is not porous.
- The heat transport in the dried layer is neglected.

Equation 3.5 can be used to study the influence of the operating parameters, namely, the product thickness and the temperature difference between the heating shelf and the sublimation interface, on the time required to complete the primary drying. In most cases, the decisive quantity influencing the time required to complete the primary drying is the heat transfer coefficient between the shelf and the sublimation front of the ice, as it has been pointed out by Tu et al. (2000) for carrot and potato slices. The heat conductivity in the product normally does not play an important part, except in the case of granulated products.

3.6   SECONDARY DRYING

When all the ice has disappeared, the first drying period is finished. If the same heat input is maintained, the dry material warms up and approaches the shelf temperature. Although all the water that had crystallized out under the form of ice has been extracted, drying is far from being completed: there remains on the internal surface of the porous dried layer a large amount of adsorbed water that would be widely sufficient to prevent any kind of safe storage at room temperature. Thus, the last stage of the FD process is a desorption period, during which the remaining water molecules are sucked out under high vacuum and moderate temperatures: this desorption period is called secondary drying.

The shelf temperature in secondary drying is much higher than that used for primary drying, so that desorption of water may occur at a practical rate. Temperature should be increased carefully because a fast ramp might cause collapse of amorphous products: the potential for collapse is very high in the first part of secondary drying because of the fairly high residual moisture content that is responsible for a low glass transition temperature. A heating rate of 0.1°C/min or 0.15°C/min for amorphous products is generally a safe and appropriate procedure. Crystalline products do not have any risk for collapse during secondary drying: a higher heating rate (0.3°C/min or 0.4°C/min) can be used (Pikal and Shah 1990).

Usually, it is better to use a high shelf temperature for a short time interval instead of a low temperature for a long period since water desorption rate decreases dramatically with time at a given temperature (Pikal et al. 1990), but the exact intermediate solution between a short or a long exposition under a high or a low temperature condition has to be determined by trial and error.

Once the product has reached the desired percentage of residual humidity, the secondary drying is stopped. The residual amount of water depends upon the type of product we are dealing with: for some foods, even a very low percentage of residual water could damage the product. The optimum secondary drying time can be determined by real-time residual moisture measurement during secondary drying: samples can be extracted from the freeze-dryer, and their moisture content can be measured using Karl Fischer titration or thermal gravimetric analysis.

3.7   COMPRESSION–LYOPHILIZATION AND LIMITED FREEZE DRYING

Freeze-dried products have the disadvantage of a very low weight/volume ratio that causes high packaging, transport, and storage costs. Studies were performed to accomplish the so-called "compression–lyophilization" for foodstuffs FD. The possibility of reducing the original volume of the product up to six times without any damage to the structure or any reduction in the rehydration ability was demonstrated. Ordinary freeze-dried foods are very brittle, but the presence of small amounts of water (5%–10% by weight) makes the cell structure plastic enough to allow compression. Generally, the freeze-dried food is first partially rehydrated (up to obtain a residual water of 12%), then compressed (with pressure ranging from 15 bar to 180 bar), and again dried (e.g., with a hot air stream) to provide storage stability. Compressed freeze-dried foods usually resume their original shape and size when they are reconstituted with hot water.

To produce compressed foods, Jones and King (1977) suggested an alternative root to rewet the product in order to obtain the desired moisture content, the so-called limited FD. In limited FD, a controlled, uniform, and predetermined amount of water is left behind in the product, usually corresponding to the requirements for compressing a freeze-dried food. Two alternative approaches to accomplish limited FD are the use of hydrating salts (Jones and King 1977) or the modification of a conventional freeze-dryer (Carn and King 1977). In the first case, hydrating salts are used as a water-uptake medium and for humidity control in the drying chamber; a circulating gas at moderate pressure flows through alternating layers. In the latter case, the relative humidity at the product surface (and thus the moisture content that will be left in the whole product) is set by means of the control of chamber pressure and plate temperature: this requires carrying out the operation under closely controlled conditions in an ordinary freeze-dryer using a higher partial pressure of water and a much lower plate temperature (e.g., are 250 Pa and 5°C).

3.8   MATHEMATICAL MODELING

The aim of this section is to propose a mathematical model to describe the dynamics of the drying process and to show how this model can be used for off-line optimization of the FD cycle. Several theoretical models describing the heat and mass transfer in a FD process have been proposed in the past (see, among the others, Pikal 1985; Millman et al. 1985; Liapis and Bruttini 1995; Lombraña et al. 1997). Multidimensional models taking into account temperature and composition profiles in radial dimension have been proposed for the simulation of FD of liquid products in vials (Sheehan and Liapis 1998), but these models are very complex, and the numerical solution is time consuming. Recently, Nam and Song (2007) have used a fixed grid method for the simulation of the FD of planar and slab-shaped food products.

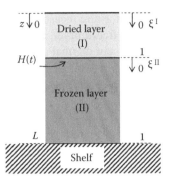

FIGURE 3.7 Schematic of the geometry of the system with the dimensional coordinate system and the dimensionless coordinate system, after the axial immobilization of the moving interface.

A simpler approach consists of assuming that radial gradients of temperature and of properties are negligible, thus lumping along the radial coordinate. Figure 3.7 shows the system under investigation: *I* refers to the dried layer, whereas *II* refers to the frozen layer; $H(t)$ is the moving interface between the dried and the frozen layer, assumed to be planar. The heat and mass balance equations describing the dynamics of the system are given in Table 3.1 with the appropriate initial and boundary conditions. This model has been proposed and validated for the drying of liquid products in vials (Velardi and Barresi 2008), but it can also be used for other products if the adequate values of the physical and transport properties are used. With respect to the equations given in Table 3.1, the following issues must be highlighted:

- The material fluxes N_w and N_{in} inside the pores of the dried layer are calculated using the dusty gas model (Mason et al. 1967; Kast and Hohenthanner 2000):

$$N_w = -\frac{M_w}{RT_I}\left(k_a \nabla p_w + k_b p_w \nabla p\right)$$

$$N_{in} = -\frac{M_{in}}{RT_I}\left(k_c \nabla p_{in} + k_b p_{in} \nabla p\right)$$

(3.6)

where

$$k_a = \left(\frac{1-y_w\left(1-\sqrt{M_w/M_{in}}\right)}{\delta_{w,in}\dfrac{\varepsilon_P}{\tau_P}} + \frac{1}{c_1\sqrt{\dfrac{RT}{M_w}}}\right)^{-1}, \quad k_b = k_a\frac{c_1\sqrt{\dfrac{RT}{M_{in}}}}{\delta_{w,in}\dfrac{\varepsilon_P}{\tau_P}} + \frac{\kappa}{\nu}p, \quad k_c = k_a\sqrt{M_w/M_{in}}.$$

(3.7)

- At time $t = 0$, no dried layer is present, and the partial pressure of water, p_w^0, and of inert gas, p_{in}^0, are those found in the freeze-dryer chamber. For

TABLE 3.1

Mass and Energy Balances for Dried and Frozen Layers of a Sample during FD

Mass Balances in the Dried Layer

$$\frac{\partial}{\partial t}\left(\frac{p_w}{T_I}\right) = -\left(\frac{R}{\varepsilon_p M_w}\right)\frac{\partial N_w}{\partial z} - \left(\frac{R}{\varepsilon_p M_w}\right)\frac{\partial \rho_{sw}}{\partial t}, \ 0 < z < H(t)$$

$$\frac{\partial}{\partial t}\left(\frac{p_{in}}{T_I}\right) = -\left(\frac{R}{\varepsilon_p M_{in}}\right)\frac{\partial N_{in}}{\partial z}, \ 0 < z < H(t)$$

$t = 0, z = 0 \qquad p_w = p_w^0, p_{in} = p_{in}^0 \qquad t > 0, z = 0 \qquad p_w = p_{w,0}, p_{in} = p_{in,0}$

$$t > 0, z = H(t) \quad p_w = p_w(T_i), \ \left.\frac{\partial p_{in}}{\partial z}\right|_{z=H(t)} = 0, \ \left.\rho_{sw}\right|_{z=H(t)} = \rho_{sw}^0$$

Energy Balance in the Dried Layer

$$\frac{\partial T_I}{\partial t} = \left(\frac{k_{I,e}}{\rho_{I,e}c_{p,I,e}}\right)\frac{\partial^2 T_I}{\partial z^2} - \left(\frac{c_{p,g}}{\rho_{I,e}c_{p,I,e}}\right)\frac{\partial}{\partial z}(T_I N_{tot}) + \left(\frac{\Delta H_v}{\rho_{I,e}c_{p,I,e}}\right)\frac{\partial \rho_{sw}}{\partial t}$$

$t = 0, z = 0 \ T_I = T^0 \qquad t > 0, z = 0 \qquad \left.-k_{I,e}\frac{\partial T_I}{\partial z}\right|_{z=0} = \sigma F\left(T_{shelf}^4 - \left.T_I^4\right|_{z=0}\right)$

$$t > 0, z = H(t) \quad \left.T_I\right|_{z=H(t)} = T_i$$

Energy Balance in the Frozen Layer

$$\frac{\partial T_{II}}{\partial t} = \left(\frac{k_{II}}{\rho_{II}c_{p,II}}\right)\frac{\partial^2 T_{II}}{\partial z^2}$$

$t = 0, 0 < z < L \ T_{II} = T^0 \qquad t > 0, z = H(t) \qquad \left.T_{II}\right|_{z=H(t)} = T_i$

$$t > 0, z = L \quad \left.k_{II}\frac{\partial T_{II}}{\partial z}\right|_{z=L} = K_v\left(T_{shelf} - \left.T_{II}\right|_{z=L}\right)$$

time $t > 0$, a moving interface is present at axial position $z = H(t)$, where the function $p_w(T_i)$ represents the thermodynamic equilibrium between frozen water and vapor; in the latter, T_i is the temperature of the moving front. While receding, the sublimation interface leaves a porous layer with an initial concentration of bound water ρ_{sw}^0.

- Water adsorption/desorption rate at the interface between the pore surface and the vapor phase can be described by the following equation:

$$\frac{\partial \rho_{sw}}{\partial t} = k_{sw,1}p_w\left(\rho_{sw}^* - \rho_{sw}\right) - k_{sw,2}\left(\rho_{sw} - \rho_{sw}^*\right) \tag{3.8}$$

where ρ_{sw}^* represents the mass concentration of adsorbed water that would be in equilibrium with the water vapor at the pore interface. The terms with $k_{sw,1}$ and $k_{sw,2}$ describe the adsorption and desorption mechanisms, respectively. Other laws have been proposed by Liapis and Bruttini (1995) to describe the desorption mechanism.

- The energy balance in the dried layer considers the porous product and the gas flowing through it as a pseudo-homogeneous system, and thus, effective properties are used taking into account the contribution from the two phases. Radiation heat transfer to the upper dried surface is taken into account in the boundary condition.
- Thermal conduction and accumulation in the frozen mass are considered in the energy balance equation for the frozen product. The overall heat transfer coefficient K_v takes into account the mechanisms of heat transfer from the heating fluid to the product (see Section 3.4).

The velocity of the sublimating moving interface, v_z, is related to the rate of sublimation and can be determined through a material balance across the interface, stating that the difference between the rate of disappearance of mass of frozen layer and the rate of formation of mass of dried layer is equal to the vapor flow rate at the interface:

$$N_w\big|_{z=H} S = -\rho_{II} v_z S + \rho_{I,e} v_z S. \tag{3.9}$$

If the interface velocity is expressed as $v_z = dH/dt$, the dynamic evolution of the sublimating interface can be written as

$$\frac{dH}{dt} = -\frac{1}{\rho_{II} - \rho_{I,e}} N_w\big|_{z=H} \tag{3.10}$$

with $z = H(t)$ and the initial condition $H = 0$ when $t = 0$.

The heat balances in the frozen product and in the porous layer are coupled at the moving front by a boundary equation that gives the enthalpy change across the interface:

$$c_{p,g} T_i N_w\big|_{z=H} S + \Delta H_s N_w\big|_{z=H} S = -Sk_{II} \frac{\partial T_{II}}{\partial z}\bigg|_{z=H} + Sk_{I,e} \frac{\partial T_I}{\partial z}\bigg|_{z=H} +$$
$$S - \rho_{II} c_{p,II} v_z T_i S + \rho_{I,e} c_{p,I,e} v_z T_i S. \tag{3.11}$$

The contributions taken into account in Equation 3.11 are the sensible heat of the water vapor leaving the interface, the latent heat associated to the change of phase, the conduction in layers I and II, and the change in the thermal capacity of the product due to the conversion of a portion of frozen mass to a dried one. Combining Equations 3.10 and 3.11, the final form of the energy balance at the moving interface is obtained:

$$-k_{II} \frac{\partial T_{II}}{\partial z}\bigg|_{z=H} + k_{I,e} \frac{\partial T_I}{\partial z}\bigg|_{z=H} - N_w\big|_{z=H} \big[\Delta c_p T_i + \Delta H_s\big] = 0 \tag{3.12}$$

with $z = H(t)$ and where

$$T_i = T_I\big|_{z=H(t)} = T_{II}\big|_{z=H(t)}$$

and

$$\Delta c_p = c_{p,g} - \frac{\rho_{II} c_{p,II} - \rho_I c_{p,I,e}}{\rho_{II} - \rho_{I,e}}.$$

The equations of the model define a moving boundary problem, whose resolution requires the use of a spatial grid evolving with time, as the position of the sublimating interface is not fixed. In order to overcome this problem, a mathematical artifice can be used, consisting of the axial immobilization of the moving interface. Following the procedure suggested by Sheehan and Liapis (1998), the immobilization can be performed introducing two nondimensional axial variables, $\xi^I, \xi^{II}, \in[0,1]$:

$$\xi^I = \frac{z}{H(t)} \text{ for } 0 < z < H(t); \; \xi^{II} = \frac{z - H(t)}{L - H(t)} \text{ for } H(t) < z < L \qquad (3.13)$$

thus obtaining a spatial grid fixed in time. In order to solve the problem, each function of (z, t), namely, the temperatures, the pressures, etc., and its derivatives are transformed accordingly to the change of coordinates defined by Equation 3.13. The numerical solution of the problem can be carried out by discretizing the spatial domains of the system, thus transforming the system of partial differential equations into a system of ordinary differential equations.

Figure 3.8 compares the measured temperatures and the model predictions for the temperature at the bottom of the sample in the experimental run of Figure 3.3: when

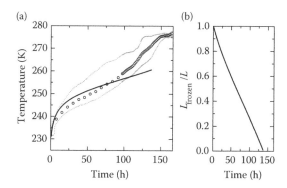

FIGURE 3.8 Time evolution of the interface temperature (a) and of the interface position (b) for the same experimental cycle of Figure 3.3. The mean value of temperature measured by the thermocouples (symbols) as well as the minimum and maximum temperatures measured (dotted lines) are also shown.

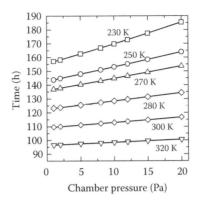

FIGURE 3.9 Model prediction of the influence of the chamber pressure and of the shelf temperature on the time required to complete the primary drying for the same batch of spinaches of the run of Figure 3.3.

primary drying approaches the end, and thus secondary drying starts occurring simultaneously, the proposed model is no more adequate to describe the dynamics of the process due to poor knowledge of the kinetic parameters of water desorption. Anyway, a fairly good estimation of the final water content and of the drying time is obtained. This allows exploiting the model to investigate the role of the chamber pressure and that of the shelf temperature on the time required to complete the primary drying, thus optimizing off-line the process. Results are given in Figure 3.9 and can be used to fix the values of the operating conditions for the system under investigation: it appears that the minimum of the time required to complete the primary drying is obtained for the lowest chamber pressure and the highest shelf temperature. Feasibility constraints on the minimum pressure that can be obtained in the chamber and on the maximum temperature allowed by the product lead to the selection of the most adequate operating conditions.

3.9 ATMOSPHERIC FREEZE DRYING

The main disadvantages of the FD process described in the previous paragraphs are the fixed and operating costs (Wolff and Gibert 1990a,b): the latter are due to the energy consumption for freezing the fresh product, for heating the frozen food at low temperature, for condensing water vapor, and for maintaining the vacuum. Actually, the presence of vacuum is not a necessary requisite for sublimation to occur: the requirement is that the partial pressure of water vapor in the drying medium is kept low enough to provide a driving force for water vapor removal from the frozen sample. On the contrary, heat transfer is hindered under vacuum, as a consequence of the reduction in the conductivity of gases at low pressure. Therefore, as an alternative to the classical FD process under vacuum conditions, several authors investigated the feasibility of FD at atmospheric pressure by using gas (air, nitrogen, or helium) as a water removal and heating medium at (or near) atmospheric pressure: this allows saving the vacuum apparatus, but very long process times are required as the mass transfer becomes rate controlling and is severely slowed down.

Particularly interesting is the use of molecular sieve particles or dehydrating salts to remove the water vapor, as suggested, for example, by Clark and King (1971) who tried a new process by placing the product in shallow beds in sandwich disposal through fixed layers of adsorbent, with gas allowed to pass through the staggered layers. The external energy source is eliminated, as the heat generated by a water absorption material is used; in this case, heat is transferred from the desiccant to the product (FD of meat was tested) by convection, as light gas is circulated at moderate pressure. This would also allow eliminating the condenser, but the process is too slow to be practically applicable.

Atmospheric FD, realized using a bed of adsorbent particles fluidized with cold air, has been proposed as an alternative process (Boeh Ocansey 1985; Lombraña and Villaran 1996; Wolff and Gibert 1990a) as the apparatus is simple, due to the absence of a vacuum chamber and of ancillary equipment, and the energy costs are significantly reduced. Moreover, no devices for supplying external energy to the product are required inside the chamber, as a consequence of the fact that the adsorbent material plays the double role of capturing the water vapor that forms by sublimation and providing the heat of adsorption to the sublimation front. Moreover, operation in a fluidized bed improves the heat transfer coefficients, which can be more than one order-of-magnitude higher than those between gas and solid (Donsì and Ferrari 1995; Donsì et al. 1998). Finally, since the process operates at atmospheric pressure, the internal heat transfer by conduction becomes more efficient (Boeh Ocansey 1988; Wolff and Gibert 1988, 1990b) due to the increase in gas thermal conductivity with pressure. The main drawback of atmospheric FD is the increase in the time required to complete the operation due to the decrease in the drying rate. This, in turn, is due to the decrease in water vapor diffusivity with increasing pressure in the chamber. A recent study by Di Matteo et al. (2003) evidences that the sample size is a key parameter in determining the contact time required by the process: if product size is reduced, the processing time can be reduced down to values that compare well with those required by the traditional process. The choice of a proper set of other variables also contributes to make atmospheric FD a technique suitable for wider application in the food industry.

Stawczyk et al. (2004) proposed to apply a mixed drying technique that combines surface freezing (about −10°C) and drying at atmospheric pressure, and then, after reaching the critical moisture content of the product (corresponding to the formation of a rigid product that reduces or eliminates its shrinkage), an increase in the process temperature of up to several or dozen degrees centigrade above zero until desired moisture content of the final product is reached. The application of surface freezing and drying under normal pressure with the use of a heat pump allows reducing significantly the energy consumption of the process beside preserving at the same time the advantages of the FD method.

3.10   CONCLUSIONS

The feasibility of a FD process for foodstuffs has been discussed in this chapter, pointing out the main advantages of this process. The various steps of a typical

FD cycle have been analyzed, and basic calculations have been proposed in order to point out the role of the operating variables and of the product characteristics on the results. Moreover, a detailed model for the primary drying phase has been described, and its effectiveness in optimizing the process has been shown.

Details about the equipment used for FD and about the main issues concerning monitoring and control of the process are discussed in Chapter 18.

LIST OF SYMBOLS

ΔH_s	Enthalpy of sublimation (J/kg)
ΔH_v	Enthalpy of vaporization (J/kg)
ΔJ	Enthalpy difference between the initial freezing point and the final temperature (J/kg)
ΔT	Temperature difference between the freezing point and the cooling medium (K)
c_1	Constant dependent only upon the structure of the porous medium (m)
c_p	Specific heat capacity (J/kg K)
d	Thickness of the product parallel to the direction of prevailing heat transfer (m)
F	View factor for radiative heat transfer
H	Position of the moving front (m)
k	Thermal conductivity (J/s m K)
k_a, k_b, k_c	Diffusivity coefficients defined by Equation 3.7
K_{su}	Heat transfer coefficient between the cooling medium and the freezing zone (J/m² s K)
$k_{sw,1}$	Kinetic rate constant for adsorption (m³/kg s)
$k_{sw,2}$	Kinetic rate constant for desorption (1/s)
K_v	Overall heat transfer coefficient between the shelf and the product bottom (J/m² s K)
L	Total product thickness (m)
L_{frozen}	Frozen layer thickness (m)
M	Molecular weight (kg/kmol)
N	Mass flux (kg/s m²)
p	Pressure (Pa)
Q	Heat flux (J/s)
Q_e	Melting heat of ice (J/kg)
Q_{fr}	Heat that has to be withdrawn for freezing (J/kg)
R	Ideal gas constant (J/kmol K)
S	Surface of sublimation (m²)
t	Time (s)
T	Temperature (K)
T_g	Glass transition temperature (K)
v_z	Moving front velocity (m/s)
x	Mass fraction

y Molar fraction
z Axial coordinate (m)

GREEKS

$\delta_{w,in}$ Free binary diffusion coefficient (m^2/s)
ε Radiant efficiency
ε_p Porosity of the dried layer
φ Permeability of water vapor through the dried product (kg/m s Pa)
κ Constant dependent only upon the structure of the porous medium (m^2)
ν Dynamic viscosity (kg/m s)
ξ Nondimensional axial coordinate
ρ Density (kg/m^3)
ρ_{sw} Mass concentration of bound water (kg/m^3)
ρ_{sw}^* Equilibrium bound water mass concentration (kg/m^3)
σ Boltzmann constant (J/m^2 s K^4)
τ_p Tortuosity factor

SUBSCRIPTS

0 At $z = 0$
e Effective
f Ending point
fp   Frozen product
fr   Freezing
g Gas
i Interface, moving front
I Layer I, dried layer
ice   Frozen water
II Layer II, frozen layer
in   Inert gas
max   Maximum allowable value
PD   Primary drying
s Solid
shelf   Heating shelf
tot   Total
w Water

SUPERSCRIPTS

0 Initial

ACKNOWLEDGMENTS

Valuable suggestions and contributions of Dr. Roberto Pisano and Prof. A. Barresi (Politecnico di Torino) are gratefully acknowledged.

REFERENCES

Boeh Ocansey, O. 1985. Some factors influencing the freeze drying of carrot discs in vacuo and at atmospheric pressure. *Journal of Food Engineering* 4: 229–243.

Boeh Ocansey, O. 1988. Freeze-drying in a fluidized bed atmospheric dryer and in a vacuum dryer: evaluation of external transfer coefficients. *Journal of Food Engineering* 7: 127–146.

Carn, R. M., and King, C. J. 1977. Modification of conventional freeze-dryers to accomplish limited freeze-drying, *AIChE Symposium Series* 163(73): 103–112.

Clark, J. P., and King, C. J. 1971. Convective freeze drying in mixed or layered beds. *Chemical Engineering Progress Symposium Series* 108(67): 102–111.

Dalgleish, J. McN. 1990. *Freeze-drying for the Food Industries*. London: Elsevier Applied Science.

D'Arsonval, M., and Bordas, F. 1906. De la distillation et de la dessiccation dans le vide à l'aide des basses températures. *Comptes Rendus de l'Académie des Sciences de Paris* 143: 567–570.

Di Matteo, P., Donsì, G., and Ferrari, G. 2003. The role of heat and mass transfer phenomena in atmospheric freeze-drying of foods in a fluidised bed. *Journal of Food Engineering* 59: 267–275.

Donsì, G., and Ferrari, G. 1995. Heat transfer coefficients between gas fluidised beds and immersed spheres: dependence on the sphere size. *Powder Technology* 82: 293–299.

Donsì, G., Ferrari, G., and De Vita, A. 1998. Experimental determination of heat and mass transfer coefficients in two component fluidised beds. In: *Proceedings of the 6th International Conference Multiphase Flow in Industrial Plants*, Milan, Italy, 219–226.

Elia, A. M., and Barresi, A. A. 1998. Intensification of transfer fluxes and control of product properties in freeze-drying. *Chemical Engineering and Processing* 37: 347–358.

Flosdorf, E. W. 1954. The development of freeze-drying. In *Biological Applications of Freezing and Drying*, ed. R. J. C. Harris. New York: Academic Press.

Ghio, S., Barresi, A. A., and Rovero, G. 2000. A Comparison of evaporative and conventional freezing prior to freeze-drying of fruits and vegetables. *Transactions IchemE Part C, Food Bioproducts Processing* 78: 187–192.

Hammami, C., and Rene, F. 1997. Determination of freeze-drying process for strawberries. *Journal of Food Engineering* 32: 298–304.

Jones, R. L., and King, C. J. 1977. Use of hydrating salts to accomplish limited freeze-drying. *AIChE Symposium Series* 163(73): 113–123.

Karmas, R., Buera, M. P., and Karel, M. 1992. Effect of glass transition on rates on nonenzymatic browning in food systems. *Journal of Agriculture Food Chemistry* 40: 873–879.

Kast, W., and Hohenthanner, C. R. 2000. Mass transfer within the gas phase of porous media. *International Journal of Heat and Mass Transfer* 43: 807–823.

Krokida, M. K., Karathanos, V. T., and Maroulis, Z. B. 1998. Effect of freeze-drying conditions on shrinkage and porosity of dehydrated agricultural products. *Journal of Food Engineering* 35: 369–381.

Levi, G., and Karel, M. 1995. Volumetric shrinkage (collapse) in freeze-dried carbohydrates above their glass transition temperature. *Food Research International* 28: 145–151.

Liapis, A. I., and Bruttini, R. 1995. Freeze drying. In *Handbook of Industrial Drying*, ed. A. S. Mujumdar. New York: Marcel Dekker.

Lombraña, J. I., and Villaran, M. C. 1996. Interactions of kinetic and quality aspects during freeze-drying in adsorbent medium. *Industrial Engineering Chemistry Research* 35: 1967–1975.

Lombraña, J. I., De Elvira, C., and Villaran, M. 1997. Analysis of operating strategies in the production of special foods in vials by freeze drying. *International Journal of Food Science and Technology* 32: 107–115.

Mason, E. A., Malinauskas, A. P., and Evans III, R. B. 1967. Flow and diffusion of gases in porous media. *The Journal of Chemical Physics* 46: 3199–3216.

Mellor, J. D. 1978. *Fundamentals of Freeze-Drying*. London: Academic Press.

Millman, M. J., Liapis, A. I., and Marchello, J. M. 1985. An analysis of the lyophilisation process using a sorption sublimation model and various operational policies. *AIChE Journal* 31: 1594–1604.

Nam, J. H., and Song, C. S. 2007. Numerical simulation of conjugate heat and mass transfer during multi-dimensional freeze drying of slab-shaped food products. *International Journal of Heat and Mass Transfer* 50: 4891–4900.

Oetjen, G. W., and Haseley, P. 2004. *Freeze-Drying*. Weinheim: Wiley-VHC.

Peleg, M. 1996. On modelling changes in foods and biosolids at and around their glass transition temperature range. *Critical Reviews in Food Science and Nutrition* 36: 49–67.

Pikal, M. J. 1985. Use of laboratory data in freeze-drying process design: heat and mass transfer coefficients and the computer simulation of freeze-drying. *Journal of Parenteral Science Technology* 39: 115–139.

Pikal, M. J., and Shah, S. 1990. The collapse temperature in freeze-drying: dependence on measurement methodology and rate of water removal from the glassy phase. *International Journal of Pharmaceutics* 62: 165–186.

Pikal, M. J., Shah, S., Roy, M. L., and Putman, R. 1990. The secondary drying stage of freeze-drying: drying kinetics as a function of temperature and chamber pressure. *International Journal of Pharmaceutics* 60: 203–217.

Ratti, C. 2001. Hot air and freeze-drying of high-value foods: a review. *Journal of Food Engineering* 49: 311–319.

Rey, L. 1975. Freezing and freeze-drying. *Proceedings of the Royal Society of London. Series B, Biological Sciences* 191: 9–19.

Riedel, L. 1972. Enthalpy–water content diagram for lean beef (also valued for other meats with aft content below 4%). In *Recommendations for the Processing and Handling of Frozen Foods*. Paris: International Institute of Refrigeration.

Roos, T. 1992. Phase transitions and transformations in food systems. In *Handbook of Food Engineering*, eds. D. R. Heldman and D. B. Lund. New York: Marcel Dekker.

Roos, Y., and Karel, M. 1991. Plasticizing effect of water on thermal behavior and crystallization of amorphous food model. *Journal of Food Science* 56: 38–43.

Roos, Y., and Himberg, M. 1994. Nonenzymatic browning behavior, as related to glass transition, of a food model at chilling temperatures. *Journal of Agriculture Food Chemistry* 42: 893–898.

Rosemberg, U., and Bögl, W. 1987. Microwave thawing, drying, and baking in the food industry. *Food Technology* 41: 85–91.

Rovero, G., Ghio, S. and Barresi, A. A. 2001. Development of a prototype capacitive balance for freeze-drying studies. *Chemical Engineering Science* 56: 3575–3584.

Searles, J. A., Carpenter J. F., and Randolph, T.W. 2001. The ice nucleation temperature determines the primary drying rate of lyophilization for samples frozen on a temperature-controlled shelf. *Journal of Pharmaceutical Sciences* 90: 860–871.

Sheehan, P., and Liapis, A. I. 1998. Modelling of the primary and secondary drying stages of the freeze-drying of pharmaceutical product in vials: numerical results obtained from the solution of a dynamic and spatially multi-dimensional lyophilisation model for different operational policies. *Biotechnology and Bioengineering* 60: 712–728.

Slade, L., and Levine, H. 1991. Beyond water activity: recent advances based on an alternative approach to the assessment of food quality and safety. *Critical Reviews in Food Science and Nutrition* 30: 115–360.

Stawczyk, J., Li, S., Zylla, R. 2004. Freeze drying of food products in a closet system. In *Proceedings of the 14th International Drying Symposium*, Sao Pulo (Brazil), 22–25 August, vol. B, pp. 949–953.

Sunderland, J. E. 1982. An economic study of microwave freeze-drying. *Food Technology* 36: 50–56.

Tu, W., Chen, M., Yang, Z., and Chen, H. 2000. A mathematical model for freeze-drying. *Chinese Journal of Chemical Engineering* 8: 118–122.

Velardi, S. A., and Barresi, A. A. 2008. Development of simplified models for the freeze-drying process and investigation of the optimal operating conditions. *Chemical Engineering Research and Design* 86(1): 9–22.

Wolff, E., and Gibert, H. 1988. Développments technologiques nouveaux en lyophilisation. *Journal of Food Engineering* 8: 91–108.

Wolff, E., and Gibert, H. 1990a. Atmospheric freeze-drying part 1: Design, Experimental investigation and energy saving advantages. *Drying Technology* 8: 385–404.

Wolff, E., and Gibert, H. 1990b. Atmospheric freeze-drying part 2: Modelling drying kinetics using adsorption isotherms. *Drying Technology* 8: 405–428.

Part II

Operations Used in Refrigeration Technologies

4 Sizing, Peeling, Cutting, and Sorting of Fruits and Vegetables

Enrique Moltó García and José Blasco Ivars

CONTENTS

4.1 INTRODUCTION

Statistics show that fruit and vegetable consumption increases as nations become wealthier and consumer habits change toward healthier nutritional diets. This has produced an important growth of the demand of fresh horticultural products in developed societies.

In parallel, consumers' lifestyle rejects employing large amounts of time to prepare meals, which has led to a continuous increase in the demand of "minimally processed" or "fresh-cut" fruits and vegetables that is substituting the traditional market of nonprocessed horticultural products. Nowadays, fresh-cut fruits and vegetables are the fastest growing segment of the fresh produce industry (Kim 2007), and it is easy to find in the marketplace commodities such as shredded lettuces, washed and trimmed spinach, fruit slices, and carrot or celery sticks.

Intact fruits and vegetables are safe to eat partly because their peel is an effective physical and chemical barrier to most microorganisms. In addition, the acidity of the pulp or juice of fruits ordinarily prevents the growth of their populations. On the other hand, the presence of soil microorganisms is common in vegetables. In these cases, microbes responsible for spoilage usually have a competitive advantage over other organisms that could potentially be harmful to humans. However, changes induced in the environmental conditions during processing may result in significant changes in this balance. For instance, film packaging produces high relative humidity and low oxygen conditions that increase the risk of pathogenic bacteria development, and the same occurs when the product is exposed to relatively high temperatures (Cantwell 2002).

The objective of the minimally processed fruit and vegetable industry is to maintain the fresh nature of these products. When processing them, it is important to preserve their natural characteristics like crispness, color, aroma, and taste. For this reason, production of minimally processed fruits and vegetables involves mostly physical operations like cleaning, washing, trimming, coring, slicing, or shredding (Cantwell 2008).

While most food processing techniques stabilize the products and lengthen their storage and shelf life, minimal processing generally increases the rates of metabolic reactions that cause deterioration of fresh products. Damages caused by common processing operations release enzymes that increase respiration and ethylene production (Abeles et al. 1992; Watada et al. 1996) and other biochemical reactions responsible for changes in color, flavor, texture, and nutritional quality. In order to reduce the impact of bruising and wounding, the product has to be maintained at low temperatures before and during processing and kept refrigerated during its shelf life.

Microbial growth on minimally processed products must also be controlled by good sanitation and adequate temperature management. Because moisture increases microbial activity, removal of wash and cleaning water is critical, but it also leads to drying of the product, which in some cases can deteriorate its quality.

Hygiene and accessibility are essential factors when constructing machines for industrial fruit and vegetable processing. All equipment must be designed to permit effective cleaning and sanitation and to prevent contamination. The equipment should also ensure that it is capable of delivering the requirements of the process (for instance, adequate refrigeration). It has to facilitate maintenance, cleaning, sanitizing, and inspection and should be designed to permit proper drainage when necessary (Canadian Food Inspection Agency 2008).

4.2 PRELIMINARY OPERATIONS

One of the first steps for preparing many fruits and vegetables is to wash them in order to remove the rest of the soil and dust attached to them. Soil removers are crucial in separating stones and mold before further processing, which prevents damages caused to the following equipment. They are the unavoidable, preliminary operation for potatoes and root crops.

Some fruits and vegetables contain inedible parts or have particular parts that have to be removed. In preliminary operations, bruised zones, undesired leaves, or organs and extraneous matter are separated. For instance, in the processing of artichokes, external leaves are removed, or in the case of lettuces, the heart is taken apart.

Undesired parts or materials can also be removed from the useful parts by flotation. Soil particles often sink, while lighter vegetal materials float. In this method, fruits or vegetables are put in a tank with a slight current of water that pushes away the floating material. Sometimes this separating action is facilitated by injecting air into the tank. The floating material is removed with water wheels, nets, or similar mechanisms.

Other machines that serve a similar purpose are drum washers. They normally have a big drum, which is partially submerged in water. The drum rotates, and the fruits or vegetables are rinsed thanks to the friction with water or between themselves. Brushes conveniently located can also help in this operation.

As it can be said, free moisture must be completely removed after washing in order to prevent microbial growth (Cantwell and Suslow 2002). For this reason, centrifuges are widely used as a complement to washers. In this case, the product is put in a drum or in a special crate that rotates at high speed, and the water is evacuated through a perforated side. In some especial, high added value products, vacuum can be used to suck the water from the skin of the product.

Other drying machines use vibration screens and air blasts. These may produce a slight desiccation that, in some commodities, can deteriorate the product, but in others, it is used to enlarge the product's postprocessing life.

In many processes, rough, preliminary sizing is crucial for subsequent operations. Some peeling machines, for instance, may require presized fruits to reach an adequate performance: the peel of a too small fruit may not reach the peeling mechanisms, and on the opposite side, a too large fruit may produce great losses due to an excessive removal of material by the cutting device. Size sorting is also especially important for food products that have to be heated or cooled in subsequent steps, as large differences in size might lead to an over-processing or under-processing of the product.

Intact fruit is often waxed in machines that spray waxes, and sometimes particular fungicides, while fruit is transported on rollers. In some commodities, like apples, the fruit can be brushed in order to make its natural waxes shine.

De-hulling is the removal of hulls from soybean, legumes, and the shells from cocoa beans and similar products. There are wet or dry de-hulling methods, depending on the hardness of the hull. Wet methods consist of soaking the legumes for some time, milling them, and then blowing them to remove the coat. Dry methods are suitable for tough seed coats. They use oil and emery-coated rollers to abrade the surface of the seed.

Some fruits need de-stemming, for instance, strawberries, blueberries, or grapes. Use of de-stemmers is very common in wineries. These machines essentially consist of a rotating, perforated drum in which the berries are separated from the stem by friction with each other or by shocks with palettes mounted on an axis concentric to the drum that rotate in the opposite direction.

Metal check is also required in many commodities in order to detect any metal particles in the raw material that can be harmful for consumers or detrimental for subsequent processes.

4.3 PEELING

The objective of peeling is to remove unwanted or inedible layers of fruit. Peeling losses need to be minimized by removing as little of the underlying product as possible. In many cases, peeling is still performed manually, in a very time consuming and laborious task. Some peeling machines are also still manually fed. There are different methods for peeling, for example, using steam, knife blades, abrasion, enzymatic reactions, or flames.

Steam peeling is carried out in batches or in a continuous process. In the first method, the products are put in a pressure vessel and exposed to high-pressure steam. The high temperature causes a rapid heating and cooking of their surface. When the pressure is released, it causes the detachment of the cooked skin. In the second method, the steam is fed directly into a pipe with a screw inside. The product is heated for a specified time in the pipe, and most of the peeled material is discharged with the steam.

In knife peeling, fruits or vegetables are placed onto a rotating disc and pressed against stationary or rotating blades to remove the skin. Many peelers use cups that transport the product through knives that are attached to flexible arms, which are automatically opened by the product and adjust to its size.

Machines must adapt to the geometrical characteristics of fruits and vegetables. For instance, there are machines specifically designed for cylindrical shaped produces, like carrots or cucumbers, or for spherical shaped fruits like citrus. There are also manual peelers in which the blade is bent to arc shape; they are easy and accurate to use for melons and watermelons.

Abrasion peeling consists of feeding the material onto abrasive rollers or feeding it into a rotating bowl, which is lined with an abrasive. The abrasive surface removes the skin, which is then washed away with water.

Enzymatic peeling of fruit is a technology that has been developed to produce segments of citrus fruits with a good texture and flavor for use as fresh, frozen, or canned products. In enzymatic peeling, rollers with prickers insert enzymes (pectinases and celluloses) under the skin of the fruit. After a certain reaction time, the skin loosens from the fruit flesh and can be easily removed manually or by water pressure. Moreover, enzymes may separate individual segments, as in the industry of canned mandarin segments (Pretel et al. 2004). Pectinases have also been used to remove the skins of stone fruits such as peaches, apricots, and nectarines (Toker and Bayindirli 2003).

Enzymatic peeling is widely used to prepare cut citrus sections for fresh consumption, which have shelf life stability of about 12 days (Rocha et al. 1996). Some juice leakage and softening are adverse side effects of this method, along with microbial contamination (Pao et al. 1997), and for this reason, many works have been devoted to finding an optimal combination of enzymes (Pretel et al. 1998; Pagan et al. 2005; Ismail et al. 2005). A machine has been developed by the Florida Department of Citrus that was found to be effective for enzymatic peeling (Ismail and Thomas 2002).

Flame peeling is used for onions. The peeler transports and rotates the onions through an oven at high temperatures (above 1000°C). The skin and small roots are burned and then removed by high-pressure water sprays.

4.4 CUTTING AND FORMING OPERATIONS

The objective of cutting and forming operations is to reduce the size or to provide a convenient shape to the material, either for further processing or to improve its edibility. As a consequence of these processes, fruits and vegetables are cut to a desired size or shape, and inedible or defective parts are removed. Although some horticultural products are still cut manually, in many current food industries, most of them are cut mechanically. This is performed by transporting them through rotating knives, blades, cleavers, or saws.

Slicing is aimed at obtaining regular pieces of material. Slicers have fixed, rotating, or oscillating blades, which cut the product when it passes underneath or is pressed against the blades. A variant of slicing is dicing; in this case, the product is first sliced and then cut into strips by rotating blades. These strips are then passed to a second set of rotating knives, which operate at right angles to the first set and cut the strips into cubes.

In some cases, other special forming of the product is required. Wedging machines cut blocks of tomatoes, potatoes, apples, fruits, beets, and other vegetables. Some machines form the product to the shape of balls (e.g., for Paris potatoes) or ovals.

Pulping is used for the size reduction and homogenization of fruits and vegetables and also for juice extraction. A moving rough surface breaks the product and squeezes the material through a gap, producing a homogeneous mass. The most common pulpers are drum pulpers and disc pulpers.

Hydrocutters are special machines that are widely used for potatoes for the production of French fries. In a typical hydrocutter, the potatoes are dropped into a tank filled with water, then pumped through a conduit, aligned, and accelerated to high speed before impinging upon a fixed array of cutter blades. These machines can also be used for making wedge cut potatoes, carrot sticks and wedges, onion rings, zucchini sticks, broccoli cubes, etc.

It is crucial for any cutting operation that cutting devices are very sharp in order to induce minimal damages to the vegetal tissues. Damage to cells near cut surfaces greatly influences the shelf life and the quality of a product due to the release of enzymes (Portela and Cantwell 2001), oxidants, and sugars, which facilitate microbial growth and browning (Artes et al. 1998). It is also important to notice that washing the cut product removes sugar and other nutrients from the cut surface.

After these operations, browning of fruits and vegetables may occur. Browning is detrimental to quality, particularly in postharvest storage of fresh fruits, fresh-cut fruits and vegetables, and juices. The control of browning is one of the most important issues in the food industry, as color is a significant attribute of food that influences consumer decision. This phenomenon occurs by the action of a family of enzymes called polyphenoloxidases.

Polyphenols are responsible for the color of many plants and are part of the taste and flavor of beverages, but are also substrates for the enzymes that cause browning. During processing and storage, many polyphenols are unstable, with their enzymatic oxidation causing browning of the product (Hodges and DeLong 2007). This reaction mostly occurs after any mechanical treatment that breaks cells because polyphenoloxidases appear to reside in the cytoplasm of senescing or ripening plant tissues.

In the presence of oxygen, these enzymes catalyze the first steps in the biochemical conversion of phenolic compounds to produce quinones, which undergo further polymerization to yield dark, insoluble polymers called melanins. These form barriers and have antimicrobial properties that naturally prevent the spread of infection or bruising in plant tissues. The reaction is dependent not only on the presence of air but also on the acidity. Several methods can be applied to avoid enzymatic browning, based on inactivating the polyphenoloxidases (Marshal et al. 2000), such as heat treatments (steam or boiling water blanching), refrigeration and chilling below 7°C, addition of citric, ascorbic, or other acids, drying, high-pressure treatments, use of inhibitors, and ultrafiltration. However, the most common solution is the use of ascorbic acid, but because its effect is temporary, other alternatives are under research (Rojas-Graü et al. 2008).

4.5 SORTING

The quality of a particular fresh or processed fruit or vegetable is defined by a series of characteristics that make it more or less attractive to the consumer, such as its ripeness, size, weight, shape, color, presence of blemishes and diseases, the presence or absence of fruit stems, the presence of seeds, and additional undesirable (leaves, stones, soil, etc.) or inappropriate (immature or rotten) material. The objectives of sorting are twofold: (1) ensuring that only good quality fruit is preserved along the process, which includes all of the factors that exert an influence on the product's appearance, on its nutritional and organoleptic qualities, or on its suitability for preservation, and (2) grouping the product in batches with similar quality to offer a more uniform product to the consumer.

The simplest method for sorting is performed manually by trained operators. This task can be facilitated by using inspection tables that make products rotate and moving forward, which helps the visual inspection by operators. These can reject units of fruits or vegetables or return them back in the production chain.

Vibration tables (Figure 4.1) are employed in order to individualize soft products (like canned fruits) before inspection. "V" belt singulators (Figure 4.2) are

FIGURE 4.1   Vibration table can be used to align soft products like satsuma segments.

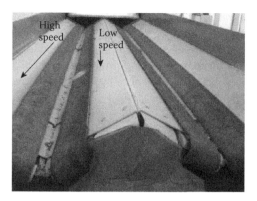

FIGURE 4.2 V-shaped belts at different velocities can be employed to align and separate fruit in packinghouses.

commonly utilized for round shaped fruits. They are designed to receive fruits in bulk and separate them into individual pieces. They consist of two inclined belts forming a V-shape lane. The belts move forward at different speeds making the fruits rotate slowly as they move forward. If two fruits move parallel to each other, the one coming into contact with the fastest belt will be forced to move ahead of the other. Both systems can be used in combination with a conveyor belt that receives the produce at a relatively higher speed, thereby increasing the separation between close objects.

For size sorting, various types of screens and sieves, with fixed or variable apertures, can be used. The screens may be stationary, rotating, or vibrating. Shape sorting can be accomplished manually or mechanically with, for example, a belt or roller sorter. Roller sorters are manufactured for spherical or cylindrical shaped produces. The product is forced to rotate and move forward by two rollers that diverge in such a way that the product falls when the separation between rollers is larger than its diameter.

Weight sorting sometimes substitutes size sorting because it is a very accurate method and is therefore used for high added value products (fresh fruits and certain vegetables).

Grading is the assessment of a number of characteristics to obtain an indication of the product's overall quality and is carried out by trained operators in many industries. The automation of fruit and vegetable grading has to deal with their intrinsic fragility, which conditions the mechanisms and technologies that can be used for the automatic separation of the product in categories. The physiological evolution that fruits and vegetables undergo during handling and storage also has to be considered.

Image processing alone or in combination with other electronic sensors can be used to grade fruits and vegetables on the basis of their size (length, diameter, etc.), color, and general appearance (presence of blemishes, rottenness, etc.) at very high speed. Image processing can provide substantial information about the nature and attributes of the objects present in a scene; moreover, it opens the possibility of studying these objects in regions of the electromagnetic spectrum where human eyes are unable to operate, as in the ultraviolet, near infrared, and infrared regions. Furthermore, computer vision has greater reliability and objectivity than human inspection, reduces inspection times, and ensures the individual quality of the products. Finally, automatic inspection allows for the generation of precise statistics on aspects related to the quality of the inspected product, which leads to greater control over the product and facilitates its traceability.

4.6   MACHINE VISION-BASED SORTERS

The appearance and external quality of the product can be evaluated automatically by machine vision systems. These have video cameras that take images of the units of the product traveling underneath and send them to a computer where they are conveniently analyzed. Normally, several images are obtained of each unit as the product translates and rotates, with the purpose of inspecting as much surface as possible. A lighting system provides uniform and diffuse light in order to reproduce the color of the objects with the highest fidelity.

Manufacturers construct modular sorters, so that the machines are composed of different elements for handling, inspecting, and sorting that are adapted to the nature of the product and the available space in the factory. However, basically all sorters have a series of elements in common: they have a feeding system that separates the product in units, a transport system, an inspection system that contains sensors that measure parameters related to the quality of each unit of product, a system that processes these measurements and assesses the overall quality, a synchronization system, and a system for separating the units into categories. Moreover, they have a user interface and software that manage the whole machine.

The performance of sorting machines is closely related to the way they are fed. The units of the product may accumulate in the initial steps, which normally complicates the separation of the product in units, hindering both the inspection and separation of these units in commercial categories. Underfeeding, on the other hand, reduces the performance of these machines.

The units of the product travel on carrying elements, which can be conveyor belts, rollers (frequently formed by two truncated cones), or a chain of receptacles (baskets) made of plastic. The size and shape of the carrying elements and the separation between them are determined by the product. Rollers formed by truncated cones (Figure 4.3) allow the product to be rotated on its major axis of inertia, enabling the inspection of most part of its surface (Leemans et al. 2004). For small or fragile products, the use of conveyor belts that allow a careful handling is more common, avoiding mechanical damages. The transport systems are complemented with damping elements (foam rollers, strips of soft materials, etc.) to reduce damages to the product.

4.6.1   Lighting Sources

The perceived color of an object in a scene basically depends on the lighting source, the reflective characteristics of its surface, and the spectral response of the observer. The success of a machine vision inspection system largely depends on the

FIGURE 4.3 Truncated cones allow transporting the fruit separately and make them rotate so the machine vision system can inspect most part of the surface of the product.

lighting system. The characteristics of the lighting source have a major influence on the system's performance and final cost and a decisive influence on the time needed for the system to process the images. The biological nature of horticultural products produces considerable variation in sizes, forms, textures, and colors, which emphasize the importance of the accurate design of illumination systems, but also make it impossible to establish a universal system of illumination that would be effective for all products and in all kinds of situations (Du and Sun 2004).

Diffuse lighting is aimed at creating a uniformly illuminated area, in which shadows disappear and where the negative effects for image analysis caused by spectral reflection are minimized. This is normally produced by focusing the light source onto a surface with certain properties of reflection or transmission. Figure 4.4 shows

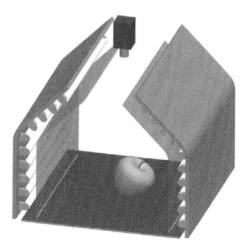

FIGURE 4.4 Inspection chamber using direct light and cross polarization to avoid bright spots.

a method for producing diffuse lighting with several fluorescent tubes, put close to each other, and increasing light scattering by placing a translucent screen beneath the tubes plus polarized filters. This is the scheme that is most commonly employed in sorting machines because it allows the construction of a wide inspection chamber on top of several lanes that carry the product.

When the object to be inspected is translucent, back-lighting can be the most appropriate technique to employ: light is applied so that the object is situated between the light source and the camera. Figure 4.5 shows a picture obtained in an inspection chamber of a mandarin segment sorter. By illuminating the segments from behind, they clearly contrast with the background, and their size and shape are easily determined. At the same time, seeds appear dark and highly contrasted, which facilitates their detection.

Near infrared radiation has been widely reported as a tool for the detection of damage in fruits. Wavelengths between 700 nm and 2200 nm have been recommended for the quality evaluation of various biological materials according to criteria of maturity, sugar concentrations, and signs of damage (Binenko et al. 1989). As a consequence of these or similar studies, some authors have explored the possibility of inspecting fruits and vegetables in several spectral bands at the same time (Blasco et al. 2007a; Gómez-Sanchís et al. 2008). The use of near infrared facilitates the discrimination between the product (with a high near-infrared (NIR) reflectance) and the background (Aleixos et al. 2002). Figure 4.6 illustrates the basics of multiband inspection systems. It shows images of the same orange with damage caused by *Penicillium digitatum* (green mold) using

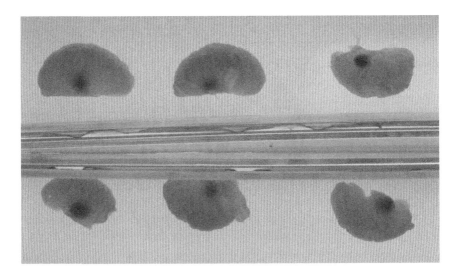

FIGURE 4.5   Back illumination of satsuma segments. This method provides a highly contrasted image that facilitates the detection of seeds and the analysis of size and shape.

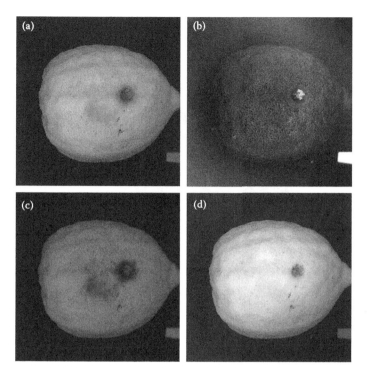

FIGURE 4.6 Images of the same fruit, affected by brown mold in (a) visible, (b) ultraviolet, (c) red, and (d) near-infrared bands.

adequate cameras and ultraviolet (200–400 nm), visible (400–700 nm), and near infrared (700–1800 nm) illumination.

4.6.2 VIDEO CAMERAS

Video cameras convert the light that they receive from the scene into electronic signals. Most widespread cameras in commercial applications acquire images by using a bi-dimensional charge-coupled device (CCD; Figure 4.7). Monochromatic cameras are suitable for estimating the size and shape of objects, as well as for detecting those defects that show a great contrast with healthy skin. Color cameras facilitate the estimation of quality parameters related to ripeness and increased precision in the detection of color and defects.

A monochromatic digital image can be considered a three-dimensional array in which the first two coordinates determine the position of the individual picture element (pixel), and the third coordinate is the intensity of the light received by the camera in this position (which is called the gray level value of the pixel). Conventional digital color images are formed by three monochromatic images, each of which represents the intensity of red, green, and blue of the scene, which commonly is referred to as an RGB image. Therefore, a color image is a five-dimensional array, in which

FIGURE 4.7   View of a matrix camera after removing the lenses. A 2/3" CCD can be observed.

the first two coordinates are the position of the pixel and the last three coordinates are the gray level of the red, green, and blue monochromatic images. Machine vision applications deal with these arrays in order to extract valuable information for the user.

4.6.3   IMAGE PROCESSING

The images acquired by the camera are transferred to the memory of the computer where they are analyzed. The first step to process an image is to label the pixels that form the image as belonging to the background and to the objects of interest (in this case, pixels that belong to the units of the product), which is known as segmentation (González and Woods 2002). The simplest technique segment of an image is thresholding. This technique consists of establishing a gray level value (threshold) and assigns to one class the pixels that have gray levels higher than the threshold and to the other the pixels that have lower gray level values (Figure 4.8). In case of the color images, the threshold can be set to only one of the three monochromatic images or can be applied to a combination of the three images.

More sophisticated methods use training. In this case, representative objects of each category are chosen by the user, and typical values of the gray levels of the pixels that compose these objects are calculated using statistical methods such as Bayesian discriminant analysis (Blasco et al. 2007b) or using other pattern recognition techniques. These methods generate models that are later used to segment new images online.

After segmentation of the image, several features of the objects are calculated in order to assign them to a class. Features calculated for the inspection of fruits and

(a) (b) (c)

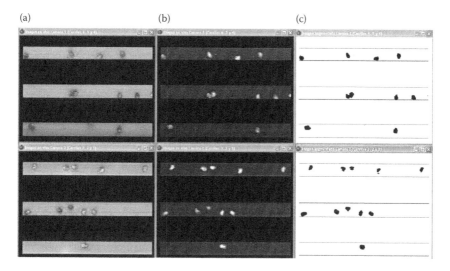

FIGURE 4.8 Images of pomegranate arils from an inspection system (a). Red band of the same image (b). Segmentation of this image by using a threshold (c).

vegetables are mostly related to size, shape, or color (Zheng et al. 2006a). Color is frequently estimated to assess the ripeness of the product or to distribute it in batches of similar color. Often the analysis of color is not enough to detect blemishes, and then the analysis of texture is required (Haralick et al. 1973). This kind of analysis has been used to classify currants, to grade apples after dehydration, or to predict the sugar content of oranges (Zheng et al. 2006b). Texture features are also important tools for describing the changes of the microstructure of food surface and have been applied to potatoes, bananas, pumpkins, carrots, and potato chips (Fernández et al. 2005).

Size and weight are primary quality attributes that most often require inspection in the industry. Machine vision systems provide morphological and appearance parameters as size, which can be measured as the length between two extreme points (Brodie et al. 1994) or can be estimated from the area (Tao et al. 1995) or from the perimeter (Sarkar and Wolfe 1985).

Figure 4.9 illustrates some image processing steps. Figure 4.9a is the original image, whereas Figure 4.9b represents the segmentation of the image in the region of interest, including the stem and defects. Figure 4.9c and d is the segmentation of the sound skin and the calculation of the size of the fruit, respectively. Figure 4.9e and f shows the detection of one blemish and the stem, respectively.

Once the features of each object have been calculated from the image, these are used to assign the unit to one category. As in the case of the segmentation of the image, the simplest classification method consists of determining thresholds of particular features (Singh et al. 1993). Sometimes these thresholds are based on national or international quality standards (Blasco et al. 2003). More complex methods of

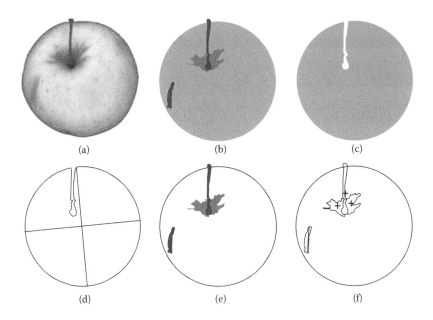

FIGURE 4.9 Image processing of a fresh apple. (a) Original image. (b) Segmentation of the fruit surface. (c) Detection of the sound skin. (d) Estimation of size. (e) Segmentation of one blemish and the stem. (f) Location of the centroid of the blemish and detected stem.

classification are also described in the literature (Blasco et al. 2007c; Díaz et al. 2004; Steinmetz et al. 1999; Guedalia and Edan 1994; Ozer et al. 1995).

4.6.4 SORTING SYSTEM

Once the computer system decides which category is assigned to each product unit, it has to be sent to an outlet. The design of the devices that separate the product units depends on the method used for transport. For example, transport systems based on rollers incorporate a pin in their lower position upon which electromagnets conveniently located above the outlets act. When the electromagnet is on standby, the pin passes below an unloading ramp, and the product is not discharged. When the tracking system detects that the unit is reaching the assigned outlet, the electromagnet is activated and a lever makes the pin pass over the ramp, which makes the rollers discharge. In the case of small products, pneumatic ejectors are commonly used (Figure 4.10). Here, the tracking system activates electrovalves, which produce the ejection of pressurized air, pushing the product out of the transportation system (Blasco et al. 2008).

Current sorting machines based on computer vision provide satisfactory measurements of quality attributes related to size, shape, color, and overall appearance. Research is focusing on the identification of particular defects, like those produced by fungal infestations or those that can favor these infestations, which may affect

FIGURE 4.10 Pneumatic systems are used for removing soft objects (i.e., mandarin segments) from a conveyor belt. Notice the air ejectors situated in front of the outlets of the different categories.

the sanitary conditions of whole batches of produce or which can evolve during refrigeration and transport to the consumer.

4.7 FINAL OPERATIONS

Depending on the particular fruit or vegetable, subsequent processing and conditioning are required prior to the conservation processes (refrigeration, chilling, etc.). These can be the following:

- Weighing and dosing are especially important if products suffer thermal treatments in order to obtain a uniform product.
- Mixing. In salads and fruit salads, several kinds of fruits and vegetables are mixed. Mixing can be mechanized with machines that have drums with edges in their sides. The products are mixed when the drum turns with the desired amount of each product.
- Packing. Normally vegetables are packed in sealed plastic bags, whereas fruits are in rigid terrines, cups, or dishes. Some machines offer the possibility of leaving a specific, controlled atmosphere inside the package that increases storage life.
- Pressing is used to extract the liquid part of the raw material. Horizontal pneumatic presses have a pneumatic membrane located in the center of the press that is inflated: berries are slowly squashed liberating the must. Hydraulic compression vertical presses use recipients that maintain the berries while a hydraulic piston squeezes them. Orange and other citrus juices are extracted with especial presses that can be used to recuperate essential oils at the same time. These oils are a valuable by-product for the cosmetic and pharmaceutical industry.

- Grinding may also improve the eating quality or the suitability of the material for further processing.
- Crushing is necessary to facilitate the yeasts' multiplication and to conduct traditional macerations before pressing in the wine and alcoholic beverage industry.
- Disinfection and hygiene are necessary for the destruction of pathogenic organisms and to prevent the spread of diseases that shorten the life of the product.

4.8 THE FUTURE: ASSESSMENT OF ORGANOLEPTIC AND INTERNAL QUALITY FEATURES

The demand for products with guaranteed organoleptic qualities has boosted the development of devices capable of estimating internal quality attributes such as sweetness. In general, these devices measure the absorption of the near infrared radiation and are becoming common.

At present, most evaluation of an agricultural product's internal quality is carried out by batch sampling and destructive techniques, so that not all of the production can be inspected. For this reason, nondestructive techniques are being studied, which are still limited in extent because of their high cost or the technological difficulties involved.

X-ray techniques have been described for inspecting the fruits' interior, which associate changes in the intensity of the image with damage to tissues. The first consists of placing the fruit between an X-ray emitter and a detector, which measures the energy absorbed by the fruit. Such systems have been employed for detecting internal defects in different kinds of fruits and vegetables (Schatzki et al. 1997). Damage caused by the dehydration of the pulp when a crop is affected by frost can be detected by such systems (Johnson 1985). However, these methods have not been effective for the detection of seeds in mandarins. Computerized tomography has also been investigated for estimating the internal quality of different fruits and vegetables (Morishima et al. 1987).

Magnetic resonance imaging (MRI) has been used for the detection of seeds in mandarins (Moltó and Blasco 2000). Difficulties arise when the object moves and special preprocessing is required (Hernández et al. 2005). This technique has also been applied to find internal damages such as that caused by freezing (Gambhir et al. 2005). Although MRI equipment is still far too expensive and thus is not used in commercial applications, application of this technology to inspection has a very promising future.

REFERENCES

Abeles, F.B., Morgan, P.W., Saltveit, M.E. Jr. (1992). *Ethylene in Plant Biology*. San Diego, CA: Academic Press. ISBN-10: 0-12-041451-1.
Aleixos, N., Blasco, J., Navarrón, F., Moltó, E. (2002). Multispectral inspection of citrus in real-time using machine vision and digital signal processors. *Computers and Electronics in Agriculture* 33(2), 121–137.

Artes, F., Castañer, M., Gil, M.I. (1998). Review: Enzymatic browning in minimally processed fruit and vegetables. *Food Science and Technology International* 4(6), 377–389.

Binenko, V.I., Voronov, N.V., Nedotsukova, D.V. (1989). Automation of quality control of some species of vegetables and citrus fruits by optical methods. *Soviet Agricultural Sciences* 1, 33–37.

Blasco, J., Aleixos, N., Moltó, E. (2003). Machine vision system for automatic quality grading of fruit. *Biosystems Engineering* 85(4), 415–423.

Blasco, J., Aleixos, N., Gómez, J., Moltó, E. (2007a). Citrus sorting by identification of the most common defects using multispectral computer vision. *Journal of Food Engineering* 83(3), 384–393.

Blasco, J., Aleixos, N., Moltó, E. (2007b). Computer vision detection of peel defects in citrus by means of a region oriented segmentation algorithm. *Journal of Food Engineering* 81(3), 535–543.

Blasco, J., Cubero, S., Arias, R., Gómez, J., Juste, F., Moltó, E. (2007c). Development of a computer vision system for the automatic quality grading of mandarin segments. *Lecture Notes in Computer Science* 4478, 460–466.

Blasco, J., Cubero, S., López, V., Gómez-Sanchís, J., Moltó, E. (2008). Machine vision system for the automatic sorting of *Punica granatum* (pomegranate) arils. *Journal of Food Engineering* 2009, 90(1): 27–34. DOI: 10.1016/j.jfoodeng.2008.05.035.

Brodie, J.R., Hansen, A.C., Reid, J.F. (1994). Size assessment of stacked logs via the Hough Transform. *Transactions of the ASAE* 37(1), 303–310.

Canadian Food Inspection Agency (2008). Code of Practice for Minimally Processed Readyto-Eat Vegetables. http://www.inspection.gc.ca/.

Cantwell, M.I. (2002). Postharvest handling systems: minimally processed fruits and vegetables. In: Kader, A.A. (ed.). *Postharvest Technology of Horticultural Crops*, 2nd ed., Davis: University of California, Division of Horticultural and Natural Resources, pp. 273–281.

Cantwell, M.I. (2008). Postharvest Handling Systems: minimally processed fruits and vegetables. Available at http://vric.ucdavis.edu/selectnewtopic.minproc.htm, accessed June 2008.

Cantwell, M.I., Suslow, T.V. (2002). Postharvest handling systems: fresh-cut fruits and vegetables. In: Kader, A.A. (ed.). *Postharvest Technology of Horticultural Crops*, 3rd ed. Publ. 3311. Oakland: University of California, Division of Agriculture and Natural Resources, pp. 445–463.

Díaz, R., Gil, L., Serrano, C., Blasco, M., Moltó, E., Blasco, J. (2004). Comparison of three algorithms in the classification of table olives by means of computer vision. *Journal of Food Engineering* 61, 101–107.

Du, C.J., Sun, D.W. (2004). Shape extraction and classification of pizza base using computer vision. *Journal of Food Engineering* 64, 489–496.

Fernández, L., Castillero, C., Aguilera, J.M. (2005). An application of image analysis to dehydration of apple discs. *Journal of Food Engineering* 67, 185–193.

Gambhir, P.N., Choi, Y.J., Slaughter, D.C., Thompson, J.F., McCarthy, M.J. (2005). Proton spin–spin relaxation time of peel and flesh of navel orange varieties exposed to freezing temperature. *Journal of the Science of Food and Agriculture* 85, 2482–2486.

Gómez-Sanchís, J. Gómez-Chova, L., Aleixos, N., Camps-Valls, G., Montesinos-Herrero, C., Moltó, E., Blasco, J. (2008). Hyperspectral system for early detection of rottenness caused by *Penicillium digitatum* in mandarins. *Journal of Food Engineering* 89(1), 80–86.

González, R.C., Woods, R.E. (2002). *Digital Image Processing*. Prentice Hall. Upper Saddle River, New Jersey, USA. ISBN 0201180758.

Guedalia, D., Edan, Y. (1994). A dynamic artificial neural network for coding and classification of multisensor quality information. ASAE Paper 94-3053.

Haralick, R.M., Shanmugam, K., Dinstein, I. (1973). Textural features for image classification. *IEEE Transactions on Systems Man and Cybernetics* 6, 610–621.

Hernández, N., Barreiro, P., Ruiz-Altisent, M., Ruiz-Cabello, J., Fernandez-Valle, M.E. (2005). Detection of seeds in citrus using MRI under motion conditions and improvement with motion correction. *Concepts in Magnetic Resonance Part B—Magnetic Resonance Engineering* 26B(1), 81–92.

Hodges, D.M., DeLong, J.M. (2007). The relationship between antioxidants and postharvest storage quality of fruits and vegetables. *Stewart Postharvest Review* 3(3), 1–9.

Ismail, M., Thomas, M. (2002). Citrus peeling technology for Florida fresh citrus. *Proceedings of the Florida State Horticultural Society* 115, 84–86.

Ismail, M.A., Chen, H., Baldwin, E.A., Plotto, A. (2005). Optimizing the use of hydrolytic enzymes to facilitate peeling of citrus fruit. *Proceedings of the Florida State Horticultural Society* 118, 400–402.

Johnson, M. (1985). Automation in citrus sorting and packing. Proceedings of Agrimation, I Conference and Exposition, Chicago, USA, pp. 63–68.

Kim, J.G. (2007). Fresh-cut market potential and challenges in far-east Asia. *Acta Horticulturae* 746, 33–38.

Leemans, V., Magein, H., Destain, M.F. (2004). A real-time grading method of apples based on features. *Journal of Food Engineering* 61, 83–89.

Marshall, M.R., Kim, J., Wei C.I. (2000). Enzymatic Browning in Fruits, Vegetables and Seafoods. Available at http://www.fao.org/ag/ags/agsi/ENZYMEFINAL/Enzymatic%20 Browning.html.

Moltó, E., Blasco, J. (2000). Detection of seeds in mandarins using magnetic resonance imaging. Proceedings of the International Society of Citriculture. Orlando, Florida, 1129–1130.

Morishima, H., Seo, Y., Sagara, Y., Yamaki, Y., Matsuura, S. (1987). Non-destructive internal quality detection of fresh fruits and vegetables by CT-scanner. Proceedings of International Symposium on Agricultural Mechanization and International Cooperation in High Technology Era, Tokyo (Japan), pp. 342–349.

Ozer, N., Engel, B., Simon, J. (1995). Fusion classification techniques for fruit quality. *Transactions of the ASAE* 38, 1927–1934.

Pagan, A., Ibarz, A., Pagan, J. (2005). Kinetics of the digestion products and effect of temperature on the enzymatic peeling process of oranges. *Journal of Food Engineering* 71(4), 361–365.

Pao, S., Widmer, W.W., Petracek, P.D. (1997). Effects of cutting on juice leakage, microbiological stability and bitter substances of peeled citrus. *Lebensmittel-Wissenschaft und-Technologie* 30, 670–675.

Portela, S.I., Cantwell, M.I. (2001). Cutting blade sharpness affects appearance and other quality attributes of fresh-cut cantaloupe melon. *Journal of Food Science* 66(9), 1265–1270.

Pretel, M.T., Fernandez, P.S., Martinez, A., Romorajo, F. (1998). Modelling design of cuts for enzymatic peeling of mandarin and optimization of different parameters of the process. *Zeitschrift fur Lebensmittel-Untersuchung und -Forschung. European Food Research and Technology* 207(4), 322–327.

Pretel, M.T., Amorós, A., Botella, M.A., Serrano, M., Romojaro, F. (2004). Study of albedo and carpelar membrane degradation for further application in enzymatic peeling of citrus fruits. *Journal of the Science of Food and Agriculture* 85(1), 86–90.

Rocha, A.M.C.N., Brochardo, C.M., Kirby, R., Morais, A.M.M.B. (1996). Shelf-life of chilled cut orange determined by sensory quality. *Food Control* 6, 317–322.

Rojas-Graü, M.A., Soliva-Fortuny, R., Martín-Belloso, O. (2008). Effect of natural anti-browning agents on color and related enzymes in fresh-cut Fuji apples as an alternative to the use of ascorbic acid. *Journal of Food Science* 73(6), S267–S272. doi: 10.1111/j.1750-3841.2008.00794.X.

Sarkar, N., Wolfe, R.R. (1985). Feature extraction techniques for sorting tomatoes by computer vision. *Transactions of the ASAE* 28(3), 970–979.

Schatzki, T.F., Haff, R.P., Young, R., Can, I., Le, L.C., Toyofuku, N. (1997). Defect detection in apples by means of X-ray imaging. *Transactions of the ASAE* 40(5), 1407–1415.

Singh, N., Delwiche, M.J., Johnson, R.S. (1993). Image analysis methods for real-time color grading of stonefruit. *Computers and Electronics in Agriculture* 9(1), 71–84.

Steinmetz, V., Roger, J.M., Moltó, E., Blasco, J. (1999). On-line fusion color camera and spectrophotometer for sugar content prediction of apples. *Journal of Agricultural Engineering Research* 73, 207–216.

Tao, Y., Heinemann, P.H., Varghese, Z., Morrow, C.T., Sommer III, H.J. (1995). Machine vision for color inspection of potatoes and apples. *Transactions of the ASAE* 38(5), 1555–1561.

Toker, I., Bayindirli A. (2003). Enzymatic peeling of apricots, nectarines and peaches. *Lebensmittel-Wissenschaft und-Technologie* 36(2), 215–221.

Watada, A.E., Ko, N.P., Minott, D.A. (1996). Factors affecting quality of fresh-cut horticultural products. *Postharvest Biology and Technology* 9, 115–125.

Zheng, C., Sun, D.W., Zheng, L. (2006a). Recent developments and applications of image features for food quality evaluation and inspection—a review. *Trends in Food Science and Technology* 17, 642–655.

Zheng, C., Sun, D.W., Zheng, L. (2006b). Recent applications of image texture for evaluation of food qualities—a review. *Trends in Food Science and Technology* 17, 113–128.

5 Blanching of Fruits and Vegetable Products

Paloma Vírseda Chamorro and
Cristina Arroqui Vidaurreta

CONTENTS

5.1 INTRODUCTION

Blanching is a low-intensive thermal treatment used mainly to destroy enzymatic activity in vegetables and some fruits. As such, it is not intended as a sole method of preservation but as a pretreatment that is normally carried out between the preparation of the raw material and later operations, particularly heat sterilization, dehydration, and freezing. Blanching is also combined with peeling and/or cleaning of food to achieve saving of energy consumption, space, and equipment costs. Blanching application determines largely the product quality. The main objective of blanching is the inactivation of the enzymes responsible for deterioration during handling and storage. Oxidative reactions produce undesirable changes of color, flavor, or texture of the product. Besides, depending on the preservation process involved, blanching can have other secondary objectives like destruction of microorganisms on the surface, cleaning the dirt, making some vegetables (e.g., broccoli, spinach) more compact, brightening of the color, retardation of loss of vitamins, softening of vegetable tissues to facilitate filling into containers, and removing of the air from intercellular spaces that assists in the formation of a head-space vacuum in cans.

The process consists of a rapid heating up of the product (usually using steam or water) to temperatures of 70°C–100°C, held during preset short time (from seconds to several minutes), followed by a rapid cooling down to near ambient temperatures. To determine the adequate thermal treatment, it is necessary to study the heat transfer process, the kinetics of enzyme inactivation, and the processes taking place during blanching that affect product quality. The factors that influence blanching time are as follows:

1. Type of fruit or vegetable
2. Size of the pieces of food
3. Blanching temperature
4. Method of heating

5.2 THEORY AND PURPOSE OF BLANCHING

5.2.1 HEAT TRANSFER DURING BLANCHING

The theory used to calculate blanching time is of unsteady-state heat transfer by conduction and convection.

Heat is transferred from the heating surroundings to the product surface by convection and within the product by conduction. The heat transfer differential equation to calculate time of blanching is the nonstationary state.

$$\frac{\delta T}{\delta t} = \alpha \cdot \left(\frac{\delta^2 T}{\delta x^2} + \frac{\delta^2 T}{\delta y^2} + \frac{\delta^2 T}{\delta z^2} \right) \qquad (5.1)$$

$t = 0;$ $\qquad T = T_0$ in all the products

$t > 0 \; x = L/2 \quad h \cdot \left(T_f - T \right) = \kappa \cdot \dfrac{\delta T}{\delta x}$

$y = M/2 \qquad h \cdot \left(T_f - T \right) = \kappa \cdot \dfrac{\delta T}{\delta y}$

$z = N/2 \qquad h \cdot \left(T_f - T \right) = \kappa \cdot \dfrac{\delta T}{\delta z}$

$t > 0 \; x = 0 \qquad \dfrac{\delta T}{\delta x} = 0$

$y = 0 \qquad \dfrac{\delta T}{\delta y} = 0$

$$z = 0 \qquad \frac{\delta T}{\delta z} = 0$$

where

$\alpha =$ thermal diffusivity of the product (m²/s), and it can be calculated by the expression $\alpha = \dfrac{k}{\rho \cdot c_p}$

$k =$ thermal conductivity (W/(m K))
$\rho =$ density of the product (kg/m³)
$c_p =$ heat capacity of the product (kJ/(kg K))
$T =$ temperature of the product (K)
$T_0 =$ initial temperature of the product (K)
$T_f =$ temperature in the medium (K)
$x, y, z =$ orthogonal coordinates
$t =$ time (s)
$L =$ product length in x direction (m)
$M =$ product length in y direction (m)
$N =$ product length in z direction (m)
$h =$ surface heat transfer coefficient of the medium (W/(m² K))

Equation 5.1 could be expressed in a different system of coordinates as a function of the product geometry (Welty et al. 1991).

Product thermal conductivity, k, and heat capacity, c_p, can be obtained using correlation equations relating them to their composition or from published values (Sweat 1974; Lamb 1976; Lamberg and Hallström 1986; Togeby et al. 1986; Andersson 1994).

$$c_p = 1.47 + 2.72\ W \quad \text{(Lamb 1976)} \tag{5.2}$$

$$k = 0.148 + 0.493 \times W \quad \text{(Sweat 1974)} \tag{5.3}$$

where $W =$ water content (kg water/kg product).

The surface heat transfer coefficient of each zone, h, depends on the water or steam flow and the spray characteristics.

5.2.2 ENZYMATIC INACTIVATION

In the case of fruits and vegetable products prior to be frozen, the time/temperature relationship used has a main objective to ensure the adequate enzyme inactivation. The enzymes that contribute to the deterioration process of vegetable products are oxidoreductases like lipoxigenase, catalase, peroxidase (POD), and polyphenol oxidase, lipases, and proteases. Those responsible for bad odors are lipoxigenases, lipases, and proteases, whereas the pectic enzymes and cellulases can affect negatively the texture of the product. On the other hand, polyphenol oxidase, chlorophyllase, and POD cause color changes, and the enzymes ascorbic acid oxidase and thiaminase can produce nutritional losses to the product (Williams and Lim 1986).

There are variabilities in the type and quantity of enzymes within the vegetable products (Table 5.1). Besides, water content and availability, pH, or presence of salts has an important effect on the enzyme stability to heat (Adams 1991).

However, lipoxigenase and POD are commonly employed as indicators of the blanching treatment, since they are the most thermoresistant within the enzyme group responsible for the product deterioration during freezing and storage.

The general equation simulating inactivation of these enzymes is the following (Palmer 1987):

$$\frac{d\left[E_A\right]}{dt} = k_1 \cdot E_A \tag{5.4}$$

where k_1 = enzymatic inactivation rate (s^{-1}) and E_A = enzymatic activity in the product (%).

The temperature dependence of k_1 for each enzyme and product may be well described by an Arrhenius-type relationship. The equations were taken from those published by different authors (Williams and Lim 1986; Togeby et al. 1986; Sarikaya and Özilgen 1991; Günes and Baymdiri 1993):

$$k_1 = A_0 e^{\frac{-E_a}{RT}} \tag{5.5}$$

where A_0 = preexponential factor (J/(m^3 K)), E_a = activation energy (J/mol), and R = universal gas constant (8.314 J/mol).

In the industry, the most commonly applied method to check the effectiveness of blanching is the one that uses guayacol to detect the total POD inactivation.

TABLE 5.1

Enzymatic Activities of Some Vegetables (per 100 g of the Product)

Vegetable	Lipoxigenase (μ atoms oxygen * min^{-1})	Peroxidase (ΔA min^{-1})	Catalase (ΔA min^{-1})
Cauliflower	5.8	150	12
Brussels sprouts	7.9	2656	59
Broccoli	8	406	40
Carrot	1.8	8	40
Green peas	44.9	289	403
Tomato	18.1	24	0
Potato	31.8	20	17

Source: Baardseth, P., and Slinde, E. *Norwegian Journal of Agricultural Sciences* 1(2), 111–117, 1987. With permission.

TABLE 5.2

Percentage of Residual Activity of POD for Green Vegetables to Obtain Good Quality after 9 Months in Freeze Storage

Product	Residual Activity of POD (%)
Green peas	0.3–1.4/2–6.3
Green beans	0.7–3.2
Cauliflower	3.8–5.7/2.9–8.2
Brussels sprouts	7.5–11.5
Spinach	2.8–3.7
Carrots	Up to 6.9%

Source: Böttcher, H., *Nahrung* 19, 173–179, 1975. With permission.

However, several works have shown that the total POD inactivation can reduce the quality of blanched vegetables, with a residual activity of the enzyme (between 3% and 10%) being more desirable for products to be frozen depending on the type of the product, its variety, or even its maturity stage (Böttcher 1975; Williams and Lim 1986; Baardseth and Slinde 1987; Günes and Baymdiri 1993; Fuster 1995a,b; Table 5.2).

On the other hand, different authors propose the total inactivation of lipoxigenase as a method to determine the effectiveness of vegetable blanching (Williams and Lim 1986; Günes and Baymdiri 1993; Barret et al. 2000; Garrote et al. 2004). Applying this method as an alternative to total POD inactivation, the process time was reduced and the quality of the product improved.

However, there is still no method to be employed industrially (Williams and Lim 1986; Günes and Baymdiri 1993). Besides, lipoxigenase is not as common as POD in vegetable products (Table 5.1).

5.2.3 FOOD SAFETY

Most of the microorganisms present in fresh-cut fruits and vegetables derive from soil, air, and water contamination. Several factors affect the microbial counts at the entry of blancher and consequently the food safety of frozen vegetables (Philippon 1992): initial raw material load, washing efficiency, and conditioning operations.

During blanching, the microbial reduction in the product was assured by an adequate sanitation of equipment and by applying the needed thermal treatment followed by quick cooling. This microorganism reduction is more or less critical in the function of the preserving treatment to be applied. When freezing, an important objective of blanching is to reduce up to a low level the surface microbial load. However, this process also kills the vegetable tissue, leaving it dead, being more sensible to the ulterior fungal or bacterial contaminations. Thus, there is a critical period between the thermal treatment and the freezer entry, with quick and effective cooling being very important, as has already been pointed out.

5.3   EFFECT OF BLANCHING ON QUALITY OF VEGETABLES

Blanching has a considerable influence on the final quality of the processed fruits and vegetables. When the thermal treatment applied on the food is not adequate, blanching can cause undesirable changes on the sensory and nutritional qualities such as softening of the tissue, cooked flavors, undesirable color changes, and water-soluble nutrient losses.

5.3.1   FLAVOR AND TASTE

Blanching inactivates oxidoreductases, which are enzymes that are mainly responsible for the undesirable odor and flavor development, and also removes part of the O_2 needed for oxidation reactions. Thus, insufficient blanching process can lead to the development of off-flavors during storage of dried or frozen foods. When correctly blanched, most foods have no significant changes of flavor or aroma.

However, excessive or inadequate treatments can produce compounds such as ethanol and acetaldehydes responsible for strange flavors and cooked tastes, as observed in green peas, green beans, carrots, and cauliflower after low-temperature long-time (LTLT) blanching treatments (Pala 1982; Fuster 1995a,b). Other undesirable changes or losses of flavors during blanching can be due to leaching of flavoring compounds or removing of the volatiles responsible for the characteristic flavors of some vegetables such as the cruciferous ones and onions. Sucrose was added to the blanching water to prevent or minimize these flavor losses in the cited vegetables.

5.3.2   NUTRIENTS

On the one hand, deterioration of vitamins by oxidation during frozen storage in a vegetable not blanched could be important. On the other hand, minerals, water-soluble vitamins, and other water-soluble components are lost during blanching. The losses of water-soluble vitamins during blanching have been specially studied due to their high solubility and their nutritional interest (Table 5.3). There are several factors that can affect the retention of vitamins during blanching:

1. Maturity of the food and variety
2. Method of blanching and cooling
3. Surface area-to-volume ratio of the food pieces
4. Time–temperature combination, etc.

Among the water-soluble vitamins, vitamin C is one of those vitamins that are present most of the time in fruits and vegetables (especially brussels sprouts, cauliflower, spinach, potatoes, etc.) and is also an important nutrient in the human diet (recommended intake of 45–90 mg/day). This vitamin also has high diffusivity in water and high sensibility to oxidation. Thus, the loss of ascorbic acid can be used as an indicator of nutritive quality of the blanched vegetable and therefore of the severity of the process.

As has already been pointed out, vitamin losses, and thus ascorbic acid losses, depend to a large extent on the blanching method used (Table 5.4), with water blanching being the one that reported the highest vitamin losses.

TABLE 5.3
Effect of Blanching on Vitamin Content of Some Vegetables

References	Product	Nutrient	Blanching Process[a]	Losses (%)
Bomben et al. (1974)	Green beans	Vitamin C	S, 2.5 min	8 mg/100 g
			IQB	11 mg/100 g
	Lime bean		S, 3.0 min	16 mg/100 g
			IQB	24 mg/100 g
	Brussels sprouts		S	47 mg/100 g
			IQB	46 mg/100 g
	Green peas		S	21 mg/100 g
			IQB	18 mg/100 g
Dietrich et al. (1970)	Brussels sprouts	Vitamin C	W, 6 min /100°C	43
			M, 1 min +	29
			W 4 min/100°C	35
			M, 3 min +	
			W 2 min/100°C	
Holmquist et al. (1954)	Green peas	Vitamin C	S	12.3
			W	25.8
Ralls et al. (1973)	Spinach	Carotene	W	5.4 mg/100 g
		Riboflavin	W	0.12 mg/100 g
		Vitamin C	W	20.8 mg/100 g
Raab et al. (1973)	Lima beans	Vitamin B_6	W, 10 min /100°C	21
			S, 10 min /100°C	14

[a] Blanching medium: S—steam; W—water; M—microwaves.

To reduce vitamin and other water-soluble compound losses, steam and microwaves may be used in the blanching process instead of the conventional water blanching (Table 5.5). Quick cooling systems after blanching that reduce the effluent production and hence the losses of solutes during the process can also be applied.

Besides this, treatments at high temperature and short time (HTST) are preferred to those at LTLT in products like peas and potatoes because the losses caused by diffusion of water solutes are reduced.

TABLE 5.4
Losses of Vitamin C for Different Blanching Methods

Blanching Technique	Average Losses (%)
Steam	32
Water	40
Microwave	3

Source: Canet W. *Alimentación, Equipos y Tecnología* June 5:75–87, 1995. With permission.

TABLE 5.5

Water Consumption and Organic Losses in Green Peas after Different Blanching and Cooling Systems

Blanching System	Green Beans Blanching		Green Beans Blanching + Cooling	
	Water Consumption (m³/t)	Organic Substances Losses (kg/t)	Water Consumption (m³/t)	Organic Substances Losses (kg/t)
Water (traditional)	0.25	0.69	5.2	2.2
Steam (traditional)	0.14	0.55	5.0	2.5

Source: Bomben, J.L., *Food Processing Waste Management*, 141–174, 1979; Bomben, J.L. et al., Integrated blanching and cooling to reduce plant effluents. *Proc. 5th. Nat. Symp. Food Proc. Wastes, Environmental Protection Agency*, pp. 120–131, 1974; Steinbuch, E., in P. Zeuthen et al. (eds.), *Thermal Processing and Quality of Foods*, Elsevier Applied Science Publishers, London, 1984.

Lately, there has been great attention paid to the importance of the antioxidant properties (ascorbic acid and total flavonoids) and the glucosinolate content of some vegetables such as cruciferous and green leafy vegetables. Thus, different studies concerning the effect of blanching treatment on these nutritive compounds were carried out (Galgano et al. 2007; Olivera et al. 2008; Rungapamestry et al. 2008).

5.3.3 Color

Blanching inactivates enzymes that can affect food pigments like chlorophyll. In order to reduce chlorophyll deterioration in green vegetables, thermal treatments have to inactivate POD. In this way, color losses of green beans during 20 days of storage at −10°C are reduced from 45% to 2% when blanching is applied correctly (Kaack 1994).

Besides, blanching brightens the color of some vegetables and fruits by removing air and dust from the surface and thus altering the wavelength of reflected light.

However, the color of fruits and vegetables can be negatively affected by blanching. Green vegetables such as spinach, green beans, broccoli, and green peas, or those sensitive to enzymatic browning, are the most affected by this color change due to the following processes:

1. Leaching of chlorophyll that can be minimized using HTST instead of LTLT treatments. Also, chemical compounds such as sodium carbonate, calcium oxide, and sodium chloride are usually added to blanching water in order to protect chlorophyll and retain the color of green vegetables.
2. Chloroplast burst that can be produced during storage at low temperatures if the tissue was affected by long-time blanching treatments. Spinach and green beans are the most sensitive vegetables to chloroplast breakage due to overblanching, having up to 30% of color loss.

3. Enzymatic browning as a consequence of phenol compound oxidation in some fruits and vegetables (apple, cauliflower, potato, mushroom, etc.; Andersson 1994, Fuster 1995a,b). The reaction is accelerated by tissue breakage prior to blanching and by the use of moderate temperature. This color deterioration is prevented in these cut fruits and vegetables by holding the food in diluted sodium phosphate, citric acid, or sulfite compound solutions before the blanching process.

5.3.4 Texture and Structure

Some changes in the texture of vegetables are desirable when blanching is applied prior to canning. In this case, making vegetables like broccoli and spinach more compact, softening the vegetable tissue and removing air from intercellular spaces to facilitate filling into containers, and forming a head-space vacuum in cans, etc., are considered as objectives of the blanching process.

Nevertheless, when the preserving treatment is freezing or drying, it is important to avoid an excessive texture loss of the vegetables. Several studies were done to analyze the effect of different blanching methods and treatment conditions on the texture of those vegetables in which texture is an important parameter of quality (Pala 1982; Hough and Alzamora 1984; Kaack 1994; Andersson 1994; Fuchigami et al. 1995; Verlinden and De Baerdemaeker 1997; Tijskens et al. 1997; Barrett et al. 2000; Canet et al. 2004; Olivera et al. 2008).

Changes in the pectin polymers of the cell wall and middle lamella during blanching can cause rapid loss of firmness and membrane integrity. This loss of turgor can be excessive in some kinds of food (e.g., certain varieties of potatoes, carrots, green beans, or cherries), in large pieces of food, or when the treatment is not the adequate one. To prevent excessive loss of firmness, LTLT treatments can be used (Pala 1982; Andersson 1994; Fuchigami et al. 1995; Verlinden and De Baerdemaeker 1997, 2000; Sanjuán et al. 2005). According to the theory of Bartolomé and Hoff (1972), low-temperature treatments cause loss of membrane selective permeability; this promotes diffusion of cations to the cell wall, causing activation of pectin methylesterase and consequently enhancing de-esterification of pectins; this facilitates the formation of divalent bridges between residues of galacturonic acid attached to adjacent pectic chains. The divalent ion–pectin complexes formed in this way provide intercellular cementation, lending firmness to the tissues.

The blanching treatment applied to enhance the firmness of vegetables consists of an initial thermal treatment at 55°C–65°C during times of 5–45 min in the function of the product (Ni et al. 2005) followed by a quick cooling and combined with a high-temperature treatment at 90°C–95°C that assures the enzymatic inactivation.

Another method to prevent excessive losses of texture is the addition of metallic salts into water (in case of water blanching), such as calcium chloride, to form insoluble calcium pectate components and thus to maintain firmness in the tissues.

5.3.5 Toxic Compounds

Blanching can favor the elimination of toxic compounds in different ways:

1. Leaching of substances such as nitrites in spinach and fertilizers in other vegetables
2. Neutralization of toxic compounds by acids or tartrates
3. Inactivation of enzymes responsible for the formation of toxic compounds

5.3.6 Weight Change

Water blanching products increase their weight by around 1%–2%, whereas steam blanching causes losses from 1.5% to 2% of the product weight. However, when water is employed as a blanching medium, there are important dry matter losses of around 5%–10%. Besides the blanching system used, there are also substantial differences in yield and nutrient retention in the blanched product due to the type of food and the method of preparation (e.g., slicing or peeling).

5.4 BLANCHING SYSTEMS

The blanching system used will affect the quality and the cost of the product; therefore, its selection has to be done carefully. Several factors have to be taken into account: product requirements, the preservation method to be applied, legal requirements, and the economical cost of the blanching system and operation (equipment, water and energy consumption, and effluent production).

The advances in the field of blanching systems are focused on the reduction of energy and water consumption, the increase in the operation yield, and the improvement of the product quality, flexibility of the product, and working capacity of the blancher. The equipment is, in general, modular and integrates the cooling system. It usually incorporates one or more systems of energy and water saving, with its working capacity also being variable.

The two most widespread commercial methods of blanching involve passing food through an atmosphere of saturated steam or a bath of hot water at temperatures near 100°C.

The main disadvantages of water blanching are the leaching losses of the product (washing of water-soluble components from the food) and the high level of effluent production. However, this system increases the product weight and has a relatively low specific cost.

Besides the water blanching system, saturated steam blanching at atmospheric pressure is traditionally used. Although this system reduces the solute losses and the effluent production, the energetic efficacy is lower than that of the water blanching one.

To select the blanching system, it is necessary to take into account the applied cooling method. In some methods, the cooling stage may result in greater losses of the product or nutrients than in the blanching stage, and it is therefore important to consider both blanching and cooling when comparing different methods. Besides this, commercial blanchers are generally integrated blancher/cooler systems.

Table 5.5 shows the differences in water consumption and organic solute losses during blanching of green beans between traditional steam and water blanching systems with and without the cooling step incorporation.

Steam blanching results in higher nutrient retention provided that cold-air or cold-water sprays were used in the cooling step. Cooling with running water (fluming) substantially increases leaching losses, but the product may gain weight and the overall yield is therefore increased. Air cooling causes weight loss of the product due to evaporation, and this may outweigh any advantages gained by nutrient retention (Bomben et al. 1975).

A brief description of different commercial blanching systems and its evolution since the industrial application of the process will be presented in the following sections.

5.4.1 Water Blanching Systems

In traditional water blanchers, the product is introduced in a hot water bath at 70°C–100°C and transported by means of a screw, a rotary drum, or by the circulated and pumped water (Figures 5.1 and 5.2). The reel blancher is a widely used method in which food is introduced into a slowly rotating cylindrical mesh drum, which is partially submerged in hot water. The food is moved through the drum by internal flights. The speed of rotation controls the heating time. Pipe blanchers consist of a continuous insulated metal pipe fitted with feed and discharge ports. Hot water is recirculated through the pipe, and food is metered in. The residence time of food in the blancher is determined by the length of the pipe and the velocity of the water. These blanchers have the advantage of a large capacity while occupying a small floor space. In some applications, they can be used to transport food simultaneously through a factory.

Developments in hot water blanchers search for the reduction of energy and water consumption and the production of effluents. The incorporation of water curtains to

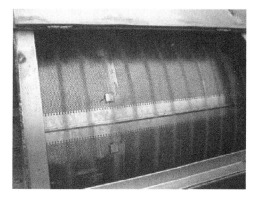

FIGURE 5.1 Rotary hot water blancher. Detail of the perforated rotary drum. (From www .heiusa.com/equipment.)

FIGURE 5.2 Screen blancher. From A.K. Robins company screw cooker. (From www .heiusa.com/equipment.)

the blancher, water fasteners, heat exchangers, or water recirculation improves the product yield and reduces the effluent production (Rumsey et al. 1981). The efficiency and the specific consumption of energy are improved when these systems are incorporated to the blancher (Table 5.6). In the case of cauliflowers, the efficiency of a conventional screw blancher is around 30%, and the specific consumption is 0.91 GJ/t. When a heat exchanger was introduced, the efficiency doubled (Rumsey et al. 1981).

Recycling of water substantially reduces the consumption of water and energy, the leaching of water-soluble nutrients, and the volume of effluents without affecting product quality or yield of the blanching process. However, it is necessary to ensure adequate hygienic standards for both the product and equipment (Swartz and Carroad 1979).

One of the most efficient systems is the integrated water blancher/cooler that includes three sections: a preheating stage, a blanching stage, and a cooling stage (Figures 5.3 and 5.4). This equipment can incorporate different systems to improve energy efficiency and reduce water consumption: water recirculation to preheating and blanching the product, heat recovery (from precooling to preheating sections), and evaporative cooling. The food remains on a single conveyor belt throughout each stage and therefore does not suffer the physical damage associated with the

TABLE 5.6
Thermal Efficiency of Some Water Blanching Systems of Vegetable Products

Blanching System	Thermal Efficiency (%)	Source
Traditional water blancher	14–32	Rammesh et al. (2000)
	26–52	Rao et al. (1986)
Screw blancher (cauliflower)	31	Rumsey et al. (1981)
Pipe blancher (lime beans)	35	Rumsey et al. (1981)
Screw blancher with heat exchanger	67.6	Rumsey et al. (1981)
Integrated blancher/cooler	>100%[a]	Togeby et al. (1986)

[a] Heat recovery up to 70% by means of heat exchangers and water recycling.

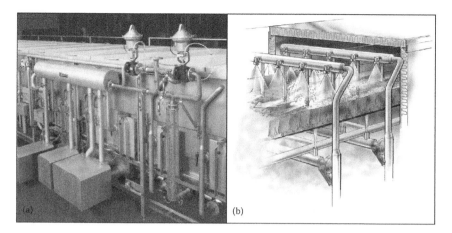

FIGURE 5.3 (a) Detail of a heat recovery system of an IBC Cabinplant integrated blancher and cooler system. (b) Detail of the spray water system. (From cabinplant.com with permission.)

turbulence of conventional hot water blanchers. The food is preheated with water that is recirculated through a heat exchanger. After blanching, a second recirculation system precools the food. Both systems pass water through the heat exchanger, where the water from the precooling section heats the preheating water and simultaneously cools. Up to 70% of heat is recovered. A recirculated water–steam mixture is used to blanch the food, and final cooling is by cold air (temperature reductions in the product from 90°C up to 20°C by evaporative cooling). Effluent production is negligible, and water consumption is reduced to approximately 1 m^3 per 10 t of product. The mass of the product blanched is 16.7–20 kg per kilogram of steam compared with 0.25–0.5 kg per kilogram in conventional hot water blanchers.

5.4.2 Steam Blanchers

As has already been mentioned, the main advantages of steam blanchers as compared to those that use water are smaller losses of water-soluble components and smaller volumes of effluent, particularly when air cooling is employed. These blanchers are also easy to clean and sterilize.

However, the energy efficiency of conventional steam blanchers is very low, with steam losses up to 95% (Table 5.7). Therefore, the use of stagnant chambers with curtains and/or water fasteners that reduce steam losses to 70% is common (Scott et al. 1981).

Hydrostatic steam blanching increases the energy efficiency applying steam at slightly higher pressure. The product is washed, blanched, and cooled in the equipment, having a thermal yield of 27% (Layhee 1975).

Likewise, in conventional steam blanching, there is often poor heating uniformity in multiple layered foods. The time–temperature combination required to ensure enzyme inactivation in the center of the bed results in overheating of food at the edges and thus in the loss of texture and other sensory characteristics.

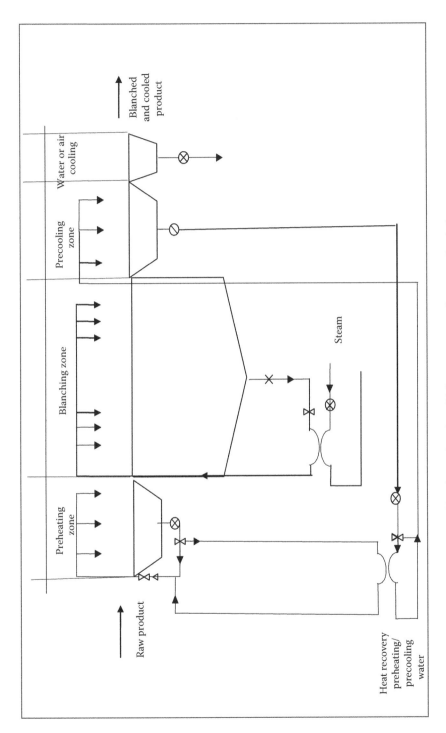

FIGURE 5.4   Scheme of an integrated water blancher and cooler with heat recovery and water recirculation.

TABLE 5.7

Efficiency of Some Steam Blanching Systems of Vegetable Products

Blanching System	Thermal Efficiency (%)	Source
Traditional steam blancher	5	Bomben (1979)
Traditional steam blancher (spinach)	13	Scott et al. (1981)
Steam blancher with water curtains (green beans)	19	Scott et al. (1981)
Blancher with hydrostatic steam (spinach)	27	Layhee (1975)
Blancher with hydrostatic steam/venture (spinach)	31	Scott et al. (1981)
Vibratory blancher/cooler	85	Bomben (1979)

To overcome the poor uniformity of conventional steam blanchers, Lazar et al. (1971) developed "individual quick blanching" (IQB; Figure 5.5). The process begins when a monolayer of the product is treated with steam for a short time (from 30 s to 300 s). After this, the product is held in an isothermal section for 30 s to 540 s in order to reach the uniform temperature in most of the product to assure the needed thermal treatment.

Lately other IQB models were developed like the ABCO Industries (Holland 1989). However, the IQB blanchers require a lot of space because vegetables are not placed in multilayer fashion.

Bomben et al. (1974) and Bomben (1979) developed the vibratory spiral blancher–cooler that improved the energetic efficiency of the IQB model. This model has a thermal yield up to 85% and has less water losses and effluent production than the steam blanching with water cooling. Timbers et al. (1984) and Cumming et al. (1984) developed a pilot prototype that combines an IQB system with thermocyclic

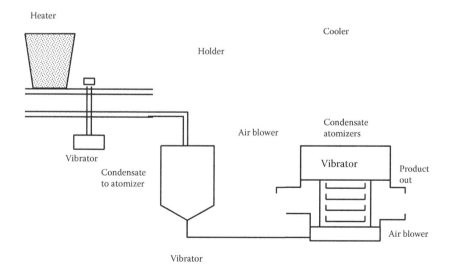

FIGURE 5.5 Schematic diagram of vibratory blancher.

blanching. This system increases the energy efficiency by the recirculation of non-condensed steam, reducing its consumption to around 0.2 t/t of the product. Besides, the amount of effluent was decreased, and the ascorbic acid retention for a complete POD inactivation was 76%–85%.

Other steam blanchers tend to increase the speed of the heat transfer searching in order to obtain a higher energetic yield in comparison to traditional systems.

This is the case of the fluidized bed blancher (Baxerres et al. 1977) that operates using an air–steam mixture that moves at approximately 4.5 m s^{-1} and in this manner fluidizes and heats simultaneously. In this model, the heat transfer improves thanks to the high exchange surface and the increase in the coefficient of convection due to the speed of the air–steam mixture.

An improved design of these blanchers is the "overpressure reel blancher" developed by Zittel (2000). A hermetically closed section allows having an internal pressure higher than the atmospheric pressure. The blanching medium can be a hot gas, steam, or a combination of both.

5.4.3 OTHER BLANCHING SYSTEMS

Other methods of blanching that have been studied for several decades are those that use dielectric heating, thermal infrared, thermal shocks, hot microwaves, or gases. Asselbergs et al. (1959) obtained satisfactory results using infrared for vegetable blanching. However, this method has the risk of causing dehydrations and burns if the product is not placed at a suitable distance from the infrared source. On the other hand, the application of a "thermal shock," that is to say, the vegetal product exposition to steam or boiling water during short periods of time (5–40 s), would suppose a remarkable power saving. High temperatures are not reached inside the product, and therefore, the residual activity of the enzymatic systems is considerable (20%–40% according to treatments in green beans). However, using "thermal shock," quality parameters such as texture of green beans are even better than those obtained with conventional blanching systems (Steinbuch 1980, 1984). Nevertheless, the equipment for the application of this heat treatment has not been developed sufficiently at the industrial level.

The use of microwaves as a blanching system can reduce the production of effluents and the necessary time for the required enzymatic inactivation. Even so, heating is not homogenous, and the color and ascorbic acid retention is smaller than that obtained using water or steam systems (Canet 1990). In addition, microwave blanching can produce product dehydrations (Varoquaux et al. 1974). Blanching with hot gas at 105°C–115°C, to which steam is injected as a humidifier, has the possibility of diminishing the amount of effluent production and reduces fuel consumption. In these systems, the use of over-warmed steam mixed with the combustion gas accelerates the heat transference by increasing the external gradient of temperature, which allows greater speed and efficiency of blanching. These new methods, in spite of the advantages that each one of them can contribute, are not actually used at the industrial level. This is due to the high substitution cost of the blanching equipment used at the present time and little knowledge on these blanching systems.

As a conclusion, the investment cost and other various costs that the use of water blanchers supposes are nowadays relatively low, taking into account that the energy efficiency of this system is superior to that of steam blanchers. Blanching by water using the new equipment obtains more uniform and effective treatments than those obtained with steam, allowing in addition the use of chemical additives, a better cleaning, and the elimination of strange flavors and odors. For these reasons, excepting products where it is important to diminish the loss of water-soluble substances like leaf vegetables, the systems that use water as a medium of heating are used traditionally in the industry.

5.5 BLANCHING OPTIMIZATION

To improve the product quality, the operational cost and environmental impact of the blanching process, and also the blanching and cooling systems, it is possible to develop models to simulate and optimize the process. POD and lipoxigenase inactivation determine the process efficacy, whereas nutrients and texture losses indicate the product quality. Likewise, the simulation models can calculate water and energy consumption and effluent production. Different works have developed blanching models that simulate the blanching process for different products and systems (Togeby et al. 1986; Arroqui et al. 2003; Garrote et al. 2004; Bevilacqua et al. 2004).

REFERENCES

Adams, J.B. 1991. Review: enzyme inactivation during heat processing of food-stuffs. *International Journal of Food Science and Technology* 26:1–20.

Andersson, A. 1994. Modelling of potato blanching. PhD diss., Lund Institute of Technology, University of Lund.153 pp.

Arroqui, C., López, A., Esnoz, A. and Vírseda, P. 2003. Mathematical model of an integrated blancher/cooler. *Journal of Food Engineering* 59(2–3):297–307.

Asselbergs, E.A., Mohr, W.P., Kemp, J.G., and Yates, A.R. 1959. Blanching of celery and apples by infra-red shows flavor, texture, appearance gains. *Quick Frozen Foods* 21(7):45–46, 150.

Baardseth, P., and Slinde, E. 1987. Enzymes and off-flavours: palmitoyl-Coa hydrolase, lipoxigenase, α-oxidation, peroxidase, catalase activities and ascorbic acid content in different vegetables. *Norwegian Journal of Agricultural Sciences* 1(2):111–117.

Barrett, D.M., Garcia, E.L., Russell, G.F., Ramirez, E., and Shirazi, A. 2000. Blanch time and cultivar effects on quality of frozen and stored corn and broccoli. *Journal of Food Science* 65(3):534–540.

Bartolomé, L.G., and Hoff, J.E. 1972. Firming of potatoes: biochemical effects of preheating. *Journal of Agricultural and Food Chemistry* 20:266–270.

Baxerres, J.L., Hauwsungcharern, A., and Gibert, H. 1977. Whirling bed: a new technique for gas fluidization of large particles. Lebensmittel-Wissenschaft Und-Technologie 10:191–197.

Bevilacqua, M., D'amore, A., and Polonara, F. 2004. A multi-criteria decision approach to choosing the optimal blanching–freezing system. *Journal of Food Engineering* 63(3):253–263.

Bomben, J.L. 1979. Waste reduction and energy use in blanching. *Food Processing Waste Management* 141–174.

Bomben, J.L., Brown, G.E., Dietrich, W.C., Hudson, J.S., and Farkas, D.F. 1974. Integrated blanching and cooling to reduce plant effluents. *Proc. 5th. Nat. Symp. Food Proc. Wastes, Environmental Protection Agency*, pp. 120–131.

Bomben, J.L., Dietrich, W.C., Hudson, J.S., Hamilton, H.K., and Farkas, O.F. 1975. Yields and solids loss in steam blanching cooling and freezing vegetables *Journal of Food Science* 40(4):660–664.

Böttcher, H. 1975. Enzyme activity and quality of frozen vegetables. I. Residual activity of peroxydase. *Nahrung* 19:173–179.

Canet, W. 1990. Effect of different blanching processes on the texture and ascorbic acid content of frozen Brussels sprouts (Efecto de diferentes escaldados en la textura y contenido de ácido ascórbico de coles de bruselas congeladas). *Alimentación, Equipos y Tecnología* June 5:46–55.

Canet, W. 1995. Stability and importance of vitamin C in frozen vegetables. (Estabilidad e importancia de la vitamina C en vegetales congelados). *Alimentación, Equipos y Tecnología* June 5:75–87.

Canet, W., Alvarez, M.D., Luna, P., Fernández, C., and Tortosa, M.E. 2004. Blanching effects on chemistry, quality and structure of green beans (cv. Moncayo). *European Food Research and Technology* 220:421–430.

Cumming, D.B., Stark, R., Timbers, G.E., and Cowmeadow, R. 1984. A new blanching system for the food industry. II. Commercial design and testing. *Journal of Food Process Engineering* 8(3–4):137–150.

Dietrich, W.C., Olson, R.L., Nutting, M., Neumann, H.J., and Boggs, M.M. 1959. Time–temperature-tolerance of frozen foods. XVIII. Effect of blanching conditions on colour stability of frozen beans. *Food Technology* 13:258.

Fuster, C. 1995a. Peroxidase activity as a quality indicator: pre-treated cauliflower preserved at different temperatures (Actividad peroxidasa como indicador de calidad: coliflor pretratada conservada a diferentes temperaturas) *Alimentaria* December:19–24.

Fuster, C. 1995b. Peroxidase activity as a quality indicator: pre-treated cauliflower preserved at different temperatures. (Actividad peroxidasa como indicador de calidad: coliflor pretratada conservada a diferentes temperaturas). *Alimentaria* December:26–28.

Galgano, F., Favati, F., Caruso, M., Pietrafesa, A., and Natella, S. 2007. The influence of processing and preservation on the retention of health-promoting compounds in broccoli. *Journal of Food Science* 72(2):130–135.

Garrote, R.L., Silva, E.R., Bertone, R.A., and Roa, R.D. 2004. Predicting the end point of a blanching process. *Lebensmittel-Wissenschaft Und-Technologie* 37(3):309–315.

Günes, B., and Baymdiri, A. 1993. Peroxidase and lipoxigenase inactivation during blanching of green beans, green peas and carrots. *Lebensmittel-Wissenschaft und -Technologie* 26:406–410.

Holland, P. 1989. Carte blanche to save cost. *Food Processing, UK* 58(10):57–58, 60.

Holmquist, J.W., Clifcorn, L.E., Heberlein, D.G., Schmidt, C.F., and Ritchell, E.C. 1954. Steam blanching of peas. *Food Technology* 8:437–445.

Hough, G., and Alzamora, S.M. 1984. Optimization of vitamin, texture and color retention during peas blanching. *Lebensmittel-Wissenschaft und -Technologie* 17:207–212.

Kaack, K. 1994. Blanching of green beans. *Plants Food for Human Nutrition* 46(4):353–360.

Lamb, J. 1976. Influence of water on the thermal properties of foods. *Chemical Industry* 24:1046–1048.

Lamberg, Y., and Hallström, B. 1986. Thermal properties of potatoes and a computer simulation model of a blanching process. *Journal of Food Technology* 21:577–585.

Layhee, P. 1975. Engineered FF line yields big production benefits. *Food Engineering* 47(2):61–62.

Lazar, M.E., Lund, D.B., and Dietrich, W.C. 1971. IQB-new concept in blanching. *Food Technology* 25:684–686.

Ni, L., Lin, D., and Barrett, D.M. 2005. Pectin methylesterase catalyzed firming effects on low temperature blanched vegetables. *Journal of Food Engineering* 70:546–556.

Olivera, D.F., Viña, S.Z., Marani, C.M., Ferreyra, R.M, Mugridge, A., Chaves, A.R., and Mascheroni, R.H. 2008. Effect of blanching on the quality of Brussels sprouts (*Brassica oleracea* l. Gemmifera dc) after frozen storage. *Journal of Food Engineering* 84(1):148–155.

Pala, M. 1982. Effect of different pretreatments on the quality of deep frozen green beans and carrots. *Refrigeration Science and Techno*logy 4:224–231.

Philippon, J. 1992. Sanitary quality of frozen vegetables. (Calidad sanitaria de vegetales congelados). *Alimentación, Equipos y Tecnología* June:105–110.

Raab, C.A., Luh, B.S., and Schweigert, B.S. 1973. Effects of heat processing on the retention of Vitamin B-6 in lima beans. *Journal of Food Science* 38:544–545.

Ralls, J.W., Maagdenberg, H.J., Yacoub, N.L., Homnick, D., Zinnecker, M., and Mercer, W.A. 1973. In-plant, continuous hot-gas blanching of spinach. *Journal of Food Science* 38(2):192–194.

Rao, M.A., Cooley, H.J., and Vitali, A.A. 1986. Thermal energy consumption for blanching and sterilization of snap beans. *Journal of Food Science* 51(2):378–380.

Ramesh, M.N. 2000. International Journal of Food Science and Technology 35 (4):377–384.

Rumsey, T.R., Scott, E.P., and Carroad, P.A. 1981. Energy consumption in water blanching. *Journal of Food Science* 47(1):295–298.

Rungapamestry, V., Duncan, A.J., Fuller, Z., and Ratcliffe, B. 2008. Influence of blanching and freezing broccoli (*Brassica oleracea* var. *italica*) prior to storage and cooking on glucosinolate concentrations and myrosinase activity. *European Food Research and Technology* 227(1):37–44.

Sanjuán, N., Hernando, I., Lluch, M.A., and Mulet, A. 2005. Effects of low temperature blanching on texture, microstructure and rehydration capacity of carrots. *Journal of the Science of Food and Agriculture* 85(12):2071–2076.

Sarikaya, A., and Özilgen, M. 1991. Kinetics of peroxidase inactivation during thermal processing of whole potatoes. *Lebensmittel-Wissenschaft und -Technologie* 24(2):159–163.

Scott, E.P., Carroad, P.A., Rumsey, T.R., Horn, J., Buhllert, J., and Rose, W.W. 1981. Energy consumption in steam blanchers. *Journal of Food Process Engineering* 5:77–88.

Steinbuch, E. 1980. The effect of heat shocks on quality retention of green beans during frozen storage. *Journal of Food Technology* 15:353–355.

Steinbuch, E. 1984. Heat shock treatment for vegetables to be frozen as an alternative for blanching. In *Thermal processing and quality of foods*. P. Zeuthen, J.C. Cheftel, C. Eriksson, M. Jul, H. Leniger, P. Linko, G. Varela, and G. Vos (eds.). London: Elsevier Applied Science Publishers.

Swartz, J., and Carroad, P.A. 1979. Recycling water in vegetable Blanching. *Food Technology* 34:54–59.

Sweat, V.E. 1974. Experimental values of thermal conductivity of selected fruits and vegetables. *Journal of Food Science* 39:1080–1083.

Tijskens, L.M.M., Waldron, K.W., Ingham, L., and Van Dijk, C. 1997. The kinetics of pectin methyl esterase in potatoes and carrots during blanching. *Journal of Food Engineering* 34(4):371–385.

Timbers, G.E., Stark, R., and Cumming, D.B. 1984. A new blanching system for the food processing industry. I: Design, construction and testing of a pilot plant prototype. *Journal of Food Processing and Preservation* 8:115–133.

Togeby, M., Hansen, N., and Mosekilde, E. 1986. Modelling energy consumption, loss of firmness and enzyme inactivation in an industrial blanching process. *Journal of Food Engineering* 4(5):251–267.

Varoquaux, P., Avisse, C., Nadal, N., and Cousin, R. 1974. The quality of canned peas. *Industries Alimentaires et Agricoles* 91(11):1415–1422.

Verlinden, B.E., and De Baerdemaeker, J. 1997. Modelling low temperature blanched carrot firmness based on heat induced processes and enzyme activity. *Journal of Food Science* 62(2):213–229.

Verlinden, B.E., Yuksel, D., Baheri, M.A., De Baerdemaeker, J., and Van Dijk, C. 2000. Low temperature blanching effect on the changes in mechanical properties during subsequent cooking of three potato cultivars. *International Journal of Food Science and Technology* 35(3):331–340.

Welty, R., Wicks, E., and Wilson, E. 1991. *Transferencia de momento, Calor y masa*. México: Ed. Limusa.

Williams, C.D., and Lim, M.H. 1986. Blanching of vegetables for freezing—which indicator to choose. *Food Technology* 40(3):130–140.

Zittel, D. 2000. Pressurized rotary blancher and method of operation. US (09/222,969) 1998-12-30.

6 Pretreatments for Meats I (Tenderization, Electrical Stimulation, and Portioning)

D. L. Hopkins

CONTENTS

6.1 INTRODUCTION

Flavor, juiciness, and tenderness influence the palatability of meat. Among these traits, tenderness is ranked as the most important for beef meat (Thompson 2002) but has lesser influence in sheep meat (Thompson et al. 2005). The three factors that determine meat tenderness are "background toughness," the toughening phase, and the tenderization phase. The toughening and tenderization phases take place during the postmortem storage or aging period. The effect of postmortem storage on tenderness is illustrated in Figure 6.1. Background toughness exists at the time of slaughter and does not change during the storage period as it is due to the connective tissue matrix that supports the muscle. Although this tissue may degrade during storage, when cooked meat toughness is determined, this degradation is not important (Purslow 2005).

The organization of the perimysium appears to affect the background toughness, since a general correlation between characteristics of the perimysium and

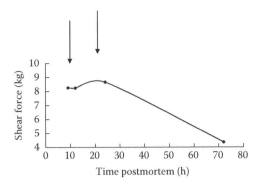

FIGURE 6.1 Proposed time course changes in shear force of ovine *longissimus* during postmortem storage. Maximum contraction occurs somewhere between the arrows as the muscle enters *rigor*. (Adapted from Wheeler, T. L. and Koohmaraie, M., *Journal of Animal Science* 72, 1232–1238, 1994. With permission of the American Society of Animal Science.)

tenderness of muscles has been found for both beef and chicken (Strandine et al. 1949). The toughening phase is caused by muscle shortening during rigor development (Koohmaraie et al. 1996; Wheeler and Koohmaraie 1994), but the degree of shortening can be manipulated by prerigor stretching, for example (Bouton et al. 1973; Hopkins et al. 2000a). The known effect of the interaction of the rate of postmortem glycolysis and temperature decline on shortening has led to industry recommendations on optimal processing conditions and procedures aimed at minimizing shortening for beef (Thompson 2002) and sheep (Thompson et al. 2005). To achieve these targets, intervention such as electrical stimulation is often required. Such intervention is very important when hot boning procedures are applied, but even with effective stimulation, this will not guarantee eating quality (Toohey and Hopkins 2006).

Electrical stimulation is defined as the application of an electric current to carcasses, where the aim is to ensure that the meat is tender under rapid carcass chilling regimes. The use of electrical stimulation for limiting meat toughness is not new (Rentschler 1951; Harsham and Deatherage 1951). The incorporation in a practical system was first used in New Zealand and then Australia to avoid toughness resulting from cold-induced shortening (Devine et al. 2004). In New Zealand, the major requirement of electrical stimulation was to accelerate the onset of *rigor mortis* so that it occurred at temperatures above those likely to result in cold-induced shortening before the meat was frozen whether taken from sheep (Chrystall and Hagyard 1976) or cattle (Davey et al. 1976).

Other potential effects of stimulation on postmortem muscle were reviewed by Cross (1979), and subsequent research has attempted to clarify the importance of these effects. These effects are described as either physical disruption of the myofibrillar matrix (Ho et al. 1997) or the acceleration of proteolysis (degradation in this case of structural proteins; Uytterhaegen et al. 1992), and these were reviewed by Hwang et al. (2003).

It has been known for a long time that meat tenderness improves during refrigerated storage (the tenderization phase), and it was suggested almost a century ago that this is due to enzymatic activity (Hoagland et al. 1917). It is now well established that postmortem proteolysis of myofibrillar and associated proteins is responsible for this process, but it is a matter of debate about which proteases are responsible for tenderization and which muscle proteins are degraded (Hopkins and Thompson 2002; Koohmaraie and Geesink 2006; Ouali et al. 2006; Hopkins and Geesink 2009).

Another approach that can be used to enhance the eating quality of red meat is to portion and retail meat not based on cuts but on eating quality. Using this approach, consumers are more likely to purchase meat that meets expectations, and this approach has emerged from the Australian system to grade meat called Meat Standards Australia (Polkinghorne 2006). There is also scope to stretch and reform meat to provide consumer acceptable portions, and several systems have emerged (O'Sullivan et al. 2003; Toohey et al. 2008d).

The purpose of this chapter is to discuss the factors that impact on aging, the principles of electrical stimulation, the latest developments in electrical stimulation technology, and how portioning can be used to improve eating quality and maximize carcass value.

6.2 MECHANISMS OF TENDERIZATION

6.2.1 Pattern and Processes of Tenderization

The processes affecting meat tenderness start at slaughter, and the endogenous enzymes responsible for proteolysis and thus tenderization will be active throughout the rigor process. While this proteolysis is taking place, significant tenderness changes are not evident until most of the muscle fibers are in rigor (Devine and Graafhuis 1995). The development of rigor and the shortening of fibers would be expected to counter early proteolysis so that the expected peak in shear force (see Figure 6.1) is eventually negated by the cumulative postrigor proteolysis resulting in tenderization. Under cooling conditions, those fibers at elevated temperatures will enter rigor early and will experience initially faster tenderization (Graafhuis et al. 1992). Thus, tenderness measured at the completion of rigor mortis (the earliest possible time) will be substantially different for electrically stimulated muscles than for nonstimulated muscles due to a difference in the rate of rigor development.

Even though there is good evidence that specific myofibrillar muscle proteins are degraded during the postmortem period (Bandman and Zdanis 1988), the total proportion of protein degraded is considered to be only small (Davey and Gilbert 1966), while there are substantial reductions in toughness. There are a growing number of reviews that summarize the biochemical changes, which occur in meat during the postmortem period (Ouali 1992; Koohmaraie 1996; Hopkins and Thompson 2002; Sentandreu et al. 2002; Hopkins and Taylor 2004; Ouali et al. 2006), and this chapter will draw on the overview provided by Hopkins and Taylor (2004).

Quantification of the contribution of individual proteins to tenderization remains a challenge, and although the rate of disappearance of individual proteins may not always match changes in toughness (Dransfield 1997), this does not diminish the strategic and cumulative contribution such degradation may have. The evidence indicates that a number of proteins associated with the thin and thick filaments and others found at the periphery of the Z-disk (Taylor et al. 1995) are degraded under normal postmortem chiller conditions. Several ultrastructural studies from the 1970s clearly show that the myofiber breaks in the I-band region and that the Z-disk is stable (Abbot et al. 1977; Davey and Graafhuis 1976; Gann and Merkel 1978).

Proteolysis can continue even in shortened muscle without the meat-becoming tender (Locker and Wild 1984; McDonagh et al. 1999). If the muscle has been stretched, then it is more tender (Hopkins and Thompson 2001a), but proteolysis may not have occurred to the same extent as nonstretched meat for the same shear force changes.

The large protein titin (up to 3700 kDa; Labeit and Kolmerer 1995), which is crucial to the orderly organization of the sarcomere, has been clearly shown to degrade into fragments during the postmortem period (Robson et al. 1991) as has another large protein nebulin (Huff-Lonergan et al. 1996) that is associated with actin. Degradation of both titin and nebulin was the suggested reason for the increased fragility of myofibrils in the I-band region (Taylor et al. 1995) and was linked to the major decrease in toughness, which occurs between 24 h and 72 h postmortem (Dransfield et al. 1992a; Wheeler and Koohmaraie 1994), but the extent of changes will be mediated by muscle types (Taylor et al. 1995).

There has been extensive work to discover which enzyme groups are responsible for protein degradation in postmortem muscle with the contenders being the cathepsins, serine proteases, proteasomes, caspases, and calpains. Of these, the most substantial evidence supports the calpains as the important proteolytic enzymatic group. A full discussion of the different enzymatic candidates is given by Hopkins and Geesink (2009), and a brief overview of the calpains will be given here.

6.2.2 CALPAINS

Murachi et al. (1981) described an enzyme (EC 3.4.22.17) that had two forms (calpain I and calpain II), subsequently called μ-calpain and m-calpain, respectively. The latter form requires much higher levels of Ca^{2+} ions for activation. An endogenous inhibitor of the enzymes was named calpastatin (Murachi et al. 1981). It was reported that less of this inhibitor was required to reduce m-calpain activity and that inhibition occurred by the binding of Ca^{2+} ions to the calpains resulting in a conformational change, thereby allowing interaction with calpastatin and inhibition. The early research also showed that the optimal condition for in vitro calpain activity was pH 7.5 at 25°C. A detailed review by Wendt et al. (2004) describes the inhibitory mode of calpastatin and discusses the importance of Ca^{2+} for calpastatin effectiveness. Other calpain forms have since been identified, and the review by Goll et al. (2003) provides an exhaustive overview of the presence of these different forms across a range of species.

Hopkins and Thompson (2001b,c) claimed using a range of indicators that proteolysis was very rapid during the first 24 h after death. This outcome was consistent with activation of μ-calpain and the significant tenderization of muscle postrigor. Indeed, autolysis of μ-calpain as early as 3 h postmortem in ovine muscle has been shown by Veiseth et al. (2001) to be indicative of proteolytic activity before the onset of rigor. More recently, Veiseth et al. (2004) have showed significant degradation of desmin at 9 h postmortem and a significant increase in myofibrillar fragmentation index (MFI) values at 6 h postmortem compared to "at death" values where greater MFI values indicate more extensive myofibrillar protein degradation.

Dransfield et al. (1992b) predicted that tenderization (defined as a decrease in shear force) commenced at pH 6.1, which even in fast glycolyzing muscle would be at least 3 h after death according to their data. Subsequently, it was suggested that μ-calpain was not fully activated until this pH (6.1) was reached (Dransfield 1993) and that this coincided with the start of tenderization. However, this does not preclude proteolysis occurring before this pH as the prediction by Dransfield (1993) sees an increase in μ-calpain activity up to a maximum at 10 h. Thus, proteolysis could proceed before tenderization commences as suggested by the results of Hopkins and Thompson (2001c). It could be speculated that since all single fibers do not enter rigor simultaneously (Jeacocke 1984), then neither would the activation of μ-calpain occur uniformly throughout fiber bundles. The temperature/pH dependency of the calpains suggests that interventions like electrical stimulation will impact on their level and duration of activity.

6.3 EFFECT OF ELECTRICAL STIMULATION

6.3.1 RATE OF GLYCOLYSIS AND RIGOR

An electric current causes muscles to contract increasing the rate of glycolysis, resulting in an immediate fall in pH (shown as ΔpH). This ranges from 0.6 pH unit at 35°C to 0.018 pH unit at 15°C. Following the ΔpH, there is a temperature-dependent acceleration of glycolysis (dpH/dt) and subsequent early rigor mortis development. Rigor mortis occurs in muscles when all supplies of adenosine triphosphate (ATP) are exhausted (Bendall 1969). This does not occur across all muscles simultaneously with a concomitant fall in pH. Jeacocke (1984) showed for single fibers that there was a contracture as the final ATP disappeared (i.e., rigor), and each fiber had its own time course depending on initial glycogen. A small temperature-dependent degree of contraction occurs for each muscle fiber as it enters rigor. The onset of the sequential progression into rigor for each muscle fiber can be tracked by measuring isometric tension and muscle shortening (Olsson et al. 1994; Devine et al. 1999), where the development of isometric tension is greatest at higher temperatures (Hertzman et al. 1993). Above 10°C–15°C, the traces reveal a steady increase in tension prerigor and continuous shortening. Such traces can be interpreted as arising from a succession of individual muscle fibers, each exhausting their energy reserves (i.e., reaching rigor) and each separately shortening (Jeacocke 1984); these sum to create tension. Stiffness is a consequence of each single fiber going into full rigor, with irreversible cross-bridge formation of the contractile components actin and myosin. With increasing numbers of fibers entering rigor, the stiffness increases and is predominant when the muscle reaches a pH of approximately 6.0.

The classical studies of Locker and Hagyard (1963) with muscle entering rigor at different temperatures showed minimal shortening at close to 15°C (for excised muscle), and this correlated with minimal meat toughness (e.g., Tornberg 1996). Above 12°C–15°C, a contracture occurs at rigor, and below this temperature, a contracture occurs before rigor. Thus, shortening effects above 15°C are a consequence of rigor shortening only and occur when muscles become depleted of glycogen. As the temperature falls below 12°C, a prerigor contracture takes place until rigor is completed due to elevated levels of cellular calcium from the sarcoplasmic reticulum that activate actomyosin ATPase. The rise of calcium in the cytoplasm under these conditions is due to the failure of the sarcoplasmic reticulum to sequester cytoplasmic calcium (Jaime et al. 1992), and this leads to "cold-induced shortening." This is different to the transitory rise in "free" calcium that occurs after electrical stimulation. The balance of calcium under the latter conditions is potentially an important factor in the activation of enzymes such as the calpains, and the tuning of the amount of stimulation with chilling rate to reach rigor mortis at 15°C resulted in optimum tenderization in several studies (e.g., Wahlgren et al. 1997).

6.3.2 STRUCTURAL AND PROTEOLYTIC CHANGES

Contracture bands have been observed in electrically stimulated muscle (Hwang and Thompson 2002; Figure 6.2) and are localized areas where significant disruption of the sarcomere is observed and therefore could contribute to the improved tenderization observed with stimulation. Such contracture bands are, however, not specific to stimulated muscle and have been observed in nonstimulated muscle (e.g., cold shortened and high temperature treated). Because such super contracture is considered related to the loss of calcium regulation by the sarcoplasmic reticulum or/and the mitochondria (Davey and Gilbert 1974; Cornforth et al. 1980), it is suggested that the resultant contracture bands are a consequence of abnormal, perhaps localized calcium release from the sarcoplasmic reticulum through a tetanic contracture.

Formation of contracture bands is dependent on current frequency (Marsh 1985; Takahashi et al. 1987) or the interaction between current frequency and voltage (Hwang and Thompson 2002). If the time interval between successive stimuli is more than approximately 0.25 s, the muscle tetanic shortening is reversible. On the other hand, when a higher frequency of current is applied, muscle may not have enough time for relaxation between succeeding twitches and forms irreversible contracture bands. Even though practical stimulation systems have focused on acceleration of glycolysis, the pulse frequency and voltage of electrical stimulation systems used in most studies appeared to be high enough to avoid this confounding effect. For example, in the study of Hwang and Thompson (2002), a 14.3 pulses/s system (800 V, for 55 s, at approximately 45 min postmortem) and a 36 pulses/s system (45 V, for 45 s, soon after stunning) resulted in 89% and 55% of the specimen blocks containing contracture bands, respectively. With some exceptions in a few studies (e.g., George et al. 1980), it appears that ultrastructural alteration takes place in electrically stimulated muscle, and that is related to improved meat tenderness to some extent. However, its relative magnitude compared to other factors such as enzymatic tenderization is a subject of debate. Further, a quantitative study of beef *M. longissimus* showed that sarcomeres adjacent to the contracture nodes had a higher frequency of I-band fracture (Ho et al. 1996; see Figure 6.2), so it is possible that physical stretching/tearing leads to an

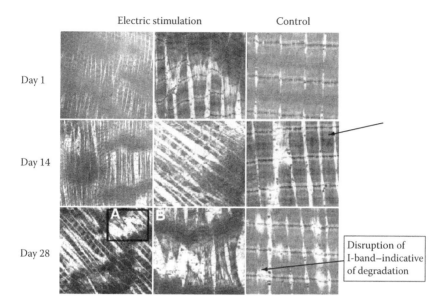

FIGURE 6.2   Characteristics of contracture bands in electrically stimulated beef *M. longissimus* compared to controls, and changes during aging. Photograph B of day 28 is a higher magnification of square A of day 28, which shows a zone of contracture bands and the major disruption that occurs in the sarcomere when this contracture occurs. The I-band is shown for the control muscle by the arrow (day 14), and the much more orderly arrangement of the sarcomere is seen in the images of control muscle. As muscle ages, degradation in the I-band occurs as shown (day 28). (From Hwang, I. H., and Thompson, J. M., *Journal of Asian Australasian Animal Science* 15, 111–116, 2002.)

acceleration of proteolysis as a result of greater exposure of proteolytic substrates within muscle fibers, in addition to the direct effect of physical tearing.

The activity of cysteine protease enzymes, most notably the calpains, has been shown to be sensitive to the prevailing pH and temperature of the meat (Dransfield 1994; Simmons et al. 1996). Since electrical stimulation alters the postmortem pH/temperature relationship, it is reasonable to expect that proteolysis would also be affected. It should be remembered that muscle types vary in their glycolytic response due to stimulation, as Devine et al. (1984) showed that red muscles like the masseter did not exhibit an accelerated rate of pH fall, whereas white muscles like the cutaneous did. This muscle specificity will flow onto differences in the rate of proteolysis. Another subtlety is the type of electricity applied, as Hopkins et al. (2006) showed increased protein degradation (as indicated by MFI values) when 14-Hz electricity was applied to carcasses compared to 10-Hz electricity with a commensurate reduction in shear force.

There are several possible explanations why stimulation would increase the activity of specific enzymes like the calpains. It may be due to some intrinsic effect associated with the rapid pH decline that results in a low pH at elevated temperatures to affect calpain/calpastatin ratios, or it could be due to a flow-on effect associated with a significant increase in "free" calcium, which leads to activation of the calpains, particularly μ-calpain. In this regard, Dransfield (1994) predicted, based on his modeling of postmortem calpain activity, that calpain activity in rapidly glycolyzing muscle would be increased by a factor of six compared to muscle with more "normal" rates of glycolysis. Under these conditions, however, μ-calpain is likely to undergo autolysis (due to high temperature and low pH), so the interplay with temperature and the levels of "free" calcium will be important. Indeed, Ducastaing et al. (1985) demonstrated that when beef carcasses were subjected to high-voltage stimulation (550 V, 60 Hz) for 2 min, the activity of μ-calpain at 4-h postmortem was reduced by 80%. These authors commented that directly after stimulation, the activity of μ-calpain increased by 10%–15% compared to unstimulated muscle. In contrast, Hwang and Thompson (2001a) reported that μ-calpain and calpastatin activity directly poststimulation was lower than prestimulation irrespective of the type of stimulation, but unfortunately, there were no control measures for comparison.

The apparent contrary results in the literature about whether stimulation results in a rise in the activity of μ-calpain are evidenced by comparing the results of Ferguson et al. (2000) and Rhee and Kim (2001). These studies highlight the importance of the interplay between sampling time and the prevailing temperature and pH, with the data from Rhee and Kim (2001) shedding some light on this interplay. Samples held at 30°C for 3-h postmortem exhibited a larger reduction in μ-calpain activity (at 3- and 9-h postmortem) than samples held at lower temperatures, but it was found that stimulation had a larger effect on μ-calpain activity than the elevated temperature. Unfortunately, the degree of interaction between pH decline and temperature was not examined in this study. This could have verified whether the conclusion of Dransfield et al. (1992a) that temperature did not affect calpain activity until the pH reached 6.2 was correct. In this regard, a previous study by Hwang and Thompson (2001b) using an experimental design aimed at unraveling the impact of pH and temperature on calpain activity reported an interaction between temperature and pH at 1.5-h postmortem. As a result, when the chilling was slow, the activity of the μ-calpain and calpastatin decreased when pH decline was rapid, whereas when chilling was rapid, the activity of the μ-calpain was largely unaffected by pH decline. This response is consistent with the interaction between chilling regime and change in MFIs reported by Pommier (1992).

Hwang et al. (2003) proposed, based on the work of Devine et al. (2002a), that electrical stimulation may confer protection to those muscle fibers that enter rigor soon after stimulation. The notion was that stimulation avoided prolonged prerigor exposure to high temperatures/low pH, thus maintaining optimal calpain levels. Limited support for this proposal was outlined in a study by Rosenvold et al. (2008), which showed that stimulated muscles that were removed prerigor, wrapped to prevent shortening, and held at 35°C prerigor exhibited a tendency to reach an acceptable shear force level quicker compared to unstimulated muscles held under the same conditions. It was argued that under such conditions where the pH dropped to 6.2, some fibers would enter rigor, and this would reduce the denaturation of muscle fiber proteins and the inhibition of aging enzymes (Rosenvold et al. 2008), but this is a speculative conclusion. In a comprehensive study to further understand the impacts of stimulation on tenderization, Martin et al. (2006) excised muscle from stimulated and nonstimulated lamb carcasses, wrapped the muscle, and allowed rigor

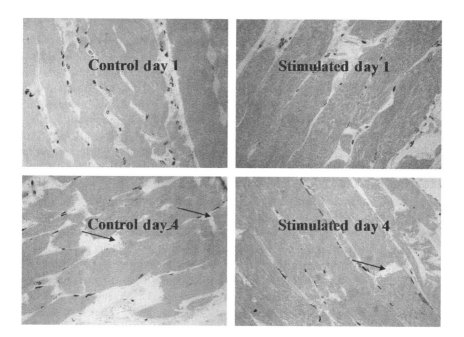

FIGURE 6.3 Arrows indicate fiber breaks, but there was no difference in this work between treatments (control versus stimulated at days 1 and 4 of aging) in the number of fiber breaks. The number of fiber breaks increases with aging indicative of degradation with none apparent in the 1-day-aged muscle. 40× magnification. (From Martin, K. M. et al., *Australian Journal of Experimental Agriculture* 46, 891–896, 2006.)

development at the 15°C optimum (Devine et al. 2002b). Unlike the study by Devine et al. (2002a), there was no difference in shear force between treatments at the various sampling points in the study of Martin et al. (2006), but it should be noted that a lower voltage was used than that in the study of Devine et al. (2002a) and so the ΔpH was likely to be much less. The study of Martin et al. (2006) did use a new novel histology method for examining breaks in fibers as shown in Figure 6.3. Although there was no support for the view that stimulation per se as applied in the study, caused physical fiber disruption, there was a strong relationship between degradation of myofibrils measured using MFIs and fiber breaks.

6.3.3 VARIATION IN RESPONSE

Electrical stimulation has been identified in many studies as a useful processing technique to improve meat quality traits (e.g., Devine et al. 2004); however, the report of Hollung et al. (2007) demonstrated considerable variation in response to stimulation between beef carcasses. This was indicated by the change in muscle pH and also manifested by variation in the relationship between stimulation and tenderness. Part of the variation was due to differing times of stimulation postdeath, with longer times likely to reduce the effectiveness of the response in pH fall, but this did not explain all of the variation. Studies by Hopkins et al. (2000b) and Devine et al. (2001) highlighted the variation that can arise. In the experiment by Hopkins et al. (2000b), results showed that there was a significant difference between slaughter days in the rate of pH decline after low voltage stimulation. This experiment was conducted under extremely controlled conditions where lambs had the same background and were exposed to identical preslaughter and preslaughter conditions. Devine et al. (2001) found a wide variability in the tenderization rate for samples held at 10°C after high-voltage electrical stimulation. However, the differences in sarcomere length were not significant compared to nonstimulated samples, and there was no detectable difference in the rate of tenderization and final tenderization between treatments. Hopkins and Toohey (2008) using data from Toohey et al. (2008a) showed that stimulation explained 62% of the variation in initial pH, indicating that other factors were contributing to the variation in initial pH. The relationship between the initial pH measurement and the shear force at 1 day of aging demonstrated that some carcasses that have low initial pH values indicative of an effective stimulation can still have high shear force values and vice versa. This same response was reported by Hollung et al. (2007). Daly et al. (2006) showed that muscle glycogen concentration at death impacts on the pH response of muscle to stimulation, and at low glycogen levels, the rate of pH decline is not elevated; but more recently, Ferguson et al. (2008) suggested that neither the form nor the concentration of glycogen influences the response. These differences can be tied up with the absolute levels of prestimulation pH, which can impact on the relationship between ΔpH and glycogen concentration. These results suggest that much of this variation may well be attributed to animal differences (Simmons et al. 1997), which is exemplified by the report of Bendall (1978) who found large variation in the rate of pH decline in beef carcasses even when the cooling rate was similar.

Although there are some carcasses that do not respond to stimulation and others that produce tender meat without stimulation, overall stimulation does reduce the mean shear force level in short aged red meat, conferring a positive effect on eating quality. In fact, recent work has shown a benefit for meat aged for up to 30 days (Pearce et al. 2009). Additionally, the evidence indicates that stimulation can also reduce the variance in eating quality traits (Hopkins and Toohey 2006), and this is an important outcome as it can eliminate very tough meat from the system.

6.4 DEVELOPMENTS IN ELECTRICAL STIMULATION

There are a number of different types of stimulation systems that have been used for both sheep and beef, and the electrical parameters for some of these systems are given by Devine et al. (2004). Traditionally high voltage electrical stimulation (HVS) systems used on sheep carcasses have applied a fixed voltage averaged across all carcasses being stimulated (Devine et al. 2004), whereas for beef, the systems have been more in the low to medium voltage range (Devine et al. 2004). Rubbing bars have been used to apply high voltage stimulation to lamb and sheep carcasses at the completion of the dressing procedure (Morton et al. 1999; Hopkins and Toohey 2006). However, this process poses concerns for work safety, gives an average stimulation effect across carcasses, and is expensive, although it can significantly reduce toughness in sheep meat (e.g., Hopkins and Toohey 2006). A new approach has been developed in Australia, where each carcass is stimulated individually using segmented electrodes to ensure that only one carcass is in contact with each electrode at a time. Thus, when a carcass contacts the first electrode, the resistance is measured and a pulse sent to the control module for that electrode, and this adjusts the voltage to maintain the desired current. This allows computer-programmed electronics to give a specified, but adjustable electrical input to each carcass to match the requirements of a particular carcass type (reflecting individual resistance) while maintaining the delivery of a predetermined level of current. The new approach reduces the installation costs with respect to occupational health and safety as the power levels and pulse widths used comply with regulations according to the Australian Standard 60479-2002 (Anon. 2002) eliminating the need for isolation of the unit. The industry uptake of the new system in Australia has been extensive, particularly among sheep abattoirs (Hopkins et al. 2008), and the system is now being extended to other countries (C. Mudford, personal communication).

The results of the study of Shaw et al. (2005) clearly showed that the approach to stimulation did achieve comparable results to an HVS system with the production of lamb meat with a similar tenderness and eating quality level. There was a clear improvement over meat, which was not subjected to any form of stimulation, and this effect was confirmed in two different muscles as shown in Table 6.1. Extensive testing and optimization of this approach has occurred, and the system has been designed so that either a predressing (Toohey et al. 2008a) or a postdressing (Pearce et al. 2006b) application of the current can be applied as shown in Figures 6.4 and 6.5. This provides greater flexibility with respect to installation of the unit in existing abattoirs. As shown in Table 6.2, the predressing system that uses different electrodes to the postdressing system can significantly increase the rate of pH decline (Toohey et al. 2008a). In this study, the predicted temperature at pH 6.0 for stimulated carcasses was 24.8°C and was 13.9°C for nonstimulated carcasses, and

---

TABLE 6.1

Mean Tenderness and Overall Liking Sensory Scores (0–100, with Highest Best) for 2-Day-Aged Loin and Rump Cuts from 18 Carcasses (6 per Treatment)

Trait	Control	New Stimulation System	Old Stimulation System (HVS)	S.E.
		Loin		
Tenderness	65.2a	74.6b	76.0b	1.6
Overall liking	65.6a	72.1b	72.4b	1.5
		Rump		
Tenderness	48.8a	63.4b	60.1b	1.60
Overall liking	53.4a	63.2b	61.1b	1.50

Source: Adapted from Shaw, F. D. et al., *Australian Journal of Experimental Agriculture* 45, 575–583, 2005.

Note: Means followed by the same letter in a row are not significantly different ($P = 0.05$).

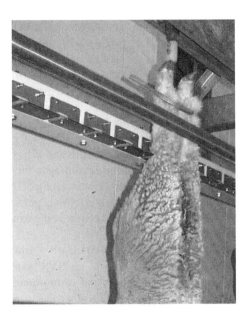

FIGURE 6.4 Photo showing a predressing electrical stimulation unit with the current administered through the skin on back legs. (Courtesy of NSW Department of Primary Industries, Cowra, Australia.)

FIGURE 6.5 Photo showing a predressing electrical stimulation unit—the separate electrodes can be seen separated by the white insulator, and only one carcass is in contact with any one electrode at any one time. (Courtesy of Murdoch University, Perth, Australia.)

TABLE 6.2
Predicted Means (S.E.D.) for Carcass Weight (kg), GR (mm), pH, Predicted Temperature at pH 6.0 (°C), Shear Force (N), Redness (a*), and Color Stability at 630/580 nm for *M. Longissimus* of Electrically Stimulated and Nonstimulated Lamb Carcasses (40 per Treatment)

Trait	Stimulation[a]	No stimulation	S.E.D.
Carcass weight (kg)	21.6a	21.4a	0.37
GR (mm)[b]	11.7a	11.0a	0.57
Initial loin pH	6.34a	6.79b	0.04
Predicted temperature at pH 6.0	24.8b	13.9a	1.50
Shear force—1 day aged (N)	36.0a	44.0b	2.40
Redness (a*)	7.70a	7.00a	0.32
Color stability (630/580 nm)	3.20a	3.00a	0.14

Source: Adapted from Toohey, E. S. et al., *Meat Science* 79, 683–691, 2008.

Note: Means followed by the same letter in a row are not significantly different ($P = 0.05$).

[a] Stimulation treatment was at a current of 800 mA with a pulse width of 0.5 ms.

[b] Adjusted to a hot carcass weight of 21.5 kg and is a measure of total tissue thickness over the 12th rib, 11 cm from the spine.

this translated into much tougher meat (Table 6.2) for the nonstimulated carcasses demonstrating the benefit of stimulation.

Likewise, the postdressing system has been shown to achieve similar results to the predressing system, and in one experiment, Pearce et al. (2006b) reported that the best combination of parameters was a current of 1000 mA, with a pulse width of 2.5 ms at 15 Hz, and showed that currents of 400 mA were much less effective. In the same paper, it was also shown by Pearce et al. (2006b) that if the frequency was altered across the electrodes, this could further increase the rate of pH fall, and this was examined in a subsequent study using a 6-electrode stimulation unit (Pearce et al. 2006a). In this case, the current was set at 1000 mA and a pulse width of 2.5 ms, and the frequency was set at the following levels: 10, 15, 25, 10, 15, and 25 Hz.

In these various studies, the impact on shear force, pH, and meat color stability has been examined, and the approach has recently been tested on goat carcasses (Toohey et al. 2008b). The general conclusion is that, irrespective of the system, there is a significant impact on the rate of pH fall, although this does vary according to animal groups. As a consequence, improvements in tenderness, particularly for product aged for less than 5 days, will be seen with no detrimental effect on color or the stability of color under simulated retail display as shown in Table 6.2.

6.5 NEW APPROACHES TO PORTIONING MEAT

6.5.1 FOCUS ON EATING QUALITY

In order to satisfy consumer demand for leaner meat of more manageable portions, there has been significant development of subprimal preparation in both the sheep and beef industries. For example, in Australia in the early 1990s, there was a strong move to change the image of lamb, and part of this program was the development of a range of boneless, low fat cuts (Hopkins et al. 1995). These cuts were designed for busy consumers who did not have the time to slow-cook meat and were looking for more versatile cuts with low waste. Swatland (2000) provides a good description of the more traditional cuts of beef and lamb, where in the past, for example, a hindleg of lamb would be marketed as such, but now the options include preparation into rump, round (knuckle), topside, and/or an easy carve leg. Preparation descriptions are available for these new cuts (Anon. 2005). Work by Farouk (2008) in New Zealand has followed the notion of value-adding to a range of subprimal cuts for both beef and lamb, and an impressive range of value-added cuts have been developed.

To maximize the value from a carcass, the inherent qualities on new cuts must be known, and this has led to extensive work in a number of countries. Extensive work in North America has been conducted to characterize the meat qualities of muscles in lower value cuts like the chuck (Von Seggern et al. 2005) so that alternative use of the cut could be made. This work has led to a comprehensive profile of the beef carcass for a range of meat quality traits (Jones et al. 2005), but these traits did not include eating quality. A contrasting approach was adopted in Australia based on development of a prediction model for the eating quality of beef (Thompson 2002), where anatomical cut descriptions were replaced with determination of the eating quality of muscles. This work highlighted the real value differences between carcasses that were of similar type and was designed to improve the "eating quality guarantee," which could be given to any piece of meat. For example, as discussed by Polkinghorne (2006), rump steak is composed of different muscles that have various levels of eating quality that will also interact with the cooking method, so the true value of the cut will shift depending on which muscles and cooking method are used. This approach challenged the concept that single muscle or carcass measures can be used to predict eating quality and thus true value. Currently, the model provides an eating quality score for 46 muscles cooked by six different methods (Polkinghorne 2006), but it has been demonstrated at the retail level that you can market products according to predicted eating quality within cooking guidelines without any mention of cuts (Polkinghorne 2006). This concept also led to the development of new value-added products, more appropriate use of particular cuts such as topsides thin sliced, and seam boning (e.g., MSA 2010).

6.5.2 Hot Boning and Portion Control

The challenge from increasing processing efficiency is to maintain or enhance eating quality, and this is a major consideration for the adoption of hot boning. Hot boning can be defined as the removal of muscle/cuts from the carcass prior to chilling. There are clear long-term economic advantages for hot boning such as a reduction in carcass weight loss, reduction in drip loss, reduction in chilling space (thus saving in refrigeration energy inputs), faster turnover of meat, labor savings, and transport costs (primal cuts versus carcasses; Pisula and Tyburcy 1996). However, there are also disadvantages to this type of system including initial costs, changes in cut shape, marketing of product, tenderness (Pisula and Tyburcy 1996), increased risk of shortening thus leading to toughening (Devine et al. 2004), and increased risk of bacterial problems (Spooncer 1993).

Research on the potential of hot boning beef has been more extensive (e.g., Taylor et al. 1980/1981) than for sheep, and several different systems of application have been proposed (Waylan and Kastner 2004). The research has been extended to the direct electrical stimulation of muscles after hot boning (White et al. 2006) with a significant reduction in toughness compared to unstimulated muscles. Hot boning offers the ability to streamline processing to cater for inherent differences in muscle characteristics. However, the adoption of hot or warm boning in the sheep meat processing industry has been limited, and in countries like Australia, only aged sheep have been processed using this approach. This process has, in some cases, been integrated with electrical stimulation with the intent of reducing muscle contraction subsequent to removal from the skeleton. The only published data on the eating quality of sheep meat processed through this system show a low level (~14%) of consumer compliance (Toohey and Hopkins 2006). Although there was some evidence that sarcomere shortening contributed to this result, sarcomere length only explained 10% of the variation in sensory-assessed tenderness. In contrast, when aged sheep were processed at the same abattoir, but conventionally chilled and cold boned, product aged for 7 days achieved a high level (86%) of consumer compliance (Hopkins and Toohey 2006). In follow-up work where loin meat was wrapped after hot boning and aged for seven days, a 14% and 24% improvement, respectively, in the overall liking and tenderness scores was achieved compared to product unwrapped and frozen at 1 day (Toohey et al. 2008c). In terms of consumer compliance, the level improved from 53% to 83%, and the reduction in shear force was dramatic at 53%. Based on sarcomere length, it appears that the benefit in this case came not from the wrapping but from the aging, and it is proposed that this was because rigor was well advanced by the time the muscle was wrapped. Devine et al. (2002b) demonstrated that excising and wrapping meat could be used to control sarcomere shortening, but it can also be reduced by stretching muscle prerigor.

An advantage of hot boning that has received little attention is the ability to reform muscles or cuts for subsequent portion cutting. O'Sullivan et al. (2003) reported on their work that utilized a technology called the PiVac to control both the contraction and shape of hot boned beef. This technology requires that meat is placed inside a chamber that is lined with a plastic wrap, which has a high degree of elasticity, and the meat is then squeezed firm under pressure and held by the wrap when removed from the machine. PiVac-treated meat had the lowest variation in shear force, indicating that this method does something different to meat structure. An alternative approach has been developed in New Zealand and is undergoing validation in both New Zealand and Australia. Recent Australian work with sheep topsides has shown that this new approach can significantly improve the tenderness of hot boned topsides (Toohey et al. 2008d), but also importantly the altered shape of the topside allows slicing of uniform portions (Figure 6.6). This offers significant potential to the industry but still requires research to ensure that the muscles within the slices remain bound on cooking. A recent review has contrasted this technology now called SmartShape™ and SmartStretch™ with other methods of stretching and shaping meat (Taylor and Hopkins 2011).

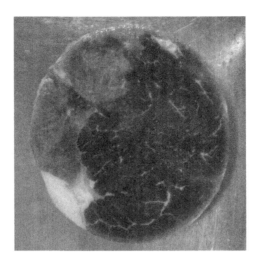

FIGURE 6.6 Photo showing a slice of topside after being taken from a prerigor stretched topside, which was frozen after treatment. (Courtesy of NSW Department of Primary Industries, Cowra, Australia.)

6.6 CONCLUSIONS

As meat ages, it becomes more tender, but the full impact of this improvement is only evidenced after rigor when tenderness can be measured. The degradation of myofibrillar proteins is responsible for tenderization, and the more important enzymatic group involved in this process are the calpains. Electrical stimulation can be used to accelerate the onset of rigor preventing "cold-induced shortening," but also accelerate the degradation of myofibrillar proteins. Recent developments in electrical technology offer the red meat processing industry new potential to improve product quality, and when combined with new approaches to preparing and grading meat, the consumer is the beneficiary.

REFERENCES

Abbot, M. T., Pearson, A. M., Price, J. F., and G. R. Hooper. 1977. Ultrastructural changes during autolysis of red and white porcine muscle. *Journal of Food Science* 42: 1185–1188.

Anon. 2002. Australian Standard AS/NZS 60479. Effects of current on human beings and livestock. Standards Australia, Standards House, North Sydney, NSW.

Anon. 2005. *Handbook of Australian Meat*, 7th edn. Brisbane: Authority for Uniform Specification Meat and Livestock.

Bandman, E., and D. Zdanis. 1988. An immunological method to assess protein degradation in *post-mortem* muscle. *Meat Science* 22:1–19.

Bendall, J. R. 1969. *Muscles, Molecules, and Movement*. London: Heinemann.

Bendall, J. R. 1978. Variability in the rates of pH fall and of lactate production in the muscles on cooling beef carcases. *Meat Science* 2:91–104.

Bouton, P. E., Fisher, A. L., Harris, P. V., and R. I. Baxter. 1973. A comparison of the effects of some post-slaughter treatments on the tenderness of beef. *Journal of Food Science* 8:39–49.

Chrystall, B. B., and C. J. Hagyard. 1976. Electrical stimulation and lamb tenderness. *New Zealand Journal of Agricultural Research* 19:7–11.

Cornforth, D. P., Pearson, A. M., and R. A. Merkel. 1980. Relationship of mitochondria and sarcoplasmic reticulum to cold shortening. *Meat Science* 4:103–121.

Cross, H. R. 1979. Effects of electrical stimulation on meat tissue and muscle properties—A review. *Journal of Food Science* 44:509–514/523.

Daly, B. L., Gardner, G. E., Ferguson, D. M., and J. M. Thompson. 2006. The effect of time of feed prior to slaughter on muscle glycogen metabolism and the rate of pH decline in three different muscles of stimulated and non-stimulated sheep carcases. *Australian Journal of Agricultural Research* 57:1229–1235.

Davey, C. L., and K. V. Gilbert. 1966. Studies in meat tenderness II. Proteolysis and the aging of beef. *Journal of Food Science* 31:135–140.

Davey, C. L., and K. V. Gilbert. 1974. The mechanism of cold-induced shortening in beef muscle. *Journal of Food Technology* 9:51–58.

Davey, C. L., and A. E. Graafhuis. 1976. Structural changes in beef muscle during ageing. *Journal of the Science of Food and Agriculture* 27:301–306.

Davey, C. L., Gilbert, K. V., and W. A. Carse. 1976. Carcass electrical stimulation to prevent cold shortening toughness in beef. *New Zealand Journal of Agricultural Research* 19:13–18.

Devine, C. E., and A. E. Graafhuis. 1995. The basal tenderness of unaged lamb. *Meat Science* 39:285–291.

Devine, C. E., Ellery, S., and S. Averill. 1984. Responses of different types of ox muscle to electrical stimulation. *Meat Science* 10:35–51.

Devine, C. E. Wahlgren, N. M., and E. Tornberg. 1999. Effect of *rigor* temperature on muscle shortening and tenderization of restrained and unrestrained beef *M. longissimus thoracis et lumborum*. *Meat Science* 51:61–72.

Devine, C. E., Wells, R., Cook, C. J., and S. R. Payne. 2001. Does high voltage electrical stimulation of sheep affect rate of tenderization? *New Zealand Journal of Agricultural Research* 44:53–58.

Devine, C. E., Lowe, T. E., Wells, R. W., Edwards, N. J., Hocking Edwards, J. E., Starbuck, T. J., and P. A. Speck. 2002a. The effect of preslaughter stress and electrical stimulation on meat tenderness. *Proceedings 48th International Congress of Meat Science and Technology*, pp. 218–219, Rome, Italy.

Devine, C. E., Payne, S. R., Peachey, B. M., Lowe, T. E., Ingram, J. R., and C. J. Cook. 2002b. High and low *rigor* temperature effects on sheep meat tenderness and ageing. *Meat Science* 60:141–146.

Devine, C. E., Hopkins, D. L., Hwang, I. H., Ferguson, D. M., and I. Richards. 2004. Electrical stimulation. In *Encyclopedia of Meat Sciences,* W. Jensen, C. Devine, and M. Dikeman (eds.), p. 413. Elsevier: Oxford.

Dransfield, E. 1993. Modeling *post-mortem* tenderization—IV: Role of calpains and calpastatin in conditioning. *Meat Science* 34:217–234.

Dransfield, E. 1994. Optimization of tenderization, aging and tenderness. *Meat Science* 36:105–121.

Dransfield, E. 1997. When the glue comes unstuck. In: *Proceedings 43rd International Congress of Meat Science and Technology*, pp. 52–63, Auckland, New Zealand.

Dransfield, E., Etherington, D. J., and M. A. J. Taylor. 1992a. Modeling post-mortem tenderization—II: Enzyme changes during storage of electrically stimulated and non-stimulated beef. *Meat Science* 31:75–84.

Dransfield, E., Wakefield, D. K., and I. D. Parkham. 1992b. Modelling *post-mortem* tenderization—I. Texture of electrically stimulated and non-stimulated beef. *Meat Science* 31:57–73.

Ducastaing, A., Valin, C., Schollmeyer, J., and R. Cross. 1985. Effects of electrical stimulation on postmortem changes in the activities of two Ca dependent neutral proteinases and their inhibitor in beef muscle. *Meat Science* 15:193–202.

Farouk, M. M. 2008. Adding value to beef or lamb. Available at http://www.agresearch.co.nz/mirinz/prod/procfoodprod.asp, accessed December 5, 2008.

Ferguson, D. M., Jiang, S. T., Hearnshaw, H., Rymill, S. R., and J. M. Thompson. 2000. Effect of electrical stimulation on protease activity and tenderness of *M. longissimus* from cattle with different proportions of *Bos indicus* content. *Meat Science* 55:265–272.

Ferguson, D. M., Daly, B. L., Gardner, G. E., and R. K. Tume. 2008. Effect of glycogen concentration and form on the response to electrical stimulation and rate of post-mortem glycolysis in ovine muscle. *Meat Science* 78:202–210.

Gann, G. L., and R. A. Merkel. 1978. Ultrastructural changes in bovine longissimus muscle during postmortem ageing. *Meat Science* 2:129–144.

George, A. R., Bendall, J. R., and R. C. Jones. 1980. The tenderizing effect of electrical stimulation of beef carcasses. *Meat Science* 4:51–68.

Goll, D. E., Thompson, V. F., Li, H., Wei, W., and J. Cong. 2003. The calpain system. *Physiology Review* 83:731–801.

Graafhuis, A. E., Lovatt, S. J., and C. E. Devine. 1992. A predictive model for lamb tenderness. *Proceedings 27th Meat Industry Research Conference*, pp. 143–147, Hamilton, New Zealand.

Harsham, A., and F. E. Deatherage. 1951. Tenderization of Meat, U.S. Patent 2,544,681.

Hertzman, C., Olsson, U., and E. Tornberg. 1993. The influence of high temperature, type of muscle and electrical stimulation on the course of *rigor*, ageing and tenderness of beef muscles. *Meat Science* 35:119–141.

Ho, C. Y., Stromer, M. H., and R. M. Robson. 1996. Effect of electrical stimulation on postmortem titin, nebulin, desmin, and troponin T degradation and ultrastructural changes in bovine longissimus muscle. *Journal of Animal Science* 74:1563–1575.

Ho, C. Y., Stromer, M. H., Rouse, G., and R. M. Robson. 1997. Effects of electrical stimulation and postmortem storage on changes in titin, nebulin, desmin, troponin-T, and muscle ultrastructure in *Bos indicus* crossbred cattle. *Journal of Animal Science* 75:366–376.

Hoagland, R., McBryde, C. N., and W. C. Powick. 1917. Changes in fresh beef during cold storage above freezing. *United States Department of Agriculture Bulletin* 433:1–100.

Hollung, K., Veiseth, E., Frøystein, T., Aass, L., Langsrud, Ø., and K. I. Hildrum. 2007. Variation in the response to manipulation of *post-mortem* glycolysis in beef muscles by low-voltage electrical stimulation and conditioning temperature. *Meat Science* 77:372–383.

Hopkins, D. L., and G. H. Geesink. 2009. Protein degradation post mortem and tenderization. In *Applied Muscle Biology and Meat Science*, M. Du and R. McCormick (eds.), pp. 149–173, USA: CRC Press, Taylor & Francis Group.

Hopkins, D. L., and R. G. Taylor. 2004. Post-mortem muscle proteolysis and meat tenderness. In *Muscle Development of Livestock Animals: Physiology, Genetics, and Meat Quality*, M. Everts, M. te Pas, and H. Haagsman (eds.), pp. 363–388. UK: CAB International.

Hopkins, D. L., and J. M. Thompson. 2001a. The relationship between tenderness, proteolysis, muscle contraction and dissociation of actomyosin. *Meat Science* 57:1–12.

Hopkins, D. L., and J. M. Thompson. 2001b. Inhibition of protease activity—Part 1: The effect on tenderness and indicators of proteolysis in ovine muscle. *Meat Science* 59:175–185.

Hopkins, D. L., and J. M. Thompson. 2001c. Inhibition of protease activity—Part 2: Degradation of myofibrillar proteins, myofibril examination and determination of free calcium levels. *Meat Science* 59:199–209.

Hopkins, D. L., and J. M. Thompson. 2002. Factors contributing to proteolysis and disruption of myofibrillar proteins and the impact on tenderization in beef and lamb meat. *Australian Journal of Agricultural Research* 53:149–166.

Hopkins, D. L., and E. S. Toohey. 2006. Eating quality of conventionally chilled sheep meat. *Australian Journal of Experimental Agriculture* 46:897–901.

Hopkins, D. L., and E. S. Toohey. 2008. Variation in the response to electrical stimulation in lamb carcases. *Proceedings of the 27th Biennial Conference of the Australian Society of Animal Production*, University of Queensland, Brisbane, Australia p. 91.

Hopkins, D. L., Wotton, J. S. A., Gamble, D. J., Atkinson, W. R., Slack-Smith, T. S., and D. G. Hall. 1995. Lamb carcass characteristics 1. The influence of carcass weight, fatness and sex on the weight of trim and traditional cuts. *Australian Journal of Experimental Agriculture* 35:33–40.

Hopkins, D. L., Garlick, P. R., and J. M. Thompson. 2000a. The effect on the sarcomere structure of super tenderstretching. *Australasian Journal of Animal Science* 13(Supplement), Vol. C, 233.

Hopkins, D. L., Littlefield, P. J. and J. M. Thompson. 2000b. The effect of low voltage stimulation under controlled conditions on the tenderness of three muscles in lamb carcasses. *Asian-Australasian Journal of Animal Sciences* 13(Supplement) July, Vol. B:362–365.

Hopkins, D. L., Shaw, F. D., Baud, S., and P. J. Walker. 2006. Electrical currents applied to lamb carcases-effects on blood release and meat quality. *Australian Journal of Experimental Agriculture* 46:885–889.

Hopkins, D. L., Toohey, E. S., Pearce, K. L., and Richards, I. 2008. Some important changes in the Australian sheep meat processing industry. *Australian Journal of Experimental Agriculture* 46:752–756.

Huff-Lonergan, E., Mitsuhashi, T., Parrish, F. C. Jr., and R. M. Robson. 1996. Sodium dodecyl sulfate-polyacrylamide gel electrophoresis and Western blotting comparisons of purified myofibrils and whole muscle preparations for evaluating titin and nebulin in postmortem muscle. *Journal of Animal Science* 74:779–785.

Hwang, I. H., and J. M. Thompson. 2001a. The effect of time and type of electrical stimulation on the calpain system and meat tenderness in beef longissimus dorsi muscle. *Meat Science* 58:135–144.

Hwang, I. H., and J. M. Thompson. 2001b. The interaction between pH and temperature decline early postmortem on the calpain system and objective tenderness in electrically stimulated beef longissimus dorsi muscle. *Meat Science* 58:167–174.

Hwang, I. H., and J. M. Thompson. 2002. A technique to quantify the extent of postmortem degradation of meat ultrastructure. *Journal of Asian Australasian Animal Science* 15:111–116.

Hwang, I. H., Devine, C. E., and D. L. Hopkins. 2003. The biochemical and physical effects of electrical stimulation on beef and sheep meat tenderness—A review. *Meat Science* 65:677–691.

Jaime, I., Beltrán, J. A., Ceña, P., López-Lorenzo, P., and P. Roncalés. 1992. Tenderization of lamb meat: Effect of rapid postmortem temperature drop on muscle conditioning and aging. *Meat Science* 32:357–366.

Jeacocke, R. E. 1984. The kinetics of *rigor* onset in beef muscle fibres. *Meat Science* 11:237–251.

Jones, S. J., Calkins, C. R., Johnson, D. D., and Gwartney, B. L. 2005. Bovine mycology. Lincoln, NE: University of Nebraska. Available at http://bovine.unl.edu.

Koohmaraie, M. 1996. Biochemical factors regulating the toughening and tenderization processes of meat. *Meat Science* 43(Suppl S):S193–S201.

Koohmaraie, M., and G. H. Geesink. 2006. Contribution of postmortem muscle biochemistry to the delivery of consistent meat quality with particular focus on the calpain system. *Meat Science* 74:34–43.

Koohmaraie, M., Seideman, S. C., Schollmeyer, J. E., Dutson, T. R., and A. S. Babiker. 1988. Factors associated with the tenderness of three bovine muscles. *Journal of Food Science* 53:407–410.

Koohmaraie, M., Doumit, M., and T. L. Wheeler. 1996. Meat toughening does not occur when rigor shortening is prevented. *Journal of Animal Science* 74:2935–2942.

Labeit, S., and B. Kolmerer. 1995. Titins: Giant proteins in charge of muscle ultrastructure and elasticity. *Science* 270:293–296.

Locker, R. H., and C. J. Hagyard. 1963. A cold shortening effect in beef muscles. *Journal of the Science of Food and Agriculture* 14:787–793.

Locker, R. H., and D. J. C. Wild. 1984. "Aging" of cold shortened meat depends on the criterion. *Meat Science* 10:235–238.

Marsh, B. B. 1985. Electrical stimulation research: Present concepts and future directions. In Advances in Meat Research—Electrical Stimulation, A. M. Pearson and T. R. Dutson (eds.), pp. 277–301, New York: Van Nostrand Reinhold Company.

Martin, K. M., Hopkins, D. L., Gardner, G. E., and J. M. Thompson. 2006. Effects of stimulation on tenderness of lamb with a focus on protein degradation. *Australian Journal of Experimental Agriculture* 46:891–896.

McDonagh, M. B., Fernandez, C., and V. H. Oddy. 1999. Hind-limb protein metabolism and calpain system activity influence post-mortem change in meat quality in lamb. *Meat Science* 52:9–18.

Morton, J. D., Bickerstaffe, R., Kent, M. P., Dransfield, E., and G. M. Keeley. 1999. Calpain–calpastatin and toughness in M. longissimus from electrically stimulated lamb and beef carcasses. *Meat Science* 52:71–79.

MSA. 2010. What next for MSA? Available at http://www.mla.com.au/files/00674356-2089-430e-bfdc-9e33010516ca/mla-producer-forum-2010-michellegorman.pdf and MSW Beef brochure http://www.mla.com.au/files/1ecd5a95-a823-4af4-b14b-9d5f007f9d59/msa-beef-brochure.pdf, accessed January 31, 2011.

Murachi, T., Tanaka, K., Hatanaka, M., and T. Murakami. 1981. Intracellular Ca^{2+}-dependent protease (calpain) and its high-molecular-weight endogenous inhibitor (calpastatin). In *Advances in Enzyme Regulation*, G. Weber (ed.), vol. 19, pp. 407–424, New York: Pergamon Press.

Olsson, U., Hertzman, C., and E. Tornberg. 1994. The influence of low temperature, type of muscle and electrical stimulation on the course of *rigor*, aging and tenderness of beef muscles. *Meat Science* 37:115–131.

O'Sullivan, A., Korzeniowska, M., White, A., and D. J. Troy. 2003. Using a novel intervention technique to reduce the variability and improve tenderness of beef *longissimus dorsi*. *Proceedings 49th International Congress of Meat Science and Technology*, pp. 513–514, Campinas, Brazil.

Ouali, A. 1992. Proteolytic and physicochemical mechanisms involved in meat texture development. *Biochimie* 74:251–265.

Ouali, A., Hernan-Mendez, C. H., Coulis, G., Beclia, S., Boudjellal, A., Aubury, L., and M. A. Sentandreu. 2006. Revisiting the conversion of muscle into meat and the underlying mechanisms. *Meat Science* 74:44–58.

Pearce, K. L., Hopkins, D. L., Pethick, D. W., Gutzke, D., Richards, I., Fuller, P., and J. K. Phillips. 2006a. Increasing the stimulation response from new generation medium voltage electrical stimulation units. *Proceedings of the 52nd International Congress of Meat Science and Technology*, pp. 625–626, Dublin, Ireland.

Pearce, K. L., Hopkins, D. L., Toohey, E. S., Pethick, D. W., and I. Richards. 2006b. Quantifying the rate of pH and temperature decline in lamb carcasses using mid voltage electrical stimulation in an Australian abattoir. *Australian Journal of Experimental Agriculture* 46:869–874.

Pearce, K. L., Hopkins, D. L., Jacob, R. H., Williams, A., Pethick, D. W., and J. K. Phillips. 2009. Alternating frequency to increase the stimulation response from medium voltage electrical stimulation and the effects on meat traits. *Meat Science* 81:188–195.

Pisula, A., and A. Tyburcy. 1996. Hot processing of meat. *Meat Science* 43:S125–S134.

Polkinghorne, R. J. 2006. Implementing palatability assured critical control point (PACCP) approach to satisfy consumer demands. *Meat Science* 74:180–187.

Pommier, S. A. 1992. Vitamin A, electrical stimulation and chilling rate effects on lysosomal enzyme activity in aging bovine muscle. *Journal of Food Science* 57:30–35.

Purslow, P. P. 2005. Intramuscular connective tissue and its role in meat quality. *Meat Science* 70:435–447.

Rentschler, H. C. 1951. Apparatus and Method for the Tenderization of Meat, U.S. Patent 2,544,724.

Rhee, M. S., and B. C. Kim. 2001. Effect of low voltage electrical stimulation and temperature conditioning on postmortem changes in glycolysis and calpains activities of Korean native cattle (hanwoo). *Meat Science* 58:231–237.

Robson, R. M., Huiatt T. W., and F. C. Parrish Jr. 1991. Biochemical and structural properties of titin, nebulin and intermediate filaments in muscle. In *Proceedings 50th Reciprocal Meat Conference,* pp. 7–20, American Meat Science Association, Chicago, IL. Manhattan, KS.

Rosenvold, K., North, M., Devine, C., Micklander, E., Hansen, P., Dobbie, P., and R. Wells. 2008. The protective effect of electrical stimulation and wrapping on beef tenderness at high *pre rigor* temperatures. *Meat Science* 79:299–306.

Sentandreu, M. A., Coulis, G., and A. Ouali. 2002. Role of muscle endopeptidases and their inhibitors in meat tenderness. *Trends in Food Science and Technology* 13:398–419.

Shaw, F. D., Baud, S. R., Richards, I., Pethick, D. W., Walker, P. J., and J. M. Thompson. 2005. New electrical stimulation technologies for sheep carcasses. *Australian Journal of Experimental Agriculture* 45:575–583.

Simmons, N. J., Singh, K., Dobbie, P., and C. E. Devine. 1996. The effect of *prerigor* holding temperature on calpain and calpastatin activity and meat tenderness. In *Proceedings 42nd International Congress of Meat Science and Technology,* pp. 414–415, Lillehammer, Norway.

Simmons, N. J, Gilbert, K. V., and J. M. Cairney. 1997. The effect of low voltage stimulation on pH fall and meat tenderness in lamb. In *Proceedings of the 43rd International Congress of Meat Science and Technology,* pp. 610–611, Auckland, New Zealand.

Spooncer, W. F. 1993. Options for hot boning. Meat 93, The Australian Meat industry Research Conference, pp. 1–6, 11–13 October 1993, Gold Coast, Australia.

Strandine, E. J., Koonz, C. H., and J. M. Ramsbottom. 1949. A study of variations in muscle of beef and chicken. *Journal of Animal Science* 8:483–494.

Swatland, H. J. 2000. Meat cuts and muscle foods. Nottingham: Nottingham University Press.

Takahashi, G., Wang, S. M., Lochner, J. V., and B. B. Marsh. 1987. Effects of 2-Hz and 60-Hz stimulation on the microstructure of beef. *Meat Science* 19:65–76.

Taylor, J. M., and D. L. Hopkins. 2011. Patents for stretching and shaping meats. *Recent Patents on Food, Nutrition and Agriculture* 3, 91–101.

Taylor, A. A., Shaw, B. G., and D. B. MacDougall. 1980–1981. Hot boning beef with and without electrical stimulation. *Meat Science* 5:109–123.

Taylor, R. G., Geesink, G. H., Thompson, V. F., Koohmaraie, M., and D. E. Goll. 1995. Is Z-disk degradation responsible for postmortem tenderization? *Journal of Animal Science* 73:1351–1367.

Thompson, J. M. 2002. Managing meat tenderness. *Meat Science* 62:295–308.

Thompson, J. M., Hopkins, D. L., D'Sousa, D., Walker, P. J., Baud, S. R., and D. W. Pethick. 2005. Sheep meat eating quality: The impact of processing on sensory and objective measurements of sheep meat eating quality. *Australian Journal of Experimental Agriculture* 45:561–573.

Toohey, E. S., and D. L. Hopkins. 2006. Eating quality of commercially processed hot boned sheep meat. *Meat Science* 72:660–665.

Toohey, E. S., Hopkins, D. L., Stanley, D. F., and S. G. Nielsen. 2008a. The impact of new generation pre-dressing medium-voltage electrical stimulation on tenderness and colour stability in lamb meat. *Meat Science* 79:683–691.

Toohey, E. S., Hopkins, D. L., and T. A. Lamb. 2008b. Preliminary studies on the impact of new generation pre-dressing medium-voltage electrical stimulation on pH decline in goat carcases. *Proceedings of the 27th Biennial Conference of the Australian Society of Animal Production*, University of Queensland, Brisbane, Australia p. 93.

Toohey, E. S., Hopkins, D. L., and T. A. Lamb. 2008c. The impact of wrapping and ageing hot boned sheep meat on eating quality. *Proceedings 54th International Congress of Meat Science and Technology,* 7B.19, pp. 1–3. Cape Town, South Africa.

Toohey, E. S., Hopkins, D. L., Lamb, T. A., Neilsen, S. G., and D. Gutzke. 2008d. Accelerated tenderness of sheep topsides using a meat stretching device. *Proceedings 54th International Congress of Meat Science and Technology,* 7B.18, pp. 1–3. Cape Town, South Africa.

Tornberg, E. 1996. Biophysical aspects of meat tenderness. *Meat Science* 43:S175–S191.

Uytterhaegen, L., Claeys, E., and Demeyer, D. (1992). The effect of electric stimulation on beef tenderness, protease activity and myofibrillar protein fragmentation. *Biochimie*, 747, 275–281.

Veiseth, E., Shackelford, S. D., Wheeler, T. L., and M. Koohmaraie. 2001. Effect of postmortem storage on μ-calpain and m-calpain in ovine muscle. *Journal of Animal Science* 79:1502–1508.

Veiseth, E., Shackelford, S. D., Wheeler, T. L., and M. Koohmaraie. 2004. Indicators of tenderization are detectable by 12 h post-mortem in ovine longissimus. *Journal of Animal Science* 82:1428–1436.

Von Seggern, D. D., Calkins, C. R., Johnson, D. D., Brickler, J. E., and Gwartney, B. L. 2005. Muscle profiling: Characterizing the muscles of the beef chuck and round. *Meat Science* 71:39–51.

Wahlgren, N. M., Devine C. E., and E. Tornberg. 1997. The influence of different pH-courses during *rigor* development on beef tenderness. In *Proceedings 43rd International Congress of Meat Science and Technology*, pp. 622–623, Auckland, New Zealand.

Waylan, A. T., and C. L. Kastner. 2004. Hot boning and chilling. In *Encyclopedia of Meat Sciences*, W. Jensen, C. Devine, and M. Dikeman (eds.), p. 606. Oxford: Elsevier.

Wendt, A., Thompson, V. F., and D. E. Goll. 2004. Interaction of calpastatin with calpain: A review. *Biological Chemistry* 385:465–472.

Wheeler, T. L., and M. Koohmaraie. 1994. *Prerigor* and *postrigor* changes in tenderness of ovine longissimus muscle. *Journal of Animal Science* 72:1232–1238.

White, A., O'Sullivan, A., Troy, D. J., and E. E. O'Neill. 2006. Effects of electrical stimulation, chilling temperature and hot-boning on the tenderness of bovine muscle. *Meat Science* 73:196–203.

7 Pretreatments for Meats II
(Curing, Smoking)

Natalia G. Graiver, Adriana N. Pinotti,
Alicia N. Califano, and Noemí E. Zaritzky

CONTENTS

7.1 MEAT CURING

Meat curing is a process in which a combination of salt, color fixing ingredients and seasoning are used to achieve unique properties in the final product. Cured meat products are characterized by addition of nitrate and/or nitrite, whereas other ingredients, particularly sodium chloride (salt), are essential parts of typical cured meat formulations. Nitrates and nitrites, either potassium or sodium salts, are used to develop cured meat color. They impart a bright reddish, pink color, which is desirable in a cured product. In addition, nitrates and nitrites have a pronounced effect on flavor; sodium nitrites also prevent the growth of food poisoning microorganisms.

7.1.1 HISTORICAL BACKGROUND

The salting of meat is an ancient practice for meat preservation, used probably in China and in the area between the Tigris and Euphrates rivers, several thousand years before the birth of Christ. Nitrate, originally present as impurity in the curing salt, was incorporated to the brine due to its conversion to nitrite by bacterial reduction. In Western Europe, the use of nitrate may have begun in the 16th century, or later, when the compound had become readily available and was widely used for medicinal purposes (Varnam and Sutherland 1995). Although the role of nitrites in cured meat was not really understood until early in the 20th century, it is clear that for thousands of years nitrite has played an important role in meat curing.

7.1.2 CURING INGREDIENTS

The two main ingredients that must be used to cure meat are salt and nitrite. However, other substances can be added to accelerate curing, stabilize color, modify flavor, and reduce shrinkage during processing.

7.1.2.1 Sodium Chloride

Salt is the primary ingredient used in meat curing. Originally, it served as a preservative by dehydration inhibiting bacterial growth; besides, it adds flavor to the cured product. Sodium chloride acts as a preservative and modifies water-holding capacity

(WHC) of the proteins. One of the functions of the addition of NaCl in meat products is to extract myofibrillar proteins. Extraction and solubilization of these muscle proteins contribute to meat particle binding, fat emulsification, and WHC, and thus, it reduces cook losses and improves quality and texture (Sofos 1986).

Salt levels are dependent on consumer taste; an acceptable level of salt has been reported to be about 2%–4% concentration in the product, although levels in the final products vary widely from less than 2% in some mild cured bacon to over 6% in some traditional types of ham.

Salt is generally used in combination with sugar, nitrite, and/or nitrate. Recently, emphasis has been placed on reducing levels of salt in meat products in view of its relationship to hypertension in about 20% of the population.

7.1.2.2 Nitrate

In the late 1800s, it was discovered that nitrate was converted to nitrite by nitrate-reducing bacteria, and that nitrite was the true curing agent. The first half of the twentieth century brought a gradual shift from nitrate to nitrite as the primary curing agent for meats due to the advantages of a faster process and a better understanding of nitrite chemistry.

Nitrates and nitrites must be used with caution because they are toxic when used in large amounts. Since 1970, it has been reported that nitrite could result in formation of carcinogenic n-nitrosamines in cured meat. Subsequent research demonstrated that a significant factor in nitrosamine formation was residual nitrite concentration (Sebranek and Bacus 2007).

7.1.2.3 Nitrite

The chemistry of nitrite in cured meat is very complex; nitrite is a highly reactive compound that can act as an oxidizing, reducing, or nitrosylating (transfer of nitric oxide) agent and can be converted to a variety of related compounds in meat, including nitrous acid, nitric oxide, and nitrate (Honikel 2004). The formation of nitric oxide (NO) from nitrite is a necessary stage for most meat curing reactions (Møller and Skibsted 2002). Research on nitric oxide in biological systems has facilitated a better understanding of nitrite and nitric oxide in cured meat (Møller and Skibsted 2002; Stamler and Meissner 2001).

Two of the most important factors governing nitrite reactions in conventionally cured meat products are pH and presence of reductants. One of the main effects of nitrite is color development of the cured meat. Nitrite does not act directly as a nitrosylating agent in meat, but first forms intermediates such as N_2O_3 (Honikel 2004) in the postmortem muscle and NOCl in the presence of salt (Fox et al. 1994; Møller and Skibsted 2002; Sebranek and Fox 1991). Formation of NO from the intermediates is facilitated by reductants such as ascorbate, and the NO reacts with the iron of both myoglobin (Fe^{+2}) and metmyoglobin (Fe^{+3}) to form cured meat pigments and curing color.

According to Varnam and Sutherland (1995), sodium nitrite is a strong oxidant and reacts with endogenous or added reductants such as ascorbic acid to produce nitric oxide (NO). Although nitrite is capable of reacting with a wide range of reducing compounds, the reaction with ascorbate (or erythorbate), if present in the cure mixture, is the most important. Nitric oxide is a gaseous molecule with an odd number of electrons. The molecule is very reactive toward radicals and oxygen and is the key to the important reactions of curing.

Nitrous acid (HNO_2) has a pKa of 3.22, and only a small proportion of added nitrite is converted to the more reactive free acid. Nitrous acid is unstable in solution and produces nitric oxide (Varnam and Sutherland 1995):

$$3HNO_2 \rightarrow H^+ + NO_3^- + 2NO + H_2O$$

In solution, nitrous acid can behave as both a reductant and an oxidant, but in meat systems, the role is primarily as an oxidant, with oxidizing capacity increased at low pH values. Part of the nitrite is, however, oxidized to nitrate during both the curing process and the subsequent storage:

$$NO_2^- + 2OH \rightarrow NO_3^- + H_2O + 2e^-$$

In reaction with ascorbic acid, nitrous acid is reduced by one equivalent to yield the free-radical product NO, while ascorbic acid is simultaneously oxidized by two equivalents to yield dehydroascorbic acid:

$$\text{ascorbic acid} + 2HNO_2 \rightarrow \text{dehydroascorbic acid} + 2NO + 2H_2O$$

Ascorbic acid is a bifunctional acid ($pKa_1 = 4.04$; $pKa_2 = 11.34$), and because of the different reactivities of the acid and base forms, the reaction with nitrous oxide is highly dependent on the pH value.

Several other nitrite reactions are involved in cured meats and contribute to nitric oxide production. The addition of nitrite to minced meat leads to the formation of metmyoglobin (brown color) because nitrite acts as a strong heme pigment oxidant and is reduced to NO. The NO reacts with metmyoglobin, and subsequent reduction reactions convert the oxidized heme to reduced nitric oxide myoglobin for typical cured color following cooking. Further, nitrite can also react with sulfhydryl groups on proteins to release nitric oxide in an oxidation–reduction reaction that results in a disulfide (Pegg and Shahidi 2000).

In addition to these reactions related to the cured color development, nitrite plays an important role in cured meat as a bacteriostatic and bacteriocidal agent. Nitrite is strongly inhibitory to anaerobic bacteria, *Clostridium botulinum*, and contributes to the control of other microorganisms such as *Listeria monocytogenes*. The effects of nitrite and the inhibitory mechanism differ in different bacterial species (Tompkin 2005). The effectiveness of nitrite as an antibotulinal agent is dependent on several environmental factors including pH, sodium chloride concentration, reductants, and iron content, among others. Some researchers have suggested that nitrous acid (HNO_2) and/or nitric oxide (NO) may be responsible for the inhibitory effects of nitrite (Tompkin 2005).

7.1.2.3.1 Nitrosamines

A wide range of nitrogen-containing substances occur in meat, which could be expected to react with nitrite under certain conditions to produce potent carcinogens called nitrosamines. Nitrosamines are formed by the nitrosation of organic compounds such as amines and other nitrogen-containing compounds; nitrite can act as the nitrosating agent (Pearson and Gilett 1996).

$$\begin{array}{c} R_1 \\ \diagdown \\ NH + HONO \\ R_2 \diagup \end{array} \longrightarrow \begin{array}{c} R_1 \\ \diagdown \\ N - NO + H_2O \\ R_2 \diagup \end{array}$$

| Secondary   Nitrous acid | N-nitroso-secondary amine |
| amine | (a nitrosamine) |

Nitrosamines can form either directly by interaction of nitrous oxide with the secondary amines in the food or indirectly by formation of nitrites, which may then react with secondary amines to form nitrosamines. The pH of meat being in the acid range may not favor the direct reaction of nitrous oxide with secondary amines to form N-nitrosamines, since this reaction occurs best under alkaline conditions.

Although acid conditions that are ideal for the reactions do not occur normally in meat or meat mixtures, nevertheless, evidence exists that suggests that under extreme conditions of cooking, for example, grilling or frying, small concentrations of these substances are produced in cured meats, in concentrations of the order of a few parts per thousand million (i.e., 10×10^{-9} to 30×10^{-9}). They could also be formed in the acid conditions within the stomach. Although no direct evidence exists for stating that consumption of cured meats causes cancer, there is a marked tendency toward reducing the use of nitrite in curing. While it might be desirable to reduce the use of nitrite in cured meats, this must be approached with caution due to the known protection that nitrite provides against botulism.

Recent proposals for legislation in the United States have suggested that bacon could be produced with low concentrations of 40 mg kg^{-1} nitrite coupled with 0% to 26% of ascorbic or erythorbic acid and 0% to 50% of potassium sorbate. Ascorbate is added in order to reduce residual levels of nitrite in the meat after curing, thereby reducing the possible nitrosamine production subsequently. Potassium sorbate is claimed to replace the antibotulinum activity of the lost nitrite; however, its action is pH dependent, and it is ineffective in this role in meat above pH 6. A gradual reduction in permitted levels of nitrite, and possible exclusion of nitrate, can be expected to continue in the future (Wilson 1981).

7.1.2.3.2 Regulations

Current regulations on use of nitrite and nitrate in the United States vary depending on the method of curing used and the product that is cured. For comminuted products, the maximum ingoing concentration of sodium or potassium nitrite is 156 parts per million (ppm), based on the fresh weight of the meat block (USDA 1995). Maximum ingoing nitrate for these products is 1718 ppm. Sodium and potassium nitrite and nitrate are limited to the same amount despite the greater molecular weight of the potassium salts, which means that less nitrite or nitrate will be included when the potassium salt is used. For immersion curing, and massaged or pumped products, maximum ingoing sodium or potassium nitrite and nitrate concentrations are 200 ppm and 700 ppm, respectively, again based on the fresh weight of the meat block. If nitrite and nitrate are both used for a single product, the ingoing limits remain the same for each, but the combination must not result in more than 200 ppm of analytically measured nitrite, calculated as sodium nitrite in the finished product.

Bacon is an exception to the general limits for curing agents because of the potential for nitrosamine formation. For pumped and/or massaged bacon without the skin, 120 ppm of sodium nitrite or 148 ppm of potassium nitrite is required along with 550 ppm of sodium ascorbate or sodium erythorbate, which is also required. To accommodate variation in pumping procedures and brine drainage from pumped products, the regulations for pumped and/or massaged bacon permit ±20% of the target concentrations at the time of injecting or massaging. For example, sodium nitrite concentrations within the range of 96–144 ppm are acceptable. Nitrate is not permitted for any bacon curing method. There are two exceptions to these regulations for pumped and/or massaged bacon: first, 100 ppm of sodium nitrite (or 123 ppm of potassium nitrite) with an "appropriate partial quality control program" is permitted, and second, 40–80 ppm of sodium nitrite or 49–99 ppm of potassium nitrite is permitted if sugar and a lactic acid starter culture are included. Immersion cured bacon is limited to 120 ppm of sodium nitrite or 148 ppm of potassium nitrite, while dry cured bacon is limited to 200 ppm or 246 ppm, respectively. It is important to note that the regulations also require a minimum of 120 ppm of ingoing nitrite for all cured "Keep Refrigerated" products "unless safety is assured by some other preservation process, such as thermal processing, pH, or moisture control." The establishment of minimum ingoing nitrite concentration is considered critical to subsequent product safety. This is a significant consideration for natural and organic cured meat products.

7.1.2.4 Phosphates

Phosphates are added to the cure to increase the water-binding capacity and the yield of finished product. The action of phosphates in improving water retention is related to both the increase in pH and the unfolding of the muscle proteins, thereby making more sites available for water binding.

Phosphates are used in most pumped meats such as ham, bacon, roast beef, cooked corned beef, pastrami, and some similar poultry products. Advantages include a reduction in cookout, improvement in sliceability, retention of flavor, and greater juiciness. Thus, tripolyphosphates must be hydrolyzed to diphosphate before they are active. No one phosphate can do everything, but combinations work best. Although tetrasodium pyrophosphate produces the best bind, it is highly caustic at pH of about 11. This results in production of soap in meat containing any fat. Sodium acid pyrophosphate has an acid pH and, if used alone, results in greatly reduced yields. The acid nature of this phosphate results in rapid development of cure color and avoids development of a green core in rapidly cooked sausages, especially those containing alkaline phosphates. More complex polyphosphates may be used in the mixtures in order to chelate metal ions, especially calcium and iron. For curing brines, a mixture of tripolyphosphate and sodium hexametaphosphate is used. They are dissolved in water and injected into hams and bacons, where they are slowly hydrolyzed to diphosphate and become active more slowly. Too high temperature may result in faster hydrolysis and cause rapid formation of the monophosphate, which is inactive in the tissues. Thus, low-temperature pumping (about 0°C) is recommended. The low temperature inhibits microbial growth and extends the shelf life.

Sodium acid pyrophosphate produces the best bind, but its acid pH results in reduced yields. On the other hand, pyrophosphate (diphosphate) increases the water-binding capacity of the muscle proteins by causing swelling. Tripolyphosphate–sodium hexametaphosphate blends are commonly used in curing brines (Pearson and Gilett 1996).

7.1.2.5 Sodium Ascorbate and Erythorbate

Ascorbate and erythorbate are isomers, with ascorbate being the L-form and erythorbate the R-isomer. Both compounds have the same functions, although erythorbate is more stable than ascorbate; thus, erythorbate is the preferred compound for use in meat curing. The salts of ascorbic acid and erythorbic acid are commonly used to stabilize the color of cured meat. In practice, only sodium erythorbate or sodium ascorbate is used in curing pickles, since ascorbic or erythorbic acid reacts with the nitrite to form nitrous oxide. Since nitrous oxide is dangerous in confined spaces and the nitrite is destroyed, ascorbates or erythorbates are always used under practical conditions. This group of compounds (referred to as ascorbates) serves three main functions: (1) ascorbates take part in the reduction of metmyoglobin to myoglobin, thereby accelerating the rate of curing; (2) ascorbates react chemically with nitrite to increase the yield of nitric oxide from nitrous acid; and (3) ascorbate excess acts as an antioxidant, therefore stabilizing both color and flavor. Under certain conditions, ascorbates have been shown to reduce nitrosamine formation.

An important use may be its effect in reducing nitrosamine formation, at a level of 350 ppm. The exact mechanism is not known, but current meat inspection regulations specify that all bacon be produced using no more than 120 ppm of nitrite in combination with 550 ppm of ascorbate/erythorbate.

The antioxidant properties of ascorbate or erythorbate prevent not only development of rancidity but also fading of sliced meats when exposed to light. Protection is closely associated with prevention of hemocatalyzed lipid oxidation, which results in both pigment degradation and development of rancidity. As long as excess ascorbate is present, the pigments are protected against breakdown. When ascorbate is depleted, however, the heme pigments are degraded and apparently catalyze lipid oxidation (Pearson and Gilett 1996).

Federal regulations permit addition of 21.26 g ascorbic acid or erythorbic acid (24.80 g sodium ascorbate or erythorbate) per 45.36 kg sausage emulsion, or addition of 2.13 kg (2.48 kg of sodium salts) per 378.54 L pickle for curing primal cuts. Pickles containing both sodium nitrite and sodium ascorbate are stable for at least 24 h when maintained at 10°C and pH 6.0 or higher.

7.1.2.6 Monosodium Glutamate

Monosodium glutamate has been used in a number of products to enhance the flavor. It has not, however, been widely used in the meat industry as there is little advantage in its use in good meat products, although mixed meat dishes may profit from its meat flavor-enhancing properties (Pearson and Gilett 1996).

7.1.2.7 Sodium and Potassium Lactate

Sodium or potassium lactate may be added to fresh or cured meats to prevent spoilage and extend storage life. Generally, sodium lactate is preferred over potassium lactate because the latter product has a bitter taste. Since sodium lactate has a slightly acid taste, salt levels in all formulations should be reduced by 0.25%–0.50% (Pearson and Gilett 1996).

7.1.3 CURING METHODS

Although there are a number of methods of curing primal or subprimal cuts of meat, they are all modifications or combinations of two fundamental procedures: dry curing and pickle curing.

In dry curing, the curing ingredients—usually salt, sugar, and nitrite and/or nitrate—are added to meat without additional water. In this method, the curing ingredients draw enough moisture from the meat to form brine, which serves to transport the ingredients into the meat by diffusion.

In pickle/wet curing, the ingredients are dissolved in water, forming a brine that acts in the same general manner as that formed by the natural meat juices and the curing ingredients. In actual practice, there are several modifications of the dry and wet curing procedures, which are a result of combining the two methods. For example, one may start with pickle curing and end with dry curing or vice versa (Pearson and Gillet 1996).

Dry curing is one of the oldest methods of meat preservation, achieved through reduction of water activity by means of salt addition and drying. Different modified techniques were developed in order to accelerate the salt transport through the product in the dry curing process, such as vacuuming, tumbling, and vacuum pulsing, that allow for more uniform and consistent products. Ionic mobility of Cl⁻ (chloride) and nitrite NO_2 is high, which facilitates diffusion, but conventional distribution of curing salts is still relatively slow and may be uneven.

The advantages of dry curing are as follows: (1) relatively high value-added products are produced; (2) the cuts are less perishable because of their dryness and firmness; and (3) they have more flavors.

The disadvantages of dry curing are as follows: (1) high cost due to poor space utilization and amount of labor required; (2) high inventory due to slowness of curing; and (3) salty flavor of the final product.

The mechanical action on meat produced by tumbling enhances brine distribution through structural changes in the muscle. Weiss (1973) stated that tumbling is used to enhance tenderness, ensure juiciness, and aid the development of a uniformly cured product.

Vacuum tumblers are used to accelerate solution penetration and homogeneity in cooked hams (Katsaras and Budras 1993). Vacuum tumbling increases cure penetration rate and color stability (Marriott et al. 1984; Solomon et al. 1980). The application of vacuum tumbling has been found to produce more extractable protein in beef than non-vacuum conditioned beef (Ghavimi et al. 1986; Wiebe and Schmidt 1982). Specifically, vacuum conditions increase the amount of crude myosin extracted to the meat surface.

The wet pickle curing procedure uses the same ingredients as dry curing, except that the cure is dissolved in water to form a brine or pickle. The cuts are submerged in the pickle until the cure has completely penetrated the meat. All cuts can be cured by the conventional pickle curing procedure. This method is used to produce corned beef in a large number of plants. In this case, the method is frequently altered by stitch-pumping the larger pieces of meat to speed up the rate of curing.

Acceleration of the curing process and improvement of the curing salts distribution have been achieved by (1) more efficient injection of the brine (single needle or multineedle stitch pumping) and (2) possible elimination of an immersion stage (Wilson 1981).

Curing times were drastically reduced over those required for more traditional cures. The need to immerse after injection was no longer necessary. Satisfactory bacon can be produced from boneless middles or backs by injection on multineedle machines and hanging on racks or stacking for 6 h to 12 h. Such bacon may then be smoked or partly heat-treated to fix the color, and also chilled.

In the slice-curing method, slices of pork are immersed for a predetermined period in a curing solution by passage on a continuous conveyor. They then pass through a smoking chamber before being vacuum-packed; the whole operation takes only minutes. The most recent development in curing, intended to reduce processing times, relates to curing at temperatures higher than the traditional (3°C–5°C). At these higher temperatures, rates of brine diffusion are faster, and curing is completed more rapidly. Temperatures of up to 15°C are used, but good control is necessary because of the acceleration in bacterial growth; consequently, meats cured under this method are normally cooked immediately afterward. It is also claimed that injection of pork prerigor mortis results in cooked cured ham having superior eating quality and yield; however, this method is not being used commercially.

7.1.4   CASE STUDY: MODELING OF WET MEAT CURING

As was mentioned, meat curing can be conducted by immersion of meat tissues in NaCl brines (wet or pickle curing procedure). It is used in several countries by small processors for certain pork cuts. Pickle cure usually gives a product with a milder flavor than dry curing and requires less labor (Pearson and Gilett 1996). An example of the application of this process is the Wiltshire bacon, a generic term for traditional tank-cured bacon in the United Kingdom. Similar processes are also used elsewhere, especially for ham curing.

Curing process traditionally utilizes sodium chloride, nitrite, and nitrate salts; mass transfer between solid and fluid is used for preservation and to modify characteristics of flavor, color, and nutritional value.

In cured meat, it is necessary to maintain microbiological safety by controlling the amounts of incorporated salts to achieve an adequate water activity, thus avoiding the growth of microorganisms, without adding amounts of chemical preservatives that exceed the allowed values of nitrite and nitrate and without exceeding the level of NaCl accepted by consumers. According to the effective regulations, the use of sodium nitrite or potassium nitrate or their combination must not exceed 200 ppm (0.2 mg/g) expressed as sodium nitrite in the final product (USDA-FSIS 1999). Besides, a NaCl concentration acceptable by the consumer is between 2% and 4% (30–40 mg/g meat; Pearson and Gilett 1996).

7.1.4.1   Effect of Brine Concentration on WHC of Meat Tissues

Mathematical models can help to a better understanding of the transport phenomena and to control the variables involved in the process, such as immersion times and suitable salt concentrations in the brine. The uptake of solutes cannot be interpreted as a simple Fickean diffusion process with a constant diffusion coefficient. The diffusion coefficient is suggested to be affected by changes in NaCl concentration, swelling, and degree of water movement (Vestergaard et al. 2005). Therefore, an additional contribution for the "salting in" process due to electrostatic repulsion forces between myofibrillar filaments in meat tissue can be represented as a pseudo-convective flux in which the driving forces are not pressure differences but electrostatic contributions.

The analysis of the wet curing process in meat immersed in brines ($NaNO_2$, KNO_3, and NaCl) was carried out. The influence of NaCl concentration on the changes in water content, salt uptake, and protein solubilization in meat tissue and the effect of brine concentration on porcine tissue microstructure and protein denaturation were taken into account. Likewise, the diffusion coefficients of sodium nitrite, potassium nitrate, and sodium chloride in pork tissue (*Longissimus dorsi*) using brine solutions were determined. A mathematical model for describing mass transfer of salts into pieces of size used in the industry was developed to simulate the operating conditions of pork curing, such as immersion times and suitable salt concentrations.

WHC is a ubiquitous term used to describe the ability of meat to retain its natural water content; closely related to WHC is the ability of meat to take up additional water at elevated salt concentrations (Offer and Trinick 1983). WHC is studied extensively because of its enormous economic importance and its influence on the sensory properties of the product such as juiciness, texture, and flavor (Trout 1988). It is generally accepted that only myosin and actin, and to some extent tropomyosin, are responsible for the WHC of meat (Morrisey et al. 1987).

In order to analyze the effect of brine concentration on meat WHC, Graiver et al. (2006) performed different experiments using small pork meat cylinders (2 cm length × 1.5 cm diameter). *Longissimus dorsi* pork tissue, free of visible fat, was used in all the experiments. The initial weight of each sample was determined (M_0), and afterward, the cylinders were immersed in stirred NaCl solutions of different concentrations (10, 30, 70, 100, 140, 200, and 330 g/L) at 4°C. Immersion times used were 15, 30, 60, 90, 120, 150, and 180 min; after immersion, samples were reweighed (M_1) and dried under vacuum at 95°C (AOAC 1984) until constant weight (M_2). The amount of NaCl present in the tissue (m_{NaCl}) after each immersion period was determined. Dry-matter content of untreated tissue samples was also determined ($m_{dry\ tissue\ 0}$). Equilibrium concentrations of NaCl and water were obtained by immersing tissue samples (2 cm length × 1.5 cm diameter) in the different brines during at least 48 h.

Mass balances were proposed in order to analyze changes in the water, proteins, and NaCl content in the tissue. Initially, muscle tissue was considered formed by water and dry matter (the initial content of NaCl was negligible), and thus the following equation was proposed:

$$M_0 = m_{dry\ tissue\ 0} + m_{water\ 0} \tag{7.1}$$

where M_0 is the initial mass of the sample, $m_{dry\ tissue\ 0}$ is the initial dry-tissue content, and $m_{water\ 0}$ is the initial water content in the tissue.

After immersion in NaCl solution, the salt was incorporated to the sample, and some soluble substances were leached to the solution. Then

$$M_1 = m_{NaCl} + m_{dry\ tissue} + m_{water} \qquad (7.2)$$

where M_1 is the mass of the sample, m_{NaCl} is the mass of NaCl that was uptaken by the tissue, $m_{dry\ tissue}$ is the mass of the dry tissue, and m_{water} is the water content of the tissue; all these magnitudes were evaluated after immersion of the sample in the brine.

The solid content after immersion (M_2) was considered as the sum of NaCl and dry-tissue content, and then the mass of dry tissue after immersion was evaluated as

$$m_{dry\ tissue} = M_2 - m_{NaCl}. \qquad (7.3)$$

During immersion in the brine solutions, part of the protein was solubilized; this amount was calculated by the following equation:

$$m_{solubilized\ tissue} = m_{dry\ tissue\ 0} - m_{dry\ tissue}. \qquad (7.4)$$

The average water content in the untreated tissue ranged between 72% and 74%.

The effect of NaCl on WHC at long contact times was calculated as the variation between the water content after immersion of the sample in the NaCl solution (m_{water}) and the initial water content ($m_{water\ 0}$) referred to this last value

$$WHC = \frac{m_{water} - m_{water\ 0}}{m_{water\ 0}}. \qquad (7.5)$$

Figure 7.1 shows the experimental values of WHC (Equation 7.5) as a function of NaCl concentration in the brine at long contact times, that is, under equilibrium

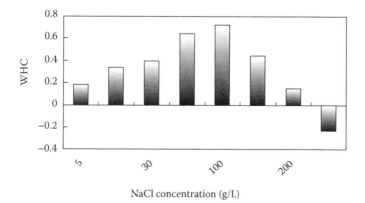

FIGURE 7.1   Water-holding capacity WHC = $((m_{water} - m_{water\ 0})/m_{water\ 0})$ as a function of the brine concentration at long contact times (equilibrium conditions).

conditions. Meat tissue treated with increasing concentrations of NaCl showed important modifications. For NaCl concentrations in the solution ranging between 5 g/L and 200 g/L, the tissue gained water and incorporated great amount of solutes ("salting in"). Salt has, in fact, been shown to cause a significant displacement of water from the outside to the inside of the myofibrillar matrix (Bertram et al. 2001).

The maximum water uptake was observed for NaCl concentrations ranging between 70 g/L and 100 g/L. Above 200 g/L, the WHC decreased showing water loss at 330 g/L ("salting out"; Graiver et al. 2006). Similar results were reported by Knight and Parsons (1988) working with rabbit muscle; they found that there was net uptake of water below 4.5 M (260 g/L) NaCl with a maximum at 1 M (58.5 g/L), but loss of water was produced above 260 g/L.

At low NaCl concentrations, swelling of the fibers and high values of WHC were observed in agreement with the results reported by Offer and Trinick (1983) and Belitz and Grosch (1997). The increase in WHC might be attributed to the lateral expansion of myofibrils, which is coupled to protein solubilization. According to Xiong et al. (2000) and Cheng and Sun (2008), an increase in water binding and hydration in salted meat and muscle fibers are generally attributed to enhanced electrostatic repulsion between myofibrillar filaments; the protein structure matrix unfolds and swelling occurs, causing the filament lattices to expand for water entrapment. Akse et al. (1993) reported that salting-in and swelling of the muscle occur at lower salt concentration (<50g/L), but salting-out of proteins is observed at higher concentrations (above 90–100g/L), which implies a strong bond between water and salt and the subsequent protein dehydration.

Using the experimental values and the mass balances, different mass ratios were determined such as m_{NaCl}/M_0, m_{water}/M_0, and $m_{solubilized\ tissue}/m_{dry\ tissue\ 0}$. Figure 7.2a and b shows that the ratio m_{water}/M_0 increased as a function of time; however, when NaCl concentrations in the brine were lower than 100 g/L, water uptake increased with salt concentration (Figure 7.2a), whereas for higher NaCl concentrations (140 g/L and 200 g/L), the opposite behavior was observed.

As expected, the ratio m_{NaCl}/M_0 increased as a function of time and increased with salt concentration. The ratio between the amount of solubilized protein and the initial dry tissue content increased during immersion time. The highest value was observed for 70 g/L NaCl in the solution for long contact times (Figure 7.3a). Figure 7.3b shows the effect of immersion time on the amount of solubilized protein for NaCl concentrations of 30 g/L and 70 g/L. At 70 g/L, protein solubilization was faster than at 30 g/L, reaching similar values at long contact times.

7.1.4.2 Effect of Sodium Chloride on Protein Denaturation

Differential scanning calorimetry (DSC) is a powerful technique for studying the thermodynamics of protein stability, and it can provide basic understanding of meat protein denaturation.

The three major transitions observed in a typical beef muscle homogenate have been attributed to muscle proteins as follows: 54°C–58°C myosin; 65°C–67°C myosin, sarcoplasmic proteins, and collagen; 71°C–83°C actin, as actomyosin and as fragments of F and G actin monomers (Stabursvik and Martens 1980; Wright and

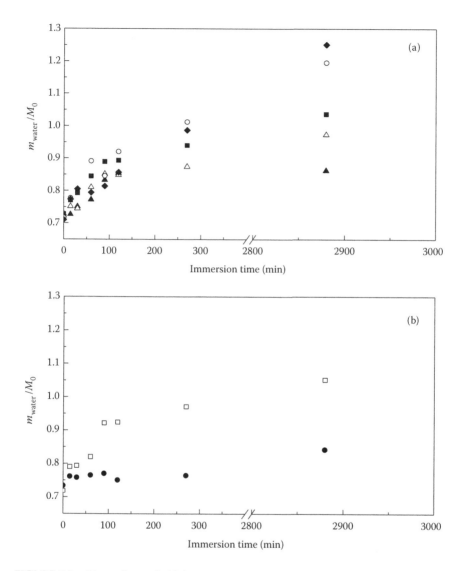

FIGURE 7.2 Mass of water/initial mass of tissue (M_0) as a function of time for different brine concentrations: (a) (▲): 5, (△): 10, (■): 30, (○): 70, (◆): 100 g NaCl/L, and (b) (□): 140, (●): 200 g NaCl/L.

Winding 1984; Findlay et al. 1986). Increasing the NaCl content of a muscle homogenate has been shown to destabilize the thermal stability of myosin and actin (Quinn et al. 1980; Kijowski and Mast 1988).

Graiver et al. (2006) carried out DSC studies using small pork meat pieces (approximately 500 mg) that were immersed in NaCl solutions of 5, 10, 20, 30, 40, and 50 g/L. DSC thermograms of pork tissue were analyzed to establish the effect of

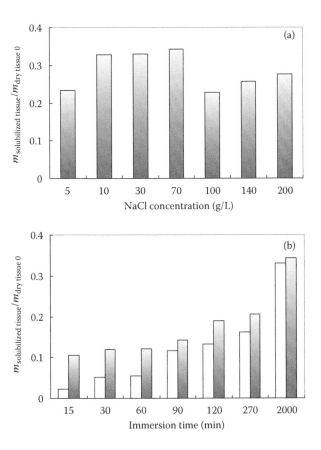

FIGURE 7.3 Ratio between the amount of solubilized protein and the initial dry tissue content: (a) as a function of NaCl concentration for long contact times; (b) as a function of immersion time for NaCl concentrations of (□) 30 g/L and (■) 70 g/L.

NaCl on protein denaturation (Figure 7.4). Untreated tissue samples showed three peaks, with T_{max} ($T_{maximum}$) values of 57.6°C, 66.2°C, and 80.3°C. The three peaks observed corresponded to myosin (I), sarcoplasmic proteins and collagen (II), and actin (III).

After soaking the samples in brine solutions of 5 and 10 g NaCl/L, these three peaks were still observed, although they were losing definition (Figure 7.4). T_I increased to 59.2°C, and T_{III} decreased to 77.4°C ($P < 0.05$) with respect to the control samples.

When NaCl concentration in the brine was increased to 20, 30, 40, and 50 g/L, only two peaks were observed in the thermograms (Figure 7.4). The maximum temperature of the first peak decreased from 56.5°C to 53.0°C and the second one from 70.3°C to 67.7°C in the assayed concentration range. Thus, peaks II and III were contracted to one peak that, according to the literature, corresponds mainly to actin thermal transition.

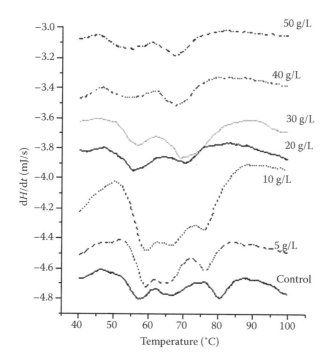

FIGURE 7.4 DSC thermograms of *Longissimus dorsi* pork muscle after different treatments with increasing concentrations of NaCl.

7.1.4.3 Analysis of Effect of NaCl on Tissue Microstructure by Scanning Electron Microscopy

Several authors (Wang et al. 2000; Pinotti et al. 2000; Graiver et al. 2005) reported that sodium chloride affects the microstructure of meat tissues. Scanning Electron Microscopy (SEM) micrographs of small pieces of meat tissue treated with different solutions showed microstructural changes; micrographs of samples treated with NaCl 5, 140, and 330 g/L for 48 h contact times are observed in Figure 7.5 (Graiver et al. 2006). Samples treated with NaCl 5 g/L (Figure 7.5a) showed slight differences with respect to the untreated sample (not shown); the essential structure of the myofibrils appeared to be intact. Fibers submitted to NaCl 140 g/L showed swelling as it can be visualized in Figure 7.5b, whereas NaCl 330 g/L (Figure 7.5c) produced fragmented and dehydrated fibers, with a granular appearance.

7.1.4.4 Mathematical Modeling of the Curing Process in Large Pieces of Pork Meat

As it was mentioned previously, Graiver et al. (2006) reported that there was a significant water uptake when NaCl concentration ranged between 5 g/L and 200 g/L. This phenomenon leads to a formulation in which Fick's law is used for the modeling of the diffusion process and "convective" terms are included to consider the global flux of brine (Graiver 2006).

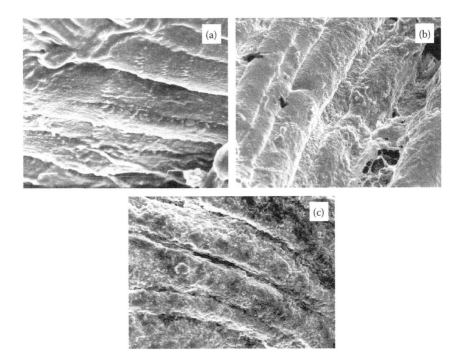

FIGURE 7.5 SEM micrographs of pork tissue treated with different solutions of NaCl 5, 140, and 330 g/L: (a) samples treated with NaCl 5 g/L; (b) samples treated with NaCl 140 g/L; (c) samples treated with NaCl 330 g/L. Scale: 100 mm between marks.

Therefore, a convective process of addition of water and solute ($C_i\, v$), as well as a diffusive process of incorporation of solutes (j_i), is proposed.

The flux $\underline{n}_i$ includes both diffusive (j_i) and convective ($C_i v$) contributions as follows:

$$\underline{n}_i = \underline{j}_i + C_i \underline{v} = -CD_m \nabla \omega_i + C_i \underline{v} \tag{7.6}$$

where i corresponds to each solute (NaCl, NaNO$_2$, or KNO$_3$), $\underline{n}_i$ is the total mass flux of a solute with respect to fixed coordinates, $\underline{j}_i$ is the diffusive mass flux of the solute that is replaced by Fick's law, $\underline{v}$ is the global average velocity of the solution, C_i is the mass concentration of the solute, D_m is the effective coefficient of diffusion of the solute in the matrix, ω_i is the mass fraction of solute i, and C is the total concentration.

Microscopic mass balances for each of the solutes were used in order to analyze the incorporation of the solutes in a piece of meat, in the absence of chemical reaction. Considering that D_m is variable with the concentration of solute, which implies a nonlinear problem and that the flux is incompressible ($\underline{\nabla} \cdot \underline{v} = 0$) with a constant total concentration, the following is obtained:

$$\frac{\partial C_i}{\partial t} + \underline{v} \cdot \left(\underline{\nabla} \cdot C_i\right) = \underline{\nabla} D_m \underline{\nabla} C_i. \tag{7.7}$$

In order to simulate the incorporation of curing salts in a finite cylindrical piece of meat, like those generally used under industrial conditions (e.g., whole *Longissimus dorsi* muscle), Equation 7.7 was written in cylindrical coordinates with radial and axial diffusive and convective fluxes. D_m was considered variable with the concentration of solute, and convective terms were added to consider v_r and v_z, with

$$\frac{\partial C_i}{\partial t} + v_r \frac{\partial C_i}{\partial r} + v_z \frac{\partial C_i}{\partial z} = \frac{1}{r} \frac{\partial}{\partial r}\left(r D_m \frac{\partial C_i}{\partial r}\right) + \frac{\partial}{\partial z}\left(D_m \frac{\partial C_i}{\partial z}\right) \tag{7.8}$$

where z is the axial coordinate and r is the radial coordinate.

The convective flux from the immersion solution takes place in opposite direction to axes r and z; therefore, $v_r = -v$ and $v_z = -v$, with v being the apparent convective contribution.

$$\frac{\partial C_i}{\partial t} = \frac{\partial D_m}{\partial r} \frac{\partial C_i}{\partial r} + \frac{D_m}{r} \frac{\partial C_i}{\partial r} + D_m \frac{\partial^2 C_i}{\partial r^2} + \frac{\partial D_m}{\partial z} \frac{\partial C_i}{\partial z} + D_m \frac{\partial^2 C_i}{\partial z^2} + v\left(\frac{\partial C_i}{\partial r} + \frac{\partial C_i}{\partial z}\right). \tag{7.9}$$

The industrial curing process is not carried out under conditions of strong agitation; then a mass-transfer coefficient k_c was considered, with the boundary conditions being

$$z = \pm\frac{H}{2}; \quad \forall r; \quad -D_m \frac{\partial C_i}{\partial z} = k_c\left(C_i - C_{if}\right) \tag{7.10}$$

$$r = R; \quad \forall z; \quad -D_m \frac{\partial C_i}{\partial r} = k_c\left(C_i - C_{if}\right) \tag{7.11}$$

where H is the cylinder height, R is the cylinder radius, C_{if} is the solute concentration of the external solution, and k_c is the mass transfer coefficient at the interface.

The initial condition was expressed as

$$C_i = 0 \quad \text{at} \quad t = 0. \tag{7.12}$$

It was considered that the initial concentration of NaCl in the tissue could be neglected since it is about 0.17 g NaCl per 100 g of fresh meat (USDA Nutrient Data Laboratory 2002).

Symmetry boundary conditions were

$$\frac{\partial C_i}{\partial r} = 0 \quad \text{at} \quad r = 0; \quad \frac{\partial C_i}{\partial z} = 0 \quad \text{at} \quad z = 0. \tag{7.13}$$

Equation 7.9 and initial and boundary conditions were discretized according to an explicit finite differences scheme, and a computer code in Fortran 90 (version 4.0 Microsoft) was developed to solve these equations.

To compare the numerical predictions of the model with experimental data, average concentrations in different regions of the cylinder (Figure 7.1) were calculated by numerical integration (Dahlquist and Björck 1974):

$$\bar{C} = \frac{2\pi \iint Cr\,dr\,dz}{2\pi \iint r\,dr\,dz}. \tag{7.14}$$

7.1.4.5 Effect of NaCl Concentration on Diffusion Coefficients of Curing Salts in Meat Tissue

Meat wet curing process requires a thorough understanding of the effects that salts ($NaNO_2$, KNO_3, and NaCl) have on the tissue microstructure, water retention, and protein solubilization (Offer and Knight 1988). Salt penetration is related to the equilibrium between the salt concentration in the interior of the meat piece and the external brine solution. Diffusion of salts in solid foods such as pork, beef, or fish has been studied by many workers (Djelveh and Gros 1988; Dussap and Gros 1980; Fox 1980; Graiver et al. 2006, 2009; Gros et al. 1984; Pinotti et al. 2001; Sabadini et al. 1998; Schwartzberg and Chao 1982; Siró et al. 2009; Turhan and Kaletunç 1992; Wang et al. 2000; Wood 1966). For the controlled manufacture of these products, it is important to know the factors influencing salt penetration and to be able to predict the diffusion rate.

Mass transport of a solute from the surface toward the center of a solid food is normally analyzed as a pseudobinary system (solute tissue; Zaritzky and Califano 1999). Different authors have determined effective diffusion coefficients in meat tissues (Wood 1966; Fox 1980; Dussap and Gros 1980; Gros et al. 1984; Djelveh and Gros 1988; Sabadini et al. 1998; Wang et al. 2000).

The process of salt uptake in the tissues is complex and produces changes in the water content and in the protein solubilization. However, in the literature, sufficient quantitative information about these phenomena has not been found; thus more studies of these changes in the system are necessary.

Meat tissue cannot be considered as formed by an insoluble matrix and an aqueous phase through which the solute diffuses, since parts of the proteins are solubilized; besides water penetration into the swelling matrix is produced.

To determine the diffusion coefficients of NaCl in the tissue, Graiver et al. (2006) used a radial (unidirectional) diffusion system. Long cylinders of *Longissimus dorsi* pork tissue (10-cm length and approximately 1.5-cm diameter) were immersed in brines of sodium chloride. The solutions were stirred to get constant solute concentration at the solid–fluid interface. Experiments were performed at 4°C in thermostatic chambers. The content of NaCl in the samples was determined at different contact times (15, 30, 60, 90, 120, 150 min) using the central zone of each cylinder (2.5-cm length). Similar procedures were used to determine sodium nitrite and potassium nitrate diffusion coefficients.

Equilibrium concentrations of salts were obtained by immersion of small tissue samples in different brines for at least 48 h.

Experimental values of D_m (m²/s) at 4°C for NaCl, NaNO$_2$, and KNO$_3$ at different NaCl concentrations (C_{NaCl}: expressed as mg NaCl/g meat; Pinotti et al. 2001; Graiver et al. 2006) were used to obtain the following linear regressions:

$$D_{m\,NaCl} = (-0.00004\,C_{NaCl}{}^2 + 0.058\,C_{NaCl} + 1.72) \times 10^{-10}\ \text{m}^2/\text{s} \quad (r^2 = 0.951) \quad (7.15)$$

$$D_{m\,NaNO2} = (0.016\,C_{NaCl} + 2.66) \times 10^{-10}\ \text{m}^2/\text{s} \quad (r^2 = 0.975) \quad (7.16)$$

$$D_{m\,KNO3} = (0.028\,C_{NaCl} + 2.63) \times 10^{-10}\ \text{m}^2/\text{s} \quad (r^2 = 0.995). \quad (7.17)$$

As it can be observed, the diffusion coefficients increased as the NaCl concentration in the brine increased. The addition of 3 g/L NaNO$_2$ and 2.5 g/L KNO$_3$ did not affect the NaCl diffusion coefficient significantly.

Different values of diffusion coefficients of salt in meat tissues were reported in literature. Wood (1966) found in *Longissimus dorsi* pork immersed in saturated salt solution (320 g/L) a coefficient of 2.2 × 10^{-10} m²/s at 12°C. Gros et al. (1984) reported a value of D_m = 2.19 × 10^{-10} m²/s at 2°C in *Longissimus dorsi* pork. Sabadini et al. (1998) obtained 2.5 × 10^{-10} m²/s for the NaCl diffusion coefficient in beef at 10°C, and Rodger et al. (1984) reported a value of 2.3 × 10^{-10} m²/s at 20°C using fish fillets in a NaCl saturated solution.

The diffusion coefficient obtained by Fox (1980) in *Longissimus dorsi* pork tissue for 180 g/L NaCl (D_m = 2.2 × 10^{-10} m²/s at 5°C) agrees with the results obtained in this work.

With reference to the effect of sodium chloride concentration on the diffusion coefficient, Wang et al. (2000) reported equations that predict salt diffusivities as a function of brine concentration in the Atlantic salmon muscle; for postrigor fish, the proposed equation was D_m = (1.08 + 0.59C') × 10^{-10} m²/s, where C' was the NaCl concentration in g/g salt-free solids.

The increase in the diffusion coefficient with salt concentration can be attributed to the effect of NaCl on the fiber microstructure as it was described in the SEM section; NaCl disrupts the fiber tissues and facilitates its penetration at higher concentrations (Pinotti et al. 2001).

Djelveh and Gros (1988) reported higher diffusion coefficients than those obtained in this work. They worked with NaCl concentrations ranging between 20 g/L and 200 g/L and obtained diffusion coefficients that increased from D_m = 7.69 × 10^{-10} to 8.73 × 10^{-10} m²/s, respectively.

It is important to remark that the use of NaCl uptake values expressed per mass of water in the tissue led to correct values of D_m. In contrast, when solute uptake values were expressed per mass of tissue without introducing the correction factors that consider the actual water content in the sample, diffusion coefficients were erroneously overestimated, turning out to be higher than the diffusion coefficient of NaCl in water, that is, 8.16 × 10^{-10} m²/s at 4°C.

7.1.4.6   Convective Contribution

As discussed in Section 7.1, the water convective flux varied with immersion time. The water macroscopic balance in the small cylinders led to

$$V \frac{dC^w}{dt} = k \left(C_{max}^w - C^w \right) A \tag{7.18}$$

where V is the volume of the small cylinder used in the mass balance, A is the cylinder area, C^w is the mass water concentration (g water/g dry tissue), k is the mass transfer coefficient (m/s), and C_{max}^w is the maximum water concentration in the tissue, reached for each NaCl concentration in the brine. Rearranging the previous equation, the following is obtained:

$$\frac{dC^w}{dt} = k' \left(C_{max}^w - C^w \right) \quad \text{where} \quad k' = \frac{kA}{V}. \tag{7.19}$$

Integrating Equation 7.18 and considering that initially $C^w = C_0^w$, where C_0^w is the initial concentration of water for each NaCl concentration, the obtained equation is

$$\ln C_w^* = \ln \left(\frac{C_{max}^w - C^w}{C_{max}^w - C_0^w} \right) = -k't \tag{7.20}$$

where C_w^* is the dimensionless water concentration.

Experimental data of water content C^w were expressed as dimensionless values C_w^*. From the linear regressions of $\ln C_w^*$ as a function of the immersion time t, the value of k' was determined. An average value of $k' = 6 \times 10^{-5}$ s^{-1} ($r^2 = 0.7431$) was calculated for the different assayed NaCl concentrations ranging between 10 and 200 g NaCl/L.

Combining Equations 7.18 and 7.20, the modulus (absolute value) of the velocity of water uptake in the tissue ($\underline{v}$) was calculated according to

$$|v| = \frac{V}{A} \frac{dC_w^*}{dt} = k e^{(-k't)}. \tag{7.21}$$

This velocity was introduced in the microscopic mass balance to represent the convective term.

7.1.4.7   Mass Transfer Coefficient in Immersion Solution

The mass transfer coefficient k_c defined in Equations 7.10 and 7.11 was determined from the correlation between j_m as a function of the dimensionless numbers Reynolds, Schmidt, and Sherwood for immersed cylinders (Sherwood et al. 1975).

The diffusion coefficients (at infinite dilution) of the salts in water at 25°C (Lide 1997–1998) were used to calculate the corresponding values at 4°C using the

Stokes–Einstein equation. The obtained values at 4°C were $D_{NaCl} = 8.5 \times 10^{-10}\,m^2/s$, $D_{NaNO2} = 8.29 \times 10^{-10}\,m^2/s$, and $D_{KNO3} = 10.18\ 10^{-10}\,m^2/s$. A low Reynolds number (Re = 3) was adopted for the stagnant solution obtaining a value of $j_m = 0.3$ that led to $k_c = 1.3 \times 10^{-7}\,m/s$ as an average value for the three curing salts.

7.1.4.8 Validation of Mathematical Model

To validate the mathematical model, different experiments were carried out using *Longissimus dorsi* pork tissue, free of visible fat. The initial water and protein contents determined according to AOAC methods (1980, 1984) were 74.10 ± 1.85 and 20.32 ± 0.21, respectively. Fat content was determined on previously dried samples by Soxhlet method using ethyl ether and petroleum ether (Bp: 35°C–60°C) in a 1:1 relationship as an extraction solvent (Andrés et al. 2006); the obtained value was 2.13 ± 0.12.

Seven cylinders of 6-cm diameter and 12-cm height cut from pork whole muscle (*Longissimus dorsi*) from different animals were used. They were immersed in a brine solution containing NaCl 140 g/L, $NaNO_2$ 3 g/L, and KNO_3 2.5 g/L at 4°C. At each immersion time (3.66, 5, 8, 13, 24, 48, and 75 h), one of the cylinders was removed from the solution and cut perpendicular to the axial axis in five cylindrical sections of 2.4-cm thickness. The center of each section was removed with a borer of 2.6-cm internal diameter, and each of the rings and internal cylinders was used to determine the concentrations of nitrite, nitrate, and chloride anions. All experiments were performed in duplicate using 14 different animals to consider biological variability. In Figure 7.6, numbers 2, 4, and 6 correspond to the external ring at different heights, whereas 1, 3, and 5 are the internal cylinders. Besides, another set of similar experiments was carried out using five cylinders of 5-cm diameter and 12-cm height of pork whole muscle from different animals to validate the mathematical model, with reference to the analysis of the effect of cylinder size on salt uptake.

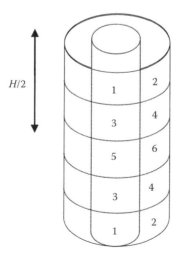

FIGURE 7.6 Scheme of the different portions of meat tissue in which the concentrations of the solutes were determined to validate the numerical model experimentally.

In order to simulate the uptake of the different solutes by the tissue, the following information has been incorporated to the mathematical model: diffusion coefficients of the solutes as functions of sodium chloride concentration, convective contribution, and mass transfer coefficient in the immersion brine. To validate the model against experimental results, the computer code was run to simulate the experimental conditions.

Figures 7.7a through c and 7.8a through c show the experimental data of the average concentrations of NaCl, $NaNO_2$, and KNO_3 corresponding to different zones of the tissue cylinders (6-cm diameter and 12-cm height) as a function of immersion

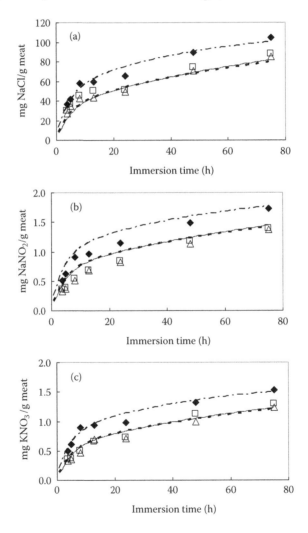

FIGURE 7.7 Comparison of the amount of (a) mg NaCl/g meat (b) mg $NaNO_2$/g meat, and (c) mg KNO_3/g meat as a function of the immersion time in external rings: experimental (◆) 2, (□) 4, (△) 6, and predicted values (- - -) 2, (———) 4, (----) 6 by the numerical model. Key numbers of the zones are shown in Figure 7.1.

time. Results for the external rings (zones 2, 4, and 6) and for the internal rings (zones 1, 3, and 5) are presented in Figures 7.7 and 7.8, respectively.

Figures 7.7 and 7.8 allow comparing experimental data with the predicted values by the numerical model (expressed in mg NaCl/g meat, mg $NaNO_2$/g meat, and mg KNO_3/g meat) as a function of immersion time. A good agreement for the three analyzed salts was observed, with an average error of 9% that includes the biological variability of the samples. The satisfactory prediction of experimental values was achieved due to the convective term introduced in the numerical model.

The effect of the cylinder diameter on sodium chloride uptake (average values in the whole meat sample) is observed in Figure 7.9. Experimental data obtained for tissue

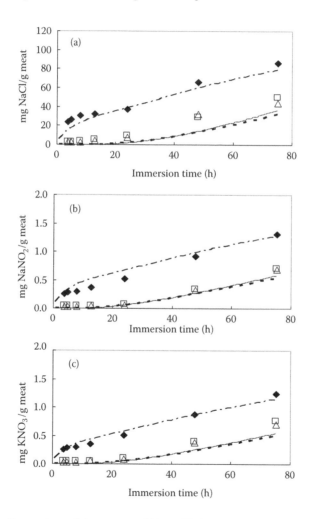

FIGURE 7.8 Comparison of the amount of (a) mg NaCl/g meat (b) mg $NaNO_2$/g meat, and (c) mg KNO_3/g meat as a function of the immersion time in internal cylinders: experimental (◆) 1, (□) 3, (△) 5, and predicted values (- - -) 1, (———) 3, (----) 5 by the numerical model. Key numbers of the zones are shown in Figure 7.6.

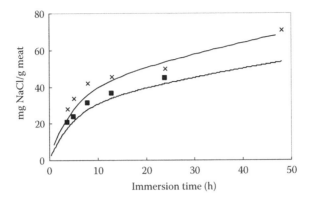

FIGURE 7.9   Total sodium chloride uptake as a function of immersion times, for two cylinders with the same height (12 cm) and different diameters (D): D = 5 cm experimental (■), predicted (———); D = 6 cm experimental (x), predicted (----). Brine concentration: 140 g NaCl/L, 3 g NaNO$_2$/L, and 2.5 g KNO$_3$/L.

samples of 5- and 6-cm diameter, both of 12-cm height, were compared with model predictions as a function of immersion times; a satisfactory agreement was observed, showing that to reach the same average sodium chloride concentrations in the cylinders, higher immersion times are necessary for the cylinder with a smaller diameter.

7.1.5   Predicted Profiles

Once the computer code was validated, it was used to predict NaCl, NaNO$_2$, and KNO$_3$ concentrations in meat tissues for different operating conditions. The input sizes of the pork meat cylinders ranged between 5- and 12-cm diameter and from 12- to 28-cm height; concentrations of sodium chloride, sodium nitrite, and potassium nitrate ranged from 20 to 80, 0 to 0.20, and 0.05 to 0.25 g/L, respectively.

Figure 7.10a shows the predicted changes in the average NaCl uptake in the tissue as a function of immersion times for cylinders 26 cm in height and diameters ranging between 9 cm and 12 cm, immersed in a brine of NaCl 70 g /L, NaNO$_2$ 0.1 g /L, and KNO$_3$ 0.15 g /L; Figure 7.11b exhibits a similar behavior for the average (NaNO$_2$ + KNO$_3$) concentrations expressed as NaNO$_2$ concentration.

In both figures, an increase in the salt concentrations in the meat tissue can be observed when the cylinder diameter increases, in agreement with the experimental data shown in Figure 7.9.

Besides, from simulations, keeping a constant diameter (10 cm) and varying the height of the cylinders between 25 cm and 28 cm, concentrations of NaCl and NaNO$_2$ + KNO$_3$ slightly decreased with decreasing height (data not shown).

Figure 7.11 shows the simulated concentration profiles of NaCl corresponding to a cylinder of 6-cm diameter and 12-cm height for 8, 24, and 72 h immersion time for a brine concentration of NaCl 140 g/L, NaNO$_2$ 3 g/L, and KNO$_3$ 2.5 g/L. Figure 7.11a corresponds to the concentration profile along the half height of the cylinder at a fixed radius of 1.45 cm, whereas Figure 7.11b shows the concentration

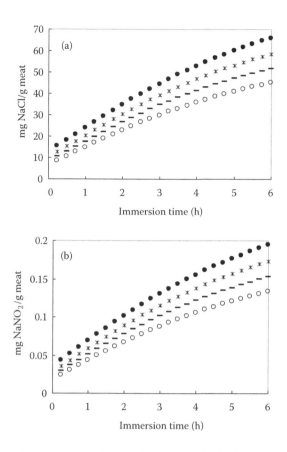

FIGURE 7.10 Predicted concentrations of the curing salts in the tissue as a function of immersion times in finite cylinders. (a) NaCl concentration (mg NaCl/g meat) (b) concentration of nitrite and nitrate (NaNO$_2$+ KNO$_3$) expressed as nitrite (mg NaNO$_2$/g meat). Pork meat cylinders 26 cm height and different diameters (o) 9 cm, (–) 10 cm, (*) 11 cm, and (♦) 12 cm. Brine concentration 70 g NaCl/L, 0.1 g NaNO$_2$/L, and 0.15 g KNO$_3$/L.

profile as a function of the radius for an axial position of 2.90 cm. Similar results were observed for NaNO$_2$ and KNO$_3$ salts.

7.1.6 APPLICATION OF THE MODEL TO DETERMINE OPERATING CONDITIONS FOR INDUSTRIAL CURING PROCESSING

After the immersion process, sodium chloride, sodium nitrate, and potassium nitrite continue the diffusion through the meat, and concentration profiles tend to become uniform. These phenomena are accomplished during the drainage and drying steps, reaching an equalization of the concentration of curing ingredients through the meat with a simultaneous loss of water. All these processes have to be considered to comply with regulations (Varnam and Sutherland 1995). To establish the average salt

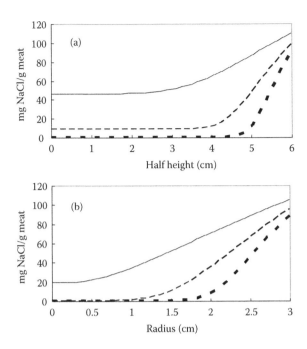

FIGURE 7.11 Concentration profiles of NaCl in a pork meat cylinder ($D = 6$ cm, $H = 12$ cm) for different immersion times as a function of (a) cylinder half height (fixed radial position: 1.45 cm) and (b) cylinder radius (fixed axial position: 2.90 cm). Immersion times: 8 h (- - - -), 24 h (----), and 72 h (————). Brine concentration: 140 g NaCl/L, 3 g NaNO$_2$/L, and 2.5 g KNO$_3$/L.

concentrations required after the immersion process, changes in water content from 75% to 50% due to the further maturation and drying process must be taken into account. The final product (water content 50%) must not exceed 0.2 mg/g of NaNO$_2$, KNO$_3$, or their mix expressed as NaNO$_2$ (Borchert and Cassens 1998), and a desirable NaCl content would be about 40 mg/g, which is the standard NaCl content in the commercial product available in Argentina. Therefore, the following target values for the wet curing process may be considered: 0.1 mg/g of NaNO$_2$, KNO$_3$, or their mix expressed as NaNO$_2$ and 20 mg/g NaCl.

The numerical code was run considering different brine concentrations (NaCl ranging between 20 and 80 g/L, NaNO$_2$ between 0 and 0.20 g/L, and KNO$_3$ between 0.05 and 0.25 g/L) for a cylinder of 10-cm diameter and 26-cm height with immersion times between 1 min and 24 h. This information is necessary to obtain a final product with the organoleptic and microbiological required conditions that complies with the regulations.

From the results computed by the code, simple regression equations were obtained using a stepwise procedure to predict the concentration of the salts in the tissue: (1) sodium chloride uptake (Equation 7.22) and (2) the sum of nitrite and nitrate concentrations (expressed as sodium nitrite; Equation 7.23) as a function of brine composition and immersion times:

$$y_{NaCl} = -1.61 + 2.64 \times 10^{-2}x_1 + 0.18x_2 - 8.89 \times 10^{-5}x_1^2$$
$$-7 \times 10^{-4}x_2^2 + 1.92 \times 10^{-3}x_1x_2 \quad (r^2 = 0.993) \tag{7.22}$$

$$y_{NaNO_2} = 0.027 + 5.16 \times 10^{-4}x_1 + 0.20x_3 - 0.14x_4 - 1.15 \times 10^{-3}x_2$$
$$-4.72 \times 10^{-7}x_1^2 + 1.86 \times 10^{-3}x_1x_3 + 1.43 \times 10^{-3}x_1x_4$$
$$+5.4 \times 10^{-3}x_2x_4 + 4.5 \times 10^{-6}x_2^2 \quad (r^2 = 0.991) \tag{7.23}$$

where y_{NaCl} is the average concentration of NaCl uptake in the tissue, y_{NaNO_2} is the average concentration of $NaNO_2$ uptake in the tissue, x_1 is the immersion time (minutes), x_2 is the NaCl concentration in the brine, x_3 is the $NaNO_2$ concentration in the brine, and x_4 is the KNO_3 concentration in the brine. y_{NaCl} and y_{NaNO_2} are expressed as g/100g tissue, and x_2, x_3, and x_4 are given in g/L brine solution.

Fixing in both equations the required conditions for the average concentrations in the wet tissue (0.1 mg/g of $NaNO_2$, KNO_3 or their mix expressed as $NaNO_2$ and 20 mg/g NaCl), the times to conform with the target values were determined for different salt concentrations in the brine. The obtained immersion times that accomplished the

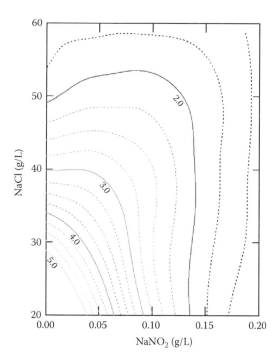

FIGURE 7.12 Contour plot to determine immersion times, expressed in hours, necessary to reach 0.1 mg/g of $NaNO_2$, KNO_3, or their mix expressed as $NaNO_2$ and 20 mg/g NaCl in meat tissue. Concentrations of NaCl and $NaNO_2$ in the brine are given in the axes for a fixed value of 0.20 g KNO_3/L.

required conditions are shown as contour lines considering a fixed value of KNO_3 0.20 g/L in the curing solution (Figure 7.12). The plot permits determining adequate immersion times (contour lines) fixing NaCl and $NaNO_2$ concentrations in the brine. As it was expected, increasing NaCl or $NaNO_2$ in the brine decreased immersion times.

The same methodology can be applied to establish optimum operating conditions for any desired geometry.

The model was applied to predict the time necessary for an industrial piece of meat to be immersed in brine without exceeding the maximum permitted nitrite value and the recommendable sodium chloride concentration; therefore, it allows determining the industrial operating conditions such as salt concentrations in the brine and immersion times in order to improve the curing process.

7.2 SMOKING

Smoking of food has been used for centuries, and its main purpose was originally to preserve the food. At first, smoking of food could be considered a side effect due to preservation by drying in the fireplace. Later on, the process was developed and changed and was combined with other processes such as salted dried, fermented, or preserved by other methods, because smoking alone does not ensure the appropriate shelf life.

Curing and smoking of meat are closely interrelated and are often practiced together, that is, cured meat is commonly smoked, and vice versa. For example, smoked ham such as Parma ham is both smoked and salted. By combining the two processes, a product is obtained that is stable against a broader spectrum of food-spoiling factors.

The use of the smoking process has also changed from being an artisanal process to being mainly an industrial process. The use of the process has also changed because there are other effective preservation methods such as cooling, heat treatment, canning, and the use of chemical preservatives. Therefore, in many countries, the main purpose of smoking food is to give the product a special taste, texture, and appearance, besides the preservation, and the antioxidant effect as a result of the smoking process is also important.

7.2.1 PURPOSES OF SMOKING

The primary purposes of smoking meat are (1) development of aroma and flavor, (2) preservation, (3) creation of new products, (4) development of color, and (5) protection from oxidation.

There are different ways to produce smoked food: (1) smoking with freshly generated smoke from wood; (2) smoking with smoke regenerated from smoke condensates; (3) flavoring with smoke flavor preparations derived from smoke condensates; and (4) flavoring with smoke flavors prepared by mixing defined chemical substances.

The first method is the traditional smoking process, and the three others are performed by the use of different kinds of smoke flavorings or "liquid smoke."

Smoking operation adds some volatile compounds to the product that inhibit bacterial growth and give a specific taste to the product (Poligné et al. 2001). The choice of the heat and smoke sources mainly depends on the desired characteristics of the product and on the economical constraints linked to the process development (Sebastian et al. 2005).

In the traditional smoking process, the first step is to make fire using dried wood. The wood used to generate the smoke should be of the hardwood species. Pine or any other resinous wood or sawdust is not recommended because the smoke from such wood will be sooty and strong smelling. It is recommended to use wood or sawdust from hickory, apple, plum, oak, maple, ash, or any nonresinous wood to obtain satisfactory results (Pearson and Gilett 1996; Poligné et al. 2001; Stołyhwo and Sikorski 2005). After having made the fire, the flames are reduced to a glow. The wood is then decomposed by the high temperature in the ignition zone. The composition of the smoke depends very much on the temperature in this area.

The period of time during which the food products are exposed to the smoke is highly important for the flavor of the food. This process can be regulated by several means including cooling of the smoke and addition of water and/or air. By this regulation, the flavor of the products can be changed, but to some degree, this regulation might also influence the formation of unwanted impurities that are transported from the smoke to the food. As changes in the composition of the smoke might also influence the taste of the food product, changes like these would often have to be the result of a compromise.

The different smoking methods are normally characterized by the temperature used:

- Cold smoking where the temperature is normally 18°C–20°C. The process is typically used for salmon, salamis, kippers, hams, and special cheeses. A cold smoking process may last several weeks. Normally the smoking process lasts for 6–24 h.
- Warm smoking by a temperature around 40°C is used for bacon, sirloin, and some types of sausage.
- Hot smoking is a combination of strong heating and smoke, which gives a temperature in the products of 70°C–90°C; therefore drying, cooking, and smoking processes occur simultaneously. Products like herring, eel, and some sausages are smoked in this way.

One of the most important properties of smoke is its effect on the bacterial population. Smoking of bacon has been shown to reduce the number of surface bacteria greatly and to extend its storage life. This is due to the bactericidal and bacteriostatic properties of smoke. These properties are attributable to certain components in the smoke, such as phenols and acids. Removal of moisture from the surface of meat during smoking also retards and reduces bacterial growth.

Smoke is also known to have a definite influence on the development of rancidity due to its antioxidant activity. This extends the shelf life of smoked meat products and helps to account for their desirability.

In a modern, industrial smoking process where the smoking generator and the smoking chamber are separated, the ability to control the density of the smoke, humidity, and temperature is quite good.

The smoke can be washed with water in order to remove some of the components with high boiling points or filtered in order to decrease the content of, for example, polycyclic aromatic hydrocarbons (PAHs), which are substances of health concern (Stołyhwo and Sikorski 2005). The temperature of smoke generally plays a very important role because the amount of PAHs in smoke, formed during pyrolysis, increases linearly with the smoking temperature within the interval 400°C–1000°C (Simko 2002).

The use of smoke flavorings often comprises atomizing of smoke flavoring and then spraying it onto the surface of the food. The result in taste, appearance, and preservation is more or less the same as that obtained by using the traditional smoking process.

7.2.2 COMPOSITION OF SMOKE

More than 300 different compounds have been isolated from wood smoke. Smoke composition depends on the conditions in the combustion chamber. The chemical components most commonly found in wood smoke include phenols, organic acids, alcohols, carbonyls, hydrocarbons, and some gaseous components such as carbon dioxide (CO_2), carbon monoxide (CO), oxygen (O_2), nitrogen (N_2), and nitrous oxide (N_2O; Pearson and Gilett 1996).

7.2.2.1 Phenols

About 20 different phenols have been isolated from wood smoke and identified. Among them are guaiacol, 4-methylguaiacol, phenol, 4-ethyl guaiacol, o-cresol, m-cresol and p-cresol, 4-propylguaiacol, eugenol (4-allylguaiacol), 4-vinylguaiacol, vanillin, 2,6-dimethoxyphenol, 2,6-dimethoxy-4-methylphenol,2,6-dimethoxy-4-propylphenol, and 2,6-dimethoxy-4-ethylphenol.

Although other phenols have been isolated from wood smoke distillates, they are apparently present in smaller amounts and are probably less important.

Phenols act as antioxidants; they contribute to the color and flavor of smoked products and have a bacteriostatic effect. The antioxidant activity of wood smoke is one of its most important attributes in smoked foods and is due to the phenols with high boiling points, especially 2,6-dimethoxyphenol, 2,6-dimethoxy-4-methylphenol, and 2,6-dimethoxy-4-ethylphenol. On the other hand, low boiling phenols have only weak antioxidant activity.

Color and flavor are the important sensory attributes that contribute to the desirability of smoked meats. Color development is caused by the interaction of the carbonyls in the vapor phase of the smoke with amino groups on the surface of the foods. Phenols also contribute to color development.

The actual color formation is believed to be due to the Maillard reaction. Maximum color formation is directly related to smoke concentration, temperature, and the moisture content at the surface of the products, with 12%–15% of moisture at the exterior surface of meat resulting in maximum color development. Some surface drying is necessary for good color formation during the smoking of meats. The characteristic flavor of smoked meats is primarily due to the phenolic compounds in the vapor phase. The phenols that are mainly responsible for the flavor and aroma of smoked meats are guaiacol, 4-methyl-guaiacol, 2,6-dimethoxyphenol, and syringol, with the first three compounds contributing most of the flavor and the latter being the primary contributor to aroma. Vanillic acid appears to be responsible for a sweet, mellow note in the aroma of wood smoke. Other acids and carbonyls probably contribute to the flavor of smoked meats; however, the flavor and aroma of whole smoke appears to be due to a more complex mixture.

The bactericidal action of smoking meat is due to the combined effects of heating, drying, and the chemical components in the smoke. When present on the surface of the meat, smoke components such as acetic acid, formaldehyde, and creosote prevent microbial growth. The phenols are known to possess strong bacteriostatic activity; in fact, this has led to the use of the phenol coefficient as a standard method for expressing the effectiveness of different germicides relative to phenol. The high boiling point phenols have the most bactericidal activity. The bacteriostatic effect is primarily on the surface since the amount of smoke penetration is limited.

Since smoke is largely concentrated on the surface of smoked-meat products, the total phenol concentration at varying depths has sometimes been used to express the depth of penetration and concentration of smoke. However, total phenol concentration is not always equivalent because the individual phenols are not equal in either color or flavor. Thus, the use of total phenols to measure the smoked flavor of meat is not necessarily closely related to sensory evaluation (Pearson and Gilett 1996).

7.2.2.2 Alcohols

A wide variety of alcohols are found in wood smoke. The most common and simplest of these is methanol. Although primary, secondary, and tertiary alcohols are all found in smoke, they are frequently oxidized to form their corresponding acids. The role of alcohols in wood smoke appears to be primarily that of a carrier for the other volatile components. Alcohols do not seem to play any major part as contributors to flavor or aroma, although they may exert a minor bactericidal effect. Thus, alcohols are probably one of the least important classes of components in smoke (Pearson and Gilett 1996).

7.2.2.3 Organic Acids

Simple organic acids ranging from 1 to 10 carbons are components of whole smoke. Only 1- to 4-carbon acids are commonly found in the vapor phase of smoke, whereas longer chain 5- to 10-carbon acids are in the particle phase of whole smoke. Thus, formic, acetic, propionic, butyric, and isobutyric acids contribute to the vapor phase of smoke, whereas valeric, isovaleric, caproic, heptylic, caprylic, nonylic, and capric acids are located in the particle phase (Pearson and Gilett 1996).

Organic acids have little or no direct influence on the aroma or flavor of smoked products; they have only a minor preservative action, which occurs as a result of greater acidity on the surface of smoked meat. They contribute to the red color of smoked products.

7.2.2.4 Carbonyls

A large number of carbonyl compounds contribute to smoke. Similar to organic acids, they occur in the steam-distillable fraction and also in the particle phase of smoke. Well over 20 compounds have been identified: 2-pentanone, valeraldehyde, 2-butanone, butanal, acetone, propanal, crotonaldehyde, ethanal, iso-valeraldehyde, acrolein, isobutyraldehyde, diacetyl, 3-methyl-2-butanone, pinacolene, 4-methyl-3-pentanone, alfa-methyl-valeraldehyde, 3-hexanone, 2-hexanone, 5-methyl furfural, methyl vinyl ketone, furfural, methacryaldehyde, methyl glyoxal, and others.

Although the largest proportion of the carbonyls is nonsteam-distillable, the steam-distillable fraction has a more characteristic smoke aroma and contains all the color from the carbonyl compounds. Short-chained simple compounds appear to be the most important to smoke color, flavor, and aroma. Many of the same carbonyls found in smoke have been isolated from a wide variety of foods. This suggests that either certain carbonyl compounds contribute to smoke flavor and aroma or more probably the level of carbonyls in smoke is much higher and thus imparts the characteristic aroma and flavor to smoked products. Regardless of the mechanisms or cause, smoke flavor and color seem to be largely due to the steam-distillable fraction of smoke (Pearson and Gilett 1996).

7.2.2.5 Hydrocarbons

A number of polycyclic hydrocarbons have been isolated from smoked foods. These include benz[a]anthracene, dibenz[a,h]anthracene, benz[a]-pyrene, benz[e]pyrene, benzo[g,h,i]perylene, pyrene, and 4-methyl pyrene.

At least two of these compounds, benz[a]pyrene and dibenz[a,h]anthracene, are recognized as being carcinogens. Both have been demonstrated to be carcinogenic in laboratory animals and have also been implicated in human cancer because Baltic Sea fishermen and Icelanders, who consume large quantities of smoked fish, have a high incidence of cancer compared to other populations.

Although the content of benz[a]pyrene and dibenz[a,h]anthracene is relatively low in most smoked foods, larger proportions have been found in smoked trout (2.1 mg/kg wet weight) and mutton (1.3 mg/kg wet weight). The concentration of benz[a] pyrene in other smoked fish is 0.5 mg for cod and 0.3 mg for red fish per kilogram of tissue. Although much higher levels of other polycyclic hydrocarbons have been found in smoked foods, none of these have been demonstrated to be carcinogens.

Fortunately, the polycyclic hydrocarbons do not appear to impart important preservative or organoleptic properties to smoked meats. Studies have shown that these compounds are removed in the particulate phase of smoke. Several liquid smoke preparations have been subjected to analysis and have been found to be free of benz[a]pyrene and dibenz[a,h]anthracene. Thus, smoke fractions can be prepared that are free of the undesirable hydrocarbons found in whole smoke. Fibrous casings prevent penetration of the hydrocarbons during smoking of meat (Pearson and Gilett 1996).

7.2.2.6 Gases

The significance of gases released in smoke is not fully understood. Most of them probably are unimportant in their contribution to smoke flavor. Both CO_2 and CO are readily absorbed on the surface of fresh meat, where they may react to produce the bright red pigments, carboxymyoglobin and carbon monoxide myoglobin, respectively. Oxygen can also combine with myoglobin to form either oxymyoglobin or metmyoglobin and makes some products take on a muddy color, especially in the presence of alkaline phosphates.

The gaseous component in smoke that is probably of the greatest significance is nitrous oxide, which has been linked to formation of nitrosamines and nitrites in smoked foods.

The cure accelerators (erythorbate and ascorbate) are known to prevent formation of N-nitrosamines during curing.

7.2.3 Methods of Smoking

Smoking as originally practiced was a relatively simple process, but as it was industrialized, it became not only more complex but also more reproducible. Old-style smokehouses had little or no control of temperature, humidity, or rate of combustion; these have largely been replaced with sophisticated equipment in which it is possible to regulate closely not only these important factors but also smoke density. Basically, there are three types of smokehouses, namely, (1) natural air circulation, (2) air-conditioned or forced air, and (3) continuous. There are, in addition, many modifications of the three types. The first type is designed so that natural ventilation will occur: regulation of the volume of air is controlled by the opening or closing of a series of dampers, thus providing natural circulation. The fire pit may be designed to use either logs or sawdust or a combination of the two.

Supplemental heat may be supplied by steam coils or gas. Sprinklers are usually installed to extinguish incipient fires. A modification of the natural air smoke house is the revolving smoke that consists of an endless chain to which the meat is attached. The endless chain keeps the meat in continuous motion. Air-conditioned or forced-ventilation smoke houses have largely replaced the natural air type. They are particularly useful where cooking or partial cooking is done and permit much more precise control of smoking. Even more important than the smoking process are the control of the temperature for cooking and the resultant control of shrinkage. Air circulation is controlled by a fan, so the air can be circulated, exhausted, or a portion of the air exhausted and part recirculated. Thus, this type of smokehouse gives uniform air movement and good control of temperature. Forced-air smokehouses usually not only control the air or smoke velocity but also regulate humidity.

The continuous smoking system offers some major advantages in space saving, speed of processing, and labor savings and allows more specific control of processing time, temperature, and relative humidity. However, the large capital investment and high output limit its usefulness.

7.2.4 Liquid Smoke

Besides the production of smoked food by using the traditional smoking process, the special taste in the smoked food can also be obtained by the use of "liquid smoke" or a combination of both. The use of liquid smoke has been developed during the twentieth century because of health concerns regarding smoked food and due to technical and economic reasons. The major advantage of using "liquid smoke" is the possibility of having a continuous process, as the addition of the smoke preparation can be done as a step in the production line. The smoke flavorings can be added, for example, by atomizing them directly on the food (Stołyhwo and Sikorski 2005).

Several liquid smoke preparations are available on the market. Liquid smoke has several advantages over natural wood smoke: it does not require the installation of a smoke generator, which usually requires a major financial outlay. The process is more repeatable, as the composition of liquid smoke is more constant. The liquid smoke can be prepared with the particle phase removed, and thereby possible problems from carcinogens can be alleviated. Liquid smoke application creates little atmospheric pollution, and its application is faster than conventional smoking.

Liquid smoke is commonly prepared by pyrolysis of hardwood sawdust. The smoke is captured in water by drawing it countercurrent to water through an absorption tower; the smoke is recycled until the desired concentration is reached. The solution is then aged to allow time for polymerization and precipitation; it is then filtered through a cellulose pulp filter, which removes any dissolved hydrocarbons that are present in the liquid smoke. These liquid smoke solutions may be used without further refinement. The final product is composed primarily of the vapor phase and contains mainly phenols, organic acids, alcohols, and carbonyl compounds.

Analyses of several liquid smoke preparations have shown that they do not contain polycyclic hydrocarbons, especially benz[a]pyrene. Animal toxicity studies have also supported the chemical analyses indicating that all carcinogenic substances in smoke are removed during production of liquid smoke (Pearson and Gilett 1996).

7.2.5 APPLICATION OF LIQUID SMOKE

There are a number of ways of adding liquid smoke to food products: (1) adding it directly to the meat emulsion; (2) dipping the product directly into the smoke solution; (3) spraying the smoke solution over the product; (4) atomizing the liquid smoke into a dense fog and injecting it into the smokehouse; (5) vaporizing the liquid by putting it on a hot surface; and (6) adding by way of smoke-treated casings. The latter four methods are commonly used for smoking meats, with the spray method most frequently being utilized for continuous meat processing.

Liquid smoke preparations are usually diluted before applying to meats. Commercially prepared liquid smoke solutions are diluted with water or frequently with vinegar or citric acid. A typical liquid smoke solution prepared and used by meat processors consists of 20 to 30 parts liquid smoke, 5 parts citric acid or acetic acid (4%), and 65 to 75 parts water. Citric or acetic acids are used to enhance skin formation on skinless frankfurters and other small sausage products. Although skin formation can be achieved by spraying with liquid smoke alone, a higher concentration of smoke is required. Thus, acid added to the smoke solution reduces costs.

Use of liquid smoke makes it much easier to keep equipment clean since deposition of residues from natural smoke requires frequent cleaning. Failure to clean regularly can result in spontaneous fires. The atomizing or spraying systems that are used to apply liquid smoke can also be adapted for use in cleaning smokehouses.

Moreover, it is necessary to provide cooking facilities even though the smoke generator is eliminated; cooking after spraying with liquid smoke preparations is essential to give good smoke color formation. This is the reason that such smoke preparations should be added just before or during cooking (Pearson and Gilett 1996).

Present trends suggest that liquid smoke will largely replace natural wood smoke. The speed with which this occurs will depend on automated procedures and on the absence of carcinogenic substances.

REFERENCES

Akse, L., Gundersen, B., Lauritzen, K., Ofstad, R., and Solberg, T. 1993. *Saltfisk: saltmodning, utproving av analysemeetoder, misfarget salfisk*. Tromso, Norway: Fiskeriforskning.

Andrés S. C., García M. E., Zaritzky N. E., and Califano, A. N. 2006. Storage stability of low-fat chicken sausages. *Journal of Food Engineering*, 72, 311–319.

AOAC 1980. *Official Methods of Analysis*, 13th ed. S. Williams (ed.), Washington, DC: Association of Official Analytical Chemists. Arlington, Virginia, USA.

AOAC 1984. *Official Methods of Analysis of the Association of Official Analytical Chemists. Met 24002. Meat and Meat Products*, 14th ed. S. Williams (ed.), Arlington, Virginia, USA.

Belitz, H. D., and Grosch, W. 1997. *Química de los alimentos*, 2nd ed. Zaragoza, España: Ed. Acribia.

Bertram, H., Karlsson, A., Rasmussen, M., Pedersen, O., Dønstrup, S., and Andersen, H. 2001. Origin of multiexponential T_2 relaxation in muscle myowater. *Journal of Agricultural and Food Chemistry*, 49(6), 3092–3100.

Borchert, L., and Cassens, R. 1998. Chemical hazard analysis for sodium nitrite in meat curing. Available at http://www.ag.ohio-state.edu/~meatsci/borca2.htm, accessed May 2, 2001.

Cheng, Q., and Sun, D.-W. 2008. Factors affecting the water holding capacity of red meat products: a review of recent research advances. *Critical Reviews in Food Science and Nutrition*, 48, 137–159.

Dahlquist, G., and Björck, A. 1974. *Numerical Methods*. Englewood Cliffs, New Jersey: Prentice-Hall, Inc., pp. 290–304.

Djelveh, G., and Gros, J. B. 1988. Measurement of effective diffusivities of ionic and non-ionic solutes through beef and pork muscles using a diffusion cell. *Meat Science*, 23, 11–20.

Djelveh, G., Petit, M., and Gros, J. B. 1988. Influence of sodium chloride concentration, potassium nitrate and temperature on the apparent diffusion coefficient of chloride ions through agar gels. *Lebensmittel Wissenschaft und Technologie*, 21, 103–107.

Dussap, C. G., and Gros, J. B. 1980. Diffusion–sorption model for the penetration of salt in pork and beef muscle. In: *Food Process Engineering*, Vol. I. P. Linko, Y. Malkii, J. Olkku, and J. Larinkari (eds.), London: Elsevier Applied Science, pp. 407–411.

Findlay, C. J., Stanley, D. W., and Gullett, E. A. 1986. Thermomechanical properties of beef muscle. *Meat Science*, 22, 57–70.

Fox, J. B. 1980. Diffusion of chloride, nitrite and nitrate in beef and pork. *Journal of Food Science*, 45, 1740–1744.

Fox, J. B., Jr., Sebranek, J. G., and Phillips, J. G. 1994. Kinetic analysis of the formation of nitrosylmyoglobin. *Journal of Muscle Foods*, 5, 15–25.

Ghavimi, B., Rogers, R. W., Althen, T. G., and Ammerman, G. R. 1986. Effects of non-vacuum, vacuum and nitrogen back-flushed tumbling on various characteristics of restructured cured beef. *Journal of Food Science*, 51, 116–1168.

Girard, J. P. 1991. *Tecnología de la Carne y de los Productos Cárnicos*. Zaragoza, España: Editorial Acribia S.A.

Graiver, N. 2006. Procesos difusionales en el curado de carne. PhD Thesis. Facultad de Ciencias Exactas. Universidad Nacional de La Plata, Argentina.

Graiver, N., Pinotti, A., Califano, A., and Zaritzky, N. 2005. Diffusion of sodium chloride in meat pork: influence on its microstructure. *The Journal of Scanning Microscopies,* 27(2), 76–77.

Graiver, N., Pinotti, A., Califano, A., and Zaritzky, N. 2006. Diffusion of sodium chloride in pork tissue. *Journal of Food Engineering,* 77, 910–918.

Graiver, N., Pinotti, A., Califano, A., and Zaritzky, N. 2009. Mathematical modeling of the uptake of curing salts in pork meat. *Journal of Food Engineering,* 95(4), 533–540.

Gros, J. B., Dussap, C. G., and González-Méndez, N. 1984. Engineering sciences in the food industry. *Engineering and Food,* Vol. 1. B. M. McKenna (ed.), London: Elsevier Applied Science, pp. 287–297.

Honikel, K. O. 2004. Curing agents. In *Encyclopedia of Meat Sciences.* W. K. Jensen, C. Devine, and M. Dikeman (eds.), Oxford, UK: Elsevier Ltd., pp. 195–201.

Katsaras, K., and Budras, K. D. 1993. The relationship of the microstructure of cooked ham to its properties and quality. *Lebensmittel Wissenschaft und Technologie,* 26(3), 229–234.

Kijowski, J. M., and Mast, M. G. E. 1988. Effect of sodium chloride and phosphates on the thermal properties of chicken meat proteins. *Journal of Food Science,* 53(2), 367–370, 387.

Knight, P. and Parsons, N. 1988. Action of NaCl and polyphosphates in meat processing: responses of myofibrils to concentrated salt solutions. *Meat Science,* 24, 275–300.

Lide, D. R. 1997–1998. *Handbook of Chemistry and Physics,* 78th ed. Florida: CRC Press.

Marriott, N. G., Graham, P. P., Boling, J. W., and Collins, W. F. 1984. Vacuum tumbling of dry-cured hams. *Journal of Animal Science,* 58(6), 1376–1381.

Møller, J. K. S., and Skibsted, L. H. 2002. Nitric oxide and myoglobins. *Chemical Reviews,* 102, 1167–1178.

Morrisey, P., Mulvihill, D., and O'Neill, E. 1987. Functional properties of muscle proteins. In *Developments in Food Proteins—5.* B. J. F. Hudson (ed.), London: Elsevier, pp. 195–256.

Offer, G., and Knight, P. 1988. In *Developments in Meat Science,* 4th ed. London: Elsevier, 63 pp.

Offer, G., and Trinick, J. 1983. On the mechanism of water holding in meat: the swelling and shrinking of myofibrils. *Meat Science,* 8, 245–281.

Pearson, A., and Gilett, T. 1996. *Curing en Processed Meats,* 3rd ed. New York: Chapman and Hall.

Pegg, R. B., and Shahidi, F. 2000. *Nitrite Curing of Meat. The n-Nitrosamine Problem and Nitrite Alternatives.* Trumbull, CT: Food and Nutrition Press, Inc.

Pinotti, A., Califano, A., and Zaritzky, N. 2000. Diffusion of curing salts in meat: effect of sodium chloride on tissue microstructure. *The Journal of Scanning Microscopies,* 22(2), 137–138.

Pinotti, A., Graiver, N., Califano, A., and Zaritzky, N. 2001. Diffusion of nitrite and nitrate salts in pork tissue in the presence of sodium chloride. *Journal of Food Science,* 67, 2165–2171.

Poligné, I., Collignan, A., and Trystram, G. 2001. Characterization of traditional processing of pork meat into boucané. *Meat Science,* 59(4), 377–389.

Quinn, J. R., Raymond, D. P., and Harwalkar, V. R. 1980. Differential scanning calorimetry of meat proteins as affected by processing treatment. *Journal of Food Science,* 45, 1146–1149.

Rodger, G., Hastings, R., Cryne, C., and Bailey, J. 1984. Diffusion properties of salt and acetic acid into herring and their subsequent effect on the muscle tissue. *Journal of Food Science and Technology,* 49, 714–720.

Sabadini, E., Carvalho, Jr. B. C., Sobral, P. J. do A., and Hubinger M. D. 1998. Mass transfer and diffusion coefficient determination in the wet and dry salting of meat. *Drying Technology,* 16, 2095–2115.

Schwartzberg, H. G., and Chao, R. Y. 1982. Solute diffusivities in leaching processes. *Food Technology*, 36, 73–86.

Sebastian, P., Bruneau, D., Collignan, A. and Rivier, M. 2005. Drying and smoking of meat: heat and mass transfer modeling and experimental analysis. *Journal of Food Engineering*, 70, 227–243.

Sebranek, J. G., and Bacus, J. N. 2007. Cured meat products without direct addition of nitrate or nitrite: what are the issues? *Meat Science,* 77, 136–147.

Sebranek, J. G., and Fox, J. B. Jr., 1991. Rate of nitric oxide formation from nitrite as affected by chloride ion concentration. *Journal of Muscle Foods*, 2, 149–163.

Sherwood, A., Pigford, R., and Wilk, C. 1975. *Mass Transfer.* New York, USA: McGraw Hill, Chemical Engineering.

Simko, P. 2002. Determination of polycyclic aromatic hydrocarbons in smoked meat products and smoke flavouring food additives. *Journal of Chromatography B*, 770, 3–18.

Siró, I., Vén, Cs., Balla, Cs., Jónas, G., Zeke, I., and Friedrich, L. 2009. Application of an ultrasonic assisted curing technique for improving the diffusion of sodium chloride in porcine meat. *Journal Food Engineering*, 91, 353–362.

Sofos, J. N. 1986. Use of phosphates in low-sodium meat products. *Food Technology*, 52–69.

Solomon, L. W., Norton, H. W., and Schmidt, G. R. 1980. Effect of vacuum and rigor condition on cure absorption in tumbled porcine muscles. *Journal of Food Science*, 45, 438–440.

Stabursvik, E., and Martens, H. 1980. Thermal denaturation of proteins in post rigor muscle tissue as studied by differential scanning calorimetry. *Journal of Science and Food Agriculture*, 31, 1034–1042.

Stamler, J. S., and Meissner, G. 2001. Physiology of nitric oxide in skeletal muscle. *Physiology Reviews*, 81, 209–237.

Stołyhwo, A., and Sikorski, Z. E. 2005. Polycyclic aromatic hydrocarbons in smoked fish—a critical review. *Food Chemistry*, 91, 303–311.

Tompkin, R. B. 2005. Nitrite. In *Antimicrobials in Food*, 3rd ed. P. M. Davidson, J. N. Sofos, and A. L. Branen (eds.), Boca Raton, FL: CRC Press, Taylor & Francis Group.

Trout, G. 1988. Techniques for measuring water-binding capacity in muscle foods—a review of methodology. *Meat Science*, 23, 235–252.

Turhan, M., and Kaletunç, G. 1992. Modelling the salt diffusion during long-term brining. *Journal of Food Science*, 57, 1082–1085.

USDA Nutrient Data Laboratory, Agricultural Research Service. Available at http://www.nal .usda.gov/fnic/foodcomp, accessed November 21, 2002.

USDA. 1995. Processing Inspector's Calculations Handbook (FSIS Directive 7620.3). Available at http://www.fsis.usda.gov/OPPDE/rdad/FSISDirectives/7620-3.pdf, accessed January 3, 2007.

USDA-FSIS. 1999. Food ingredients and sources of radiation listed or approved for use in the production of meat and poultry products; final rule. Subpart C—424.21. Use of Food Ingredients and sources of radiation. Code of Federal Regulations, Title 9, Vol. 64, Ch. III, Part. 424, pp. 72185–72186. Office of Federal Register, National Archives and Records, GSA, Washington, DC.

Varnam, A., and Sutherland, J. P. 1995. Cured meat. Chapter 4. In *Meat and Meat Products. Technology, Chemistry and Microbiology*, 1st ed. London, UK: Chapman & Hall, pp. 167–210.

Vestergaard, C., Risum, J., and Adler-Nissen, J. 2005. Na-MRI quantification of sodium and water mobility in pork during brine curing. *Meat Science*, 69, 663–672.

Wang, D., Tang, J., and Correia, L. 2000. Salt diffusivities and salt diffusion in farmed Atlantic salmon muscle as influenced by rigor mortis. *Journal of Food Engineering*, 43, 115–123.

Weiss, J. M. 1973. Ham tumbling and massaging. *Western Meat Industry*, 23.

Wiebe, W. R. Jr., and Schmidt, G. R. 1982. Effects of vacuum mixing and precooking on restructured steaks. *Journal of Food Science*, 47, 386–387, 396.

Wilson, N. R. P. 1981. *Curing in Meat and Meat Products—Factors Affecting Quality Control*. London: Applied Science Publishers.

Wood, F. W. 1966. The diffusion of salt in pork muscle and fat tissue. *Journal of Science and Food Agriculture,* 17, 138–140.

Wright, D., and Winding, P. 1984. Differential scanning calorimetric study of muscle and its proteins: myosin and its subfragments. *Journal of Science and Food Agriculture*, 357–371.

Xiong, Y., Lou, X., Harmon, R., Wang, C., and Moody, W. 2000. Salt- and pyrophosphate-induced structural changes in myofibrils from chicken red and white muscle. *Journal of Science and Food Agriculture,* 80, 1176–1182.

Zaritzky, N., and Califano, A. 1999. Effective diffusion coefficients of chemical preservatives in food tissues. In *Trends in Heat, Mass and Momentum Transfer*, Vol. 5. Trivandrum, India: Research Trends, pp. 127–139.

8 Pretreatments for Fish and Seafood in General

A. Javier Borderías

CONTENTS

8.1 INTRODUCTION

Røra et al. (2001) used the term "primary processing" to denote the first steps of fish processing in which fish is converted to "fish-as-food." These processes are more or less the same for wild and farmed fish, although fish farming has certain advantages over traditional fisheries in that the processor has more control and can influence postmortem biochemistry and various quality parameters like color (Christiansen et al. 1995; Nickell and Bromage 1998) and fat, which influence flavor and texture.

The main purpose of these steps is to enhance the hygiene and so extend the shelf life. In addition, primary processing reduces the mass of the raw material, thus making transportation more cost-effective. Also, by filleting, producers add product value and enhance product differentiation, which is important to secure current markets and to penetrate new ones. With the edible and nonedible parts, separation is also possible, to minimize waste and to try and valorize by-products, for which new process lines have to be created.

8.2 FISH HANDLING

Wild fish are harvested by a large variety of methods such as different kinds of nets, hooks, pots, etc. Incorrect handling at this point will be detrimental to the quality during ice storage. From the point of view of contamination, Shewan (1949) reported that in general, trawled fish carry microbial loads 10–100 times higher than line-caught fish because of mud stirring contamination and gut contamination produced by the pressure of the fish in the net. Moreover, in prolonged trawls, the cod end becomes very full, and the fish there can die of the pressure and become bruised (Costakes et al. 1982). Furthermore, larger catches take more time to stow and ice properly.

On the other hand, fish that have been trawled are subject to more stress from fighting the net for hours, and this stress has been shown to affect ice storage quality. In the case of some species, like tuna, when they are caught in a highly stressed state, the buildup of lactic acid in the muscle, combined with high muscle temperatures, results in a dull muscle and acidic and metallic aftertaste (Goodrick 1987). This has been reported in other species; for instance, wild salmon caught by gill netting die after from stress exhaustion. In these conditions, rigor mortis is faster, and quality is poorer during icing (Dassow 1976). With harvesting by hook and line, when the fish is killed faster, there is less stress, and the quality is better preserved during icing. In Atlantic cod (*Gadus morhua* L.), Botta et al. (1987) reported that the fishing method is more significant than the season of capture.

It is important that fish be handled carefully as soon as it is on board. Breaks in the skin and bruising introduce microorganisms in the flesh, which is sterile just after death, thus accelerating spoilage. Fish should not be exposed to the sun or the wind but should be carefully cleaned and cooled down as soon as possible (FAO 1973).

Handling of farmed fish has certain peculiarities with respect to wild fish, and there are differences depending on the species. According to Smart (2001), it is vital that techniques used in harvesting of sea bass (*Dicentrarchus labrax*) and sea bream (*Sparus aurata*) do not result in any loss of quality after a lot of care has been taken during 18 months of growing.

The first operation for farmed fish is to carefully separate fish from the main cages into smaller holding units without causing more stress than necessary. At this stage, the fish are kept at around 5–10 kg/m^3 until ready for cropping. In live tilapia (*Oreochomis* sp.), ozone pretreatment prolongs shelf life by 12 days and improves the quality characteristics during storage at 0°C for 30 days but has little effect at 5°C (Gelman et al. 2005; Glatman et al. 2006). The improvement they report is probably the result of an initial reduction of spoilage bacteria and subsequent prevention of growth.

A second very important operation is starvation for as long as is necessary to ensure that gut contents are evacuated; this will be 1 or 2 days depending on temperature. One problem is that around 1% of weight can be lost due to starvation at high temperature (>20°C). Mechanical properties of sea bream muscle change as the starvation time progresses, so that the flesh is firmer when they are starved for up to 8 days, owing to changes in protein solubility and pH (Gines et al. 2002). Starvation is very important to ensure a long iced storage life and to prevent feces trailing from the anus, which is off-putting for consumers.

The third operation for farmed fish is killing. The method most farmers use is to plunge the fish directly into iced water. It is very important to check that the water is kept close to 0°C at all depths; if the temperature should rise to 8°C, the fish will die not of thermal shock but of asphyxia, and this will adversely affect their appearance, color, and texture. In the case of sea bass, it is very important that crowding prior to cropping be kept to a minimum and that fish are swiftly killed (Smart 2001); otherwise, there can be considerable damage, such as descaling and other skin lesions, and bleeding around the belly.

The last step of harvesting is packaging. Fish must not warm up from the slaughter tank to the polystyrene boxes, which are filled with sufficient ice to maintain the temperature. Fish are placed in boxes with the belly cavity upward and fitted in so as to avoid unnecessary movement while they are in circulation, which can be up to a maximum of 10–15 days.

In other species, like salmon (*Salmo salar*), there are other problems that cause downgrading of the product due to the characteristics shown by the muscle. The causes are linked to stock characteristics, season, and maturation, and they can also be caused by handling fish from the harvest end in the processing chain (Michie 2001). The most important problem is lack of color, or paleness, which is a seasonal phenomenon linked to maturation. It is caused by an increase in fat deposition in the muscle prior to spawning, which if heavy enough can mask the pigmentation of muscle. Flesh color is also affected by the rapid drop of pH in muscle just after the slaughter of stressed salmonids (Robb and Frost 1999); this is caused by insolubilization of muscle proteins as a result of low pH and subsequent drip loss, which influence the reflection of incident light so that the flesh looks more opaque. Erikson and Misimi (2008) report that perimortem handling stress initially significantly affects several color parameters of skin and fillets, and significant transient fillet color changes also occurred in the prerigor and development of rigor mortis. Afterward, during chilled storage, other factors such as lipid oxidation and enzyme activity can cause discoloring, mainly in fatty fish.

Another important problem in salmon is gaping. This occurs when myocommata fibers break under tension. Gaps appear in salmon both along the length of the fillet and between the myotomes; they are mainly a consequence of rigor mortis and are linked to season and growth cycle. A similar effect can occur after smoking, especially if the temperature is too high.

Similarly, another problem in salmon is blood spotting. This occurs when blood vessels are ruptured during evisceration and filleting and inadequate drainage results in the appearance of small petechia in the midline of the fillets. All these spots are more visible after smoking. This phenomenon typically occurs on individual slaughter days when market demand is higher. Michie (2001) suggested that either stress or season (higher water temperature) could be influencing factors.

8.2.1 Slaughtering Stress Influence on Quality

Depending on the method used, for example, net and hook, the capture of wild fish involves various degrees of desperate struggle followed by a period of asphyxiation once the fish is on board. To control stress produced by these conditions, it is necessary to control mainly the fishing method and time; however, the method is often dictated by commercial considerations, and it is difficult to interfere. It is easier to try and reduce stress in farmed fish.

Stress in farm fish that are very active before slaughter can affect the quality of the fish in a physical and a biochemical way (Robb 2001). From the physical point of view, when fish are swimming very actively, during crowding, they can collide with other fish or solid objects, causing loss of scales or bruising. Also, killing by electrical stunning can produce enough active movements to break vertebrae and rupture blood vessels (Kestin et al. 1995). Further rough handling also causes physical damage.

From a biochemical point of view, if the fish is killed after muscle activity, its cells will contain more lactic acid from anaerobic respiration, so that adenosine triphosphate (ATP) synthesis is stopped and rigor mortis is established because of the low pH (Korhonen et al. 1990; Lowe et al. 1993). Spiking to instantaneously destroy the brain and so prevent muscle activity delays the onset of rigor mortis as compared to a slower death such as immersion in chilled water (Boyd et al. 1984). However, there is no difference in the final postslaughter pH of stressed and unstressed fish of the same species, despite differences immediately postmortem (Robb 1998).

Robb et al. (2002) do not recommend stunning salmon with carbon dioxide as these cause an earlier onset and resolution of rigor mortis; they recommend electricity as a good method for stunning prior to slaughtering. Sigholt et al. (1997) reported that a panel test differentiated between stressed and unstressed salmon killed by carbon dioxide stunning. They found that the texture of the stressed fish was softer during storage, which is detrimental, especially when slicing smoked salmon. Roth et al. (2007) report that percussive stunning is the optimal choice for meat quality of turbot (*Scophthalamus maximus*), but electric stunning by prolonged electric exposure is also good; no negative effects on injuries and texture were observed after 7 days of ice storage.

On the other hand, Poli et al. (2005) report that asphyxiated and electrically stunned fish were more stressed than spiked, knocked, or live-chilled fish.

The degree of muscle activity prior to slaughter also affects how firm the flesh becomes during rigor mortis. Mishima et al. (2005) report that lactic acid concentration is lowest when horse mackerel (*Trachurus japonicus*) are killed by spinal cord destruction as opposed to other killing methods such as struggling and temperature shock; this slow rigor mortis onset results in slower muscle degradation during the course of iced storage, as measured by the K-value. Stien et al. (2005) report that a high storage temperature masks most of the effects produced by preslaughter stress; however, it is important to follow the stress management protocols when fillets are kept at the common storage temperatures under 4°C.

The temperature of the fish just after death will affect the course of various biochemical reactions during storage. This is caused by the reduction in ATPase activity as the temperature decreases and by a reduction in the uptake of Ca^{++} (Robb 2001). Also, there are species-related differences; for instance, Watabe et al. (1989) found that the rate of onset of rigor mortis was temperature dependent in plaice (*Paralichthys oliaceus*), whereas in horse mackerel, Mochizuki and Sato (1994) found almost no differences between 0°C and 8°C.

In highly stressed fish, all muscles enter rigor very quickly and at the same time. As a result, the whole fish is very stiff and difficult to process. In fish with a low level of activity, only some muscles have been used, and these are the ones that first enter rigor mortis, while the others do so later. Because of this difference in timing, not all the muscles are in rigor at the same time, so that the fish as a whole is less stiff (Robb

2001). Although it is difficult to process fish in rigor, it is sometimes necessary to do so, especially when the onset is rapid; however, if preslaughter stress is reduced, it will take longer to go into rigor.

8.2.2 GAPING

Rigor mortis has other consequences for muscle quality. For instance, fish that have been stressed before death present a considerable amount of gaping, that is, when myotomes separate from one another, because the intervening threads of connective tissue break, causing slits or holes to appear in the fillet; in severe cases, the fillet may even fall apart when skinned. This makes it more difficult to process the flesh, especially in the case of smoked salmon, where thin slices are required.

Rough handling of the fish, for example, by throwing them against the washer or standing on them as they lie in the pound, can cause damage that may result in gaping. The processing temperature is also important with regard to gaping. The connective tissue of newly caught fish is very sensitive to small rises in temperature, so when fish are warm, any handling such as gutting, washing, or moving can result in gaping. However, when warm fish are cooled again in ice, the connective tissue recovers most of its strength, unless the temperature has risen to about 30°C, in which case the connective damage is irreversible.

Size also influences gaping; smaller fish seem to gape more because the connective tissue is thicker in larger fish.

The season of capture is also important with regard to gaping; for instance, when fish begin to feed heavily again after spawning, there is a general alteration of their biochemistry so that the myocommata are weakened, and the fish are very liable to gape.

Some types of species are more susceptible to gaping than others; for example, round fish generally gape more than flat fish, and some species, for example, catfish, never gape.

8.2.3 GUTTING AND WASHING

Where possible, fish should be bled, gutted, and gilled on board and the more valuable parts such as roes and livers collected, depending on the species. The main reason for gutting is to prevent autolytic spoilage rather than bacterial spoilage (Shewan 1961). Cakli et al. (2006) found psychrophilic counts in sea bream (*S. aurata*) and sea bass (*Dicentrarchus labrax*) a little lower in ungutted than in gutted fish; nevertheless, the quality as gauged by taking chemical, sensory, and microbial tests was similar throughout iced storage. And again, Erkan (2007) reported that shelf life was similar for gutted and whole sea bream (*S. aurata*) on the basis of overall acceptability scores in sensory evaluation. Tejada and Huidobro (2002) reached a similar conclusion in the same species, reporting that gutting reduced the intensity of rigor mortis and the microbial load, although none of the other quality parameters were affected. On the other hand, the results reported by Papadopoulos et al. (2003), based on sensory and microbiological analyses, suggest that gutted sea bass have a shorter shelf life than ungutted specimens.

Another reason for gutting is that gutted albacore (*Thunnus alalunga*; Price et al. 1991) and most other species chill more rapidly. On the other hand, during gutting, the belly area is exposed to air, which makes it more susceptible to oxidation (Røra et al. 2001).

Fed fish spoil more quickly than starved ones when ungutted (Meyer et al. 1986) because in the former, proteolytic activity in the viscera will cause autolysis after death, possibly producing off-flavors or causing rib separation or belly burst. Fatty fish like herring, sardines, sprat, and mackerel are not gutted at sea because their small size and the large numbers in which they are caught make this impracticable in the time available on typically short voyages to grounds not far from the port of landing. The spoilage of cephalopods is influenced more by autolysis (Venugopal 2006). The presence of cathepsin D proteinase of lysosomal origin and other enzymes from digested glands can play a vital role in the degradation of squid mantle muscle, resulting in an increase in the level of muscle-derived nitrogen; this favors proliferation of degenerative microflora and hence a shorter shelf life. Scott et al. (1986) compared microbiological and sensory assessment of whole and headed and gutted orange roughy (*Hoplostethus atlanticus*) during iced storage after washing with seawater. They found no significant differences in bacterial count between the two lots, but they did note a slight increase in shelf life in the gutted fish due to reduced autolysis. Working with Atlantic croaker (*Micropogon undulatus*) and grey trout (*Cynoscion regalis*), Townley and Lanier (1981) reported microbiological advantages in evisceration just after landing.

Industrial gutting and beheading are mechanized in developed countries today, but on board, this operation is traditionally done by hand with a knife, and only in large ships are machines used. Gutting is usually done by cutting, but there are machines that perform gutting by sucking the viscera out and cleaning the belly part through the mouth. This method obviates the need to open the belly, but it makes it difficult to be sure how well the fish is cleaned.

When gutting is performed, fish should be thoroughly washed to remove traces of blood and debris and to wash bacteria and gut juices out of the gut cavity, skin, and gills of the fish. Erkan (2007) reports advantages when washing is included in the processing of sea bream. Other authors, for example, Samuels et al. (1984), have reported that the practice of washing after gutting is more effective in removing remnants than in eliminating bacterial contamination.

The washing equipment on small boats may consist simply of a hose and an open mesh basket, but large trawlers usually have a more sophisticated washing tank with circulating water. In these washers, the fish are discharged over a weir and down a chute to the fish room below deck. It is not clear that using large tanks is an advantage, for although they are effective in removing blood and debris, they can introduce a source of bacterial contamination. A study by Meyer et al. (1986) on Atlantic mackerel (*Scomber scombrus*) reported that pressure washing was much better; similar results were achieved with dressed sea bass (*Dicentrarchus labrax*) and dressed sea bream (*S. aurata*). The efficiency of pressure washing with surfactants in reducing surface flora was also confirmed by Meyer et al. (1986). Kosak and Toledo (1981) reported that microbial decontamination with chlorine was highly effective in terms of the storage stability of iced finfish. On the other hand, Samuels et al. (1984) found no advantage in dipping fish in a hypochlorite solution.

Another possible reason for gutting is to eliminate the possibility of parasite migration from the intestines into the muscle. In some countries, intensive eviscera-tion of different species has been established to minimize invasion of muscle (Chory 1988). On the other hand, there are studies (Roepstorf et al. 1993) that show that Anisakis larvae are already present in the flesh of herring at the time of capture; however, no significant postmortem migration into the flesh has been demonstrated during iced storage. Immediate gutting on board cannot then eliminate or even reduce muscle infestation. Again, Karl et al. (2002) studied the possibility of migra-tion of nematodes from the intestines into the muscle of ungutted pollack (*Pollachius pollachius*), haddock (*Melanogrammus aeglefinus*), and redfish (*Sebastes marinus*); their results show that Anisakis larvae were already present in the flesh of all three species at capture, but no postmortem migration into the flesh was observed during 6 days of ice storage. Herreras et al. (2000) also found Anisakis in the muscle of hake (*Merluccius hubbsi*) that had been gutted just after the capture, indicating that worms had migrated into the muscle before capture. They also found that the density of Anisakis was significantly higher in the hypaxial muscles than in the epiaxial muscles, which means that the removal of hypaxial muscle can reduce the risk of Anisakis intake.

8.2.4 BLEEDING

Bleeding is recommended by several sources (Strom and Lien 1984; Valdimarsson et al. 1984). Robb (2001) reported that large farm fish needed to have the blood removed from the muscle and recommended cutting the gills with a sharp knife; this allows the fish to swim and so die from anoxia caused by blood loss. Robb et al. (2003) concluded that exsanguinations generally reduce blood spots, but they could not say which methodology was better; they also concluded that although bleeding affects the number of spots in smoked salmon, other factors can play an important role. On the other hand, other authors (Meyer et al. 1986; Moser 1986) have concluded that bleeding before gutting and gilling have no effect on such parameters as sensory, color, trimethylamine and hypoxanthine concentration, and surface bacterial load test. Some authors (Botta et al. 1986) have reported that bleeding is effective only if conducted within 1–2 hours of capture. Sohn et al. (2007) report that simply remov-ing a portion of the blood from live yellowtail by bleeding is not sufficient to prevent lipid oxidation in the early stages of ice storage. In farmed halibut (*Hippoglossus hipoglossus*), Aske and Midling (2001) reported only small differences in heme iron muscle residues between bled and unbled lots of halibut, and they added that halibut killed by a blow to the head bled better than specimens anesthetized with CO_2. All these discrepancies suggest that bleeding is not universally applicable and that there are many factors to be considered, such as the type of species, the size, the season, and in farmed fish, also the farming area and harvesting method.

Olsen et al. (2006) reported that the amount of residual blood was influenced by the anesthetizing and killing procedures. Fish that were chilled alive and anes-thetized then directly gutted had less residual blood in the fillet as compared to the standard industrial procedure of gill cutting and bleeding before gutting. Use of anesthesia on live-chilled fish killed by gill cutting did not result in a reduction of residual blood as compared to live-chilled unanesthetized fish killed by gill cutting. This was because blood coagulation time was prolonged at low temperature, possibly improving bleeding. Sakai et al. (2006) reported that skipjack (*Katsuwonu pelamos*) killed while struggling in iced sea water bleeds better than when killed instantly by mechanical bleeding, and the lot that bled better presented less oxidation during ice storage. Roth et al. (2005) reported that the bleeding method was of less importance in trout and salmon, in which it is the timing that is important. They observed no significant difference in blood spotting between fish that were bled alive by a gill cut or percussively killed and bled by gutting, so the industry would do well to gut directly. They also reported that drainage of blood in the fish muscle seemed to occur within the first hour postmortem, so that rigor mortis was of little importance.

8.2.5 FILLETING

Many species are filleted to satisfy consumer demand. In general, filleting adds value to the product, although this depends very much on the type of market. Most companies that commercialize fillets use machines. Basically these machines cut along the upper and lower appendices on the spine, cutting the ribs and vertebrae with a pair of symmetrical knives. Different standards of trimming are used, from removing only the backbone to removing visible fat, pin bones, and skin, and these differences produce different yields. Filleting should in principle be performed after the onset of rigor mortis, but this should be weighed against loss of freshness and the cost of storage. If fish are processed in rigor, the yield will be poor, and it may cause gaping (Laverty 1984; Huss 1995). Large farmed species like salmon are usually filleted once rigor has been resolved, normally 3–4 days after death. On the other hand, Shaw et al. (1984) reported that filleting after 7 days rendered the longest shelf life. In any case, it is difficult to industrially control the onset of rigor in large catches of wild species; fish farming makes it easier to control all the parameters that culminate in rigor.

Prerigor filleting has several advantages, one being that it ensures very fresh processed fish with little or no fillet gaping (Andersen et al. 1994), although it changes the shape (Sørensen et al. 1997; Skjervold et al. 2001; Kristoffersen et al. 2006), and prerigor fillets are significantly thicker (Skjervold et al. 2001). Also, the texture is different, and certain operations such as the removal of pin bones are more difficult.

The texture of prerigor fillets of farmed Atlantic cod (*G. morhua*) depends in part on dietary content; for instance, dietary inclusion of soybean oil has been found to result in faster reduction of breaking strength and lower L^* values (Morkore 2006). However, there may be extensive loss of weight and proteins during subsequent storage if fish are filleted prerigor (Kristoffersen et al. 2007). Prerigor processing then presents problems, but there are obvious advantages to processing immediately after slaughter, in that the products can be shipped to markets 3–5 days earlier, and a prolonged shelf life is a major economic benefit (Rosnes et al. 2003; Tobiassen et al. 2006).

Fillet yield depends on the species and its structural anatomy (Røra et al. 2001). Fish with large heads and frames relative to their musculature give a lower yield than those with smaller heads and frames. Filleting yields for farmed fish depend on species, sex, size, and shape, so these parameters can be selected to improve fillet yields. Yields can also be affected by farming conditions. Of commercially farmed fish species, tilapia (*Oreochromis* sp.) has the lowest fillet yield (33%) as compared to salmon (*S. salar*; >50%), channel catfish (*Ictalurus punctatis*; >38%), and striped bass (*Morone saxatilis*; >40%). Sea bream and sea bass also give higher fillet yields than tilapia. Freshwater eel gives the highest fillet yield (60%).

8.3 HANDLING OF CRUSTACEA

Care of shrimps, prawns, and Norway lobsters should begin before the cod end comes on board; appropriate trawling gear should be used as unnecessarily long hauls can damage the shrimp in the net. Once the crustaceans are on board, they must be handled quickly and carefully, without exposure to sun and wind. The catch is first sorted by hand or by sieve to separate it from fish or damaged crustaceans and to grade it into different sizes. After sorting and washing, shrimps, prawns, and Norway lobsters are drained, treated with antimelanotics, and packed in ice in shallow boxes. Depending on the product requirements, crustaceans are cooked or peeled, or both before further refrigeration. In some cases, depending on the trade practice, crustaceans are dipped in salt, citric acid, polyphosphates, or the like after peeling in order to improve flavor, color, or yield.

Crabs should be handled as little as possible after capture (Edwards and Early 1978). They should not be removed from traps by the claws as this can cause mutilation. Crabs must not be exposed to the wind and sun on deck as this weakens them, and they may die. Afterward, they should be packed upright in ventilated boxes in such a way as to prevent fighting. Wet materials in the bottom of the box can help in prolonging life in live storage or until killing.

In crustaceans, melanosis—or *black spot*, as it is commonly called—is a progressive blackening of the skin. It is largely species dependent and is caused by the action of prophenol oxidase in the hemolymph and crustacean cuticula. Melanosis commences when enzymes are released after death, contributing to tissue degradation and the release of hemolymph and, consequently, prophenol oxidase in the cephalothorax. Enzymatic hydrolysis also released amino acids and peptides, which form a polyphenol oxidase substrate. The photolytic enzymes in contact with the hemolymph then activate the prophenyl oxidase to produce active polyphenoloxidase, which oxidizes phenolic compounds; these in turn produce melanins, which give the shells a dark color (Diaz-Lopez et al. 2003). Various compounds such as antioxidants, acidulants, chelants, and enzymatic inhibitors can be used more or less successfully to inhibit the enzymatic process. In practice, the most commonly used are sulfite-based substances (sodium bisulfite) accompanied by acidulants or chelants. The safety of sulfites has recently been questioned in the literature as possibly causing respiratory and allergic disorders in consumers (Ruiz-Capillas and Jiménez-Colmenero 2009). Laboratory experiments have shown that 4-hexyl-resorcinol can be an efficient inhibitor of melanosis (McEvily et al. 1991; Montero et al. 2004), but this compound has not yet been approved everywhere (it has been approved in the United States, Australia, Canada, and some Latin American countries). Modified atmosphere can also be applied during further chilled storage to minimize blackening (Gonçalves et al. 2003).

8.4   HANDLING OF MOLLUSKS

Mollusks, with the exception of cephalopods, have external shells that afford them some protection against rough handling, but they spoil very rapidly after death. These two characteristics and their ability to live out of water account for their being kept alive after harvest, in most cases allowing them to survive for several days. They can therefore be delivered live to the processing plant and to consumers. As a result, the requirements for handling these organisms onboard a ship are significantly different from those of dead fish. Most mollusks are harvested by dredges, hand rakes, tongs, or similar equipment. Once on board, usable animals are separated from shells and other debris. The animals are thrown into a pile or container and remain there until they reach the dock or processing plant. Although it is possible to harvest many mollusks in temperate climates without the need to control the animal's temperature on the harvesting boat, product quality will be better if more care is exercised. In the case of live animals, gentle cooling lowers the metabolic rate and reduces their oxygen needs. Depending on the market, the flesh may be removed and sold either raw or cooked. In the case of live animals, purification is desirable and is in fact compulsory in most countries. This is a simple process that takes advantage of the fact that bivalves cleanse themselves of polluting bacteria if they are kept in clean (or ozonized) sea water for a number of hours. In bivalves, attention should be paid to the National Sanitary Authority's red tide alert.

REFERENCES

Andersen, U. B., Strømsnes, A., Steinsholt, K. and Thomassen, M. S. 1994. Fillet gaping in farmed Atlantic salmon (*Salmo salar*). *Norwegian Journal of Aquaculture Sciences*, 8: 165–179.

Aske, L. and Midling, K. 2001. Slaughtering of Atlantic halibut (*Hippoglossus hippoglossus*): effect on quality and storing capacity. In: *Farmed Fish Quality*, S. C. Kestin and P. D. Warriss (eds.), p. 381. Fishing News Book. Oxford, U.K.: Blackwell Science Ltd.

Botta, J. R., Squires, B. E. and Johnson, J. 1986. Effect of bleeding/gutting procedures on the sensory quality of fresh raw Atlantic cod (*Gadus morhua*). *Canadian Institute of Food Technology Journal*, 19(4):186–190.

Botta, J. R., Bonell, G. and Squires, B. E. 1987. Effect of method of catching and time of season on sensory quality of fresh raw Atlantic cod (*Gadus morhua* L.). *Journal of Food Science*, 52(4):928–931.

Boyd, N., Wilson, N. D., Jerrett, A. R. and Hall, B. Y. 1984. Effects of brain destruction on post harvest muscle metabolism in the fish Kahawai (*Arripis-Trutta*). *Journal of Food Science*, 49(1):177–179.

Cakli, S., Kilinc, B., Cadun, A., Dincer, T. and Tolasa, S. 2006. Effects of gutting and ungutting on microbiological, chemical, and sensory properties of aqua cultured sea bream (*Sparus aurata*) and sea bass (*Dicentrarchus labrax*) stored in ice. *Critical Reviews in Food Science and Nutrition*, 46(7):519–527.

Chory, H. 1988. Regulations regarding the wholesomeness of fish and shellfish. *Bundesgesetzlatt*, Part I:1570–1571.

Christiansen, R., Struksnaes, G., Estermann, R. and Torrisen, O. J. 1995. Assessment of colour in Atlantic salmon (*Salmo salar* L.). *Aquaculture Research*, 25:311–321.

Costakes, J., Connors, E. and Paquette, G. 1982. *Quality at Sea—Recommendations for On Board Quality Improvement Procedures*. Gloucester, Massachusetts: New Bedford Seafood Producers Association and New England Fisheries Development Foundation.

Dassow, J. A. 1976. Handling fresh fish. In: *Industrial Fishery Technology,* ed. M. E. Stansby, 45–64. Robert New Cork: E. Krieger.

Diaz-Lopez, M., Martinez-Diaz, I. and Martinez-Moya, T. 2003. Evaluación de agentes conservantes en pesquerías de crustáceos pesqueros y acuículas. In: *Estudio de los agentes conservantes e inhibidores de la melanosis en crustáceos*. Sevilla, Spain: Junta de Andalucía Press.

Edwards, E. and Early, J. C. 1978. *Catching, Handling and Processing Crabs*. Torry Advisory Note No 26. Edinburgh: HMSO Press.

Erikson, U. and Misimi, E. 2008. Atlantic salmon skin and fillet color changes effected by perimortem handling stress, rigor mortis, and ice storage. *Journal of Food Science*, 73(2):C50–C59.

Erkan, N. 2007. Sensory, chemical and microbiological attributes of sea bream (*Sparus aurata*): effect of washing and ice storage. *International Journal of Food Properties*, 10(3):421–434.

FAO. 1973. *Code of Practice for Fresh Fish*. FAO Fisheries Circular C318. Rome, Italy: Food and Agricultural Organization.

Gelman, A., Sach, O., Khanin, Y., Drabkin, V. and Glatman, L. 2005. Effect of ozone pretreatment on fish storage life at low temperatures. *Journal of Food Protection*, 68(4):778–784.

Gines, R., Palicio, M., Zamorano, M. J., Arguello, A., Lopez, J. L. and Alfonso, J. M. 2002. Starvation before slaughtering as a tool to keep freshness attributes on gilthead sea bream (*Sparus aurata*). *Aquaculture Internacional*, 10(5):379–389.

Glatman, L., Sach, O., Khanin, Y., Drabkin, V. and Gelman, A. 2006. Ozone action on survival and storage life on live and chilled tilapia. *Israeli Journal of Aquaculture-Bamidgeh*, 58(3):147–156.

Gonçalves, A., López-Caballero, M. E., Nunes, M. L. 2003. Quality changes of deepwater pink shrimp (*Parapenaeus longirostris*) packed in modified atmospheres. *Journal of Food Science*, 68(8):2586–2590.

Goodrick, B. 1987. Postharvest quality of tuna meat, a question of technique. *Food Technology of Australia*, 39:343–345.

Herreras, M. V., Aznar, F. J., Balbuena, J. A. and Raga, J. A. 2000. Anisakid larvae in the musculature of the Argentinean hake (*Merluccius hubbsi*). *Journal of Food Protection*, 63(8):1141–1143.

Hurtado, J. L., Borderías, A. J., Montero, P., An, H. 1999. Characterization of proteolytic activity in octopus (*Octopus vulgaris*) arm muscle. *Journal of Food Biochemistry*, 23:469–483.

Huss, H. H. 1995. *Quality and Quality Changes in Fresh Fish*. FAO Fisheries Technical Paper 348. Rome: FAO Press.

Karl, H., Meyer, C., Banneke, S., Sipos, G., Bartelt, E., Lagrange, F., Jark, U. and Feldhusen, F. 2002. The abundance of nematode larvae *Anisakis* sp. in the flesh of fishes and possible post-mortem migration. *Archiv fur Lebensmittelhygiene* 53(5):118–120.

Kestin, S., Wotton, S. and Adams, S. 1995. The effect of CO_2, concussion or electrical stunning of rainbow trout (*Onchoryncus mykiss*) on fish welfare. In: *Quality in Aquaculture,* Special Publication 23, pp. 380–381. Ghent, Begium: European Aquaculture Society.

Korhonen, R. W., Lanier, T. C. and Giesbrecht, F. 1990. An evaluation of simple methods for following rigor development in fish. *Journal of Food Science*, 55(2):346–349.

Kosak, P. H. and Toledo, R. T. 1981. Effects of microbial decontamination on the storage stability of fresh fish. *Journal of Food Science*, 46(4):1012–1014.

Kristoffersen, S., Tobiassen, T., Esaiassen, M., Olsson, G. B., Godvik, L. A., Seppola, M. A. and Olsen, T. L. 2006. Effect of pre-rigor filleting on quality aspects of Atlantic cod (*Gadus morhua* L.). *Aquaculture Research*, 37(15):1556–1564.

Kristoffersen, S., Vang, B., Larsen, R. and Olsen, R. 2007. Pre-rigor filleting and drip loss from fillets of farmed Atlantic cod (*Gadus morhua* L.). *Aquaculture Research*, 38(16):1721–1731.

Laverty, J. 1984. *Gaping in Farmed Salmon and Trout*. Torry Advisory Note No 90. Edimburgh: HMSO Press.

Lowe, T. E., Ryder, J. M., Carragher, J. F. and Wells, R. M. G. 1993. Flesh quality of snapper, *Pagrus auratus,* affected by capture stress. *Journal of Food Science*, 58(4):771–773.

McEvily, A. J., Iyengar, R. and Otwell, W. S. 1991. Sulfite alternative prevents shrimp melanosis. *Food Technology*, 45(9):80–86.

Meyer, B., Samuels, R. and Flick, G. 1986. A seafood quality program for the mid-Atlantic region, Part II. A report submitted to the Mid-Atlantic Fisheries Development Foundation. Virginia Polytechnic Institute and State University, Sea Grant, Blackburg.

Michie, I. 2001. Causes of downgrading in the salmon farming industry. In: *Farmed Fish Quality*, S. C. Kestin and P. D. Warriss (eds.), pp. 129–136. Oxford, U.K.: Fishing News Book. Blackwell Science Ltd.

Mishima, T., Nonaka, T., Okamoto, A., Tsuchimoto, M., Ishiya, T., Tachibana, K. and Tsuchimoto, M. 2005. Influence of storage temperatures and killing procedures on post-mortem changes in the muscle of horse mackerel caught near Nagasaki Prefecture, Japan. *Fisheries Science*, 71(1):187–194.

Mochizuki, S. and Sato, A. 1994. Effects of various killing procedures and storage temperatures on postmortem changes in the muscle of horse mackerel. *Nippon Suisan Gakkaishi*, 60(1):125–130.

Montero, P., Martinez-Álvarez, O. and Gómez-Guillén, M. C. 2004. Effectiveness of onboard application of 4-hexylresorcinol in inhibiting melanosis in shrimp (*Parapenaus longirostris*). *Journal of Food Science*, 69(8):643–647.

Morkore, T. 2006. Relevance of dietary oil source for concentration and quality of *pre-rigor* filleted Atlantic cod, *Gadus morhua*. *Aquaculture*, 251(1):56–65.

Moser, M. D. 1986. Maine Groundfish Association vessel quality handling project. A report submitted to the New England Fisheries Development Foundation.

Nickell, D. C. and Bromage, N. R. 1998. The effect of dietary lipid level on variation of flesh pigmentation in rainbow trout (*Oncorhynchus mykiss*). *Aquaculture*, 161:237–251.

Olsen, S. H., Sorensen, N. K., Stonno, S. K. and Elvevoll, E. O. 2006. Effect of slaughter methods on blood spotting and residual blood in fillets of Atlantic salmon (*Salmo salar*). *Aquaculture*, 258(1–4):462–469.

Papadopoulos, V., Chouliara, I., Badeka, A., Savvaidis, I. N. and Kontominas, M. G. 2003. Effect of gutting on microbiological, chemical and sensory properties of aquacultured sea bass (*Dicentrarchus labrax*) stored in ice. *Food Microbiology*, 20(4):411–420.

Poli, B. M., Parisi, G., Scappini, F. and Zampacavallo, G. 2005. Fish welfare quality as affected by pre-slaughter and slaughter management. *Aquaculture International*, 13(1–2):29–49.

Price, R. J., Melvin, E. F. and Bell, J. W. 1991. Post-mortem changes in chilled round, bled and dressed albacore. *Journal of Food Science*, 56(2):318–321.

Robb, D. H. F. 1998. Some factors affecting the fish quality of salmonids: pigmentation, composition and eating quality. Ph.D. Thesis, University of Bristol, U.K.

Robb, D. H. F. 2001. The relationship between killing methods and quality. In: *Farmed Fish Quality*, S. C. Kestin and P. D. Warriss (eds.), pp. 220–233. Oxford, U.K.: Fishing News Book. Blackwell Science Ltd.

Robb, D. H. F. and Frost, S. 1999. Welfare and quality. What is the relationship? Presentation at Innovation for Seafood'99, Surfer's Paradise, Queensland, Australia, 21–23 April.

Robb, D. H. F., Philips, A. J. and Kestin, S. C. 2003. Evaluation of methods for determining the prevalence of blood spots in smoked Atlantic salmon and the effect of exsanguinations method on prevalence of blood spots. *Aquaculture*, 217(1–4):125–138.

Roepstorf, A., Karl, H., Bloesma, B. and Huss, H. H. 1993. Catch handling and the possible migration of Anisakis larvae in herring (*Clupea harengus*). *Journal of Food Protection*, 56(3):783–787.

Røra, A. M. B., Morkore, T. and Einen, R. 2001. Primary processing (evisceration and filleting). In: *Farmed Fish Quality*, S. C. Kestin and P. D. Warriss (eds.), pp. 249–260. Oxford, U.K.: Fishing News Book. Blackwell Science Ltd.

Rosnes, J. R., Vorre, A., Folkvord, L., Hovda, M., Fjaera, S. O. and Skjervold, P. O. 2003. Effect of pre-, in-, and post-rigor filleted Atlantic salmon (*Salmo salar*) on microbial spoilage and quality characteristics. *Journal of Aquatic Food Products Technology*, 12:17–31.

Roth, B., Torrissen, O. J. and Slinde, E. 2005. The effect of slaughtering procedures on blood spotting in rainbow trout (*Oncorthynchus mykiss*) and Atlantic salmon (*Salmo salar*). *Aquaculture*, 250(3–4):796–803.

Roth, B., Imsland, A., Gunnarsson S., Foss, A. and Schelvis-Smith, R. 2007. Slaughter quality and rigor contraction in fanned turbot (*Scophthalmus maximus*); a comparison between different stunning methods. *Aquaculture*, 272(1–4):754–761.

Ruiz-Capillas, C. and Jiménez-Colmenero, F. 2009. Application of flow injection analysis for determining sulfites in food and beverages: A review. *Food Chemistry*, 112:487–493.

Sakai, T., Ohtsubo, S., Minami, T. and Terayama, M. 2006. Effect of bleeding on hemoglobin contents and lipid oxidation in the skipjack muscle. *Bioscience Biotechnology and Biochemistry*, 70(4):1006–1008.

Samuels, R. D., DeFeo, A. and Flick, G. J. 1984. Demonstrations of quality maintenance program for fresh fish products. A report submitted to Mid-Atlantic Fisheries Development Foundation. Virginia Polytechnic Institute and State University, Sea Grant, Blacksburg.

Scott, D. N., Fletcher, G. C., Hogg, M. G. and Ryder, J. M. 1986. Comparison of whole with headed and gutted Orange Roughy stored in ice—sensory, microbiology and chemical assessment. *Journal of Food Science*, 51(1):79–83.

Shaw, S. J., Bligh, E. G. and Woyewoda, A. D. 1984. Effect of delay in filleting on quality of cod fish. *Journal of Food Science*, 48(3):979–980.

Shewan, J. M. 1949. Some bacteriological aspects of handling, processing and distribution of fish. *Journal of Royal Sanitary Institute*, 59:394–402.

Shewan, J. M. 1961. The microbiology of seawater fish. In: *Fish as Food*, G. Borgstrom (ed.), Vol. 1, pp. 487–560. New York: Academic Press.

Sigholt, T., Erikson, U., Rustad, T., Johansen, S., Nordvedt, T. and Seland, A. 1997. Handling stress and storage temperature affect meat quality of farm-raised Atlantic salmon (*Salmo salar*). *Journal of Food Science*, 62(4):898–905.

Skjervold, P. O., Røra, A. M. B., Fjaera, S. O., Vegusdal, A., Vorre, A. and Einen, O. 2001. Effect of pre-, in- or post-rigor filleting of live chilled Atlantic Salmon. *Aquaculture*, 194(3–4):315–326.

Smart, G. 2001. Problems of sea bass and seabream quality in the Mediterranean. In: *Farmed Fish Quality*, S. C. Kestin and P. D. Warriss (eds.), pp. 121–128. Oxford, U.K.: Fishing News Book. Blackwell Science Ltd.

Sohn, J. H., Ushio, H., Ishida, N., Yamashita, M., Terayama, M. and Ohshima, T. 2007. Effect of bleeding treatment and perfusion of yellowtail on lipid oxidation in post-mortem muscle. *Food Chemistry*, 104(3):962–970.

Sørensen, N. K., Brataas, R., Nyvold, T. E. and Lauritsen, K. 1997. Influence of early processing (pre-rigor) on fish quality. In: *Seafood from Producer to Consume, Integrated Approach to Quality*, J. B. Luten, T. Børressen and J. Oehlenschläger (eds.), pp. 253–263. Amsterdam: Elsevier.

Stien, L. H., Hirmas, E., Bjornevik, M., Rarlsten, O., Nortvedt, R., Rora, A. M. B., Sunde, J. and Kiessling, A. 2005. The effect of stress and storage temperature on the colour and texture of pre-rigor filleted farmed cod (*Gadus morhua* L.). *Aquaculture Research*, 36(12):1197–1206.

Strom, T. and Lien, K. 1984. Fish handling on board Norwegian fishing vessels. In: *Fifty Years of Fisheries Research in Iceland*. Jubilee Seminar of the Icelandic Fisheries Lab., A. Moller (ed.), p. 15. Reykjavic, Iceland, Sep. 23–24.

Tejada, M. and Huidobro, A. 2002. Quality of farmed gilthead seabream (*Sparus aurata*) during iced storage related to the slaughter methods and gutting. *European Food Research and Technology*, 215(1):1–7.

Tobiassen, T., Akse, L., Midling, K., As, K., Dahl, R. and Eilertsen, G. 2006. The effect or pre-rigor processing of cod (*Gadus morhua* L.) on quality and shelf life. In: *Seafood Research from Sea to Dish*, J. B. Luten, C. Jacobsen, K, Bekaert, A. Saebo and J. Oehlenschläger (eds.), pp. 149–159. Wageningen: Wageningen Academic Publishers.

Townley, R. R. and Lanier, T. C. 1981. Effect or early evisceration on the keeping quality of Atlantic croaker (*Micropogon undulatus*) and grey trout (*Cynoscion regalis*) as determined by subjective and objective methodology. *Journal of Food Science*, 46(3):863–867.

Valdimarsson, G., Matthiansson, A. and Stefansson, G. 1984. The effect of bleeding and gutting on the quality of fresh, quick frozen, and salted products. In: *Fifty Years of Fisheries Research in Iceland*. Jubilee Seminar of the Icelandic Fisheries Lab., A. Moller (ed.), p. 61. Reykjavic, Iceland, Sep. 23–24.

Venugopal, V. 2006. Postharvest quality and safety hazards. In: *Seafood Processing: Adding Value through Quick Freezing, Cook-chilling and Other Methods,* V. Venugopal (ed.), USA: CRC Press, Taylor & Francis Group.

Watabe, S., Ushio, H., Iwamoto, M., Yamanaka, H. and Hashimoto, H. 1989. Temperature-dependency of rigor-mortis of fish muscle—myofibrillar mg-2+-atpase activity and ca-2+ uptake by sarcoplasmic-reticulum. *Journal of Food Science*, 54(5):1107–1110.

9 Processing of Poultry

Casey M. Owens

CONTENTS

9.1 INTRODUCTION

Poultry meat is a popular choice of animal protein for consumers. The success of poultry meat over the recent years can be mainly ascribed to the innovative preparation and presentation of different cuts from broilers and turkeys. Consumers have led the way in demanding products that require minimal preparation and are nutritionally adequate. Over the years, the whole-bird market has steadily decreased while the cut market and further-processed market have increased. Now, cut-up and further-processed items represent over 90% of marketed poultry in the United States, which reflects consumer interests. In other countries, the same trend of increased parts and further-processed products exists, though the market percentages may differ. Furthermore, in recent years, the further-processed market has become the predominant market segment in the United States, surpassing not only the whole-bird market but the cut-up market as well. This is a trend that will likely be observed in other countries as well. This chapter will focus on the production of various poultry products. Primary, secondary, and further processing will be discussed.

9.2   PRIMARY POULTRY PROCESSING

Primary poultry processing can be divided into several unit operations including preslaughter handling, immobilization, defeathering, evisceration, and chilling of poultry carcasses (Figure 9.1). The discussion of these steps will be brief.

9.2.1   PRESLAUGHTER HANDLING AND IMMOBILIZATION

During preslaughter handling, feed and water are withdrawn from birds, which are transported to the processing plant and then unloaded at the hanging area (initial stage of processing). The next operation is immobilization, where the birds are hung onto shackles, where they then proceed to the stunner, which renders the birds unconscious. Typically, birds are stunned with electrical currents or gas. The stunning procedure is beneficial in that it is humane to the birds, it provides a uniform heartbeat for better bleedout, and it assists with positioning the bird correctly as it enters the cutting machine. The cutting machine cuts the carotid artery and jugular vein, which allows the birds to bleed out for up to 3 minutes; this is also known as exsanguination.

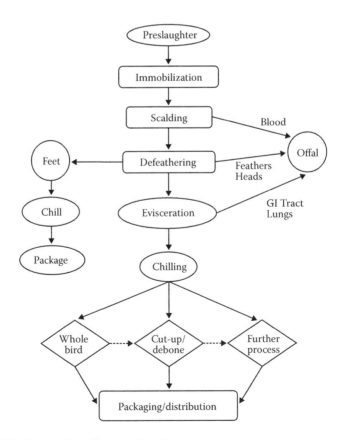

FIGURE 9.1   Process flow diagram of poultry processing.

9.2.2 ELECTRICAL STIMULATION

Postmortem electrical stimulation has been recently commercially implemented after approximately 10 to 12 years of development in broilers. Electrical stimulation uses a saline bath to deliver high-voltage and high-amperage electricity to the bird immediately after bleedout. Electrical stimulation speeds up rigor development [depletes adenosine triphosphate (ATP)] in broilers so that carcasses can be deboned in the early postmortem, between 1 and 2 hours postmortem, without negatively affecting tenderness, rather than the carcasses being aged 4 to 6 hours postmortem, which would allow rigor completion in nonstimulated birds. The advantages of using electrical stimulation are an increase in yield, the reduction of toughness due to pre-rigor deboning, a decrease in labor costs associated with aging, and an improvement in processing efficiency.

9.2.3 DEFEATHERING—SCALDING AND PICKING

The next unit operation in primary processing is defeathering, which consists of scalding and picking. After exsanguination, birds enter the scalder and then the picking machine. The purpose of scalding is to denature the proteins in the feather follicle so that feather release is much easier when the birds are picked in the picking machine. Birds are scalded at temperatures ranging from approximately 50°C to 60°C from 2 to 4 minutes. The scalding process is a time- and temperature-dependent process, so generally, the longer the birds are in the scalder, the lower the temperatures are. Lower temperatures and longer periods are used to obtain a soft scald, which leaves the cuticle (waxy layer that contains pigments) on the bird. Higher temperatures for shorter periods result in a hard scald, which removes the cuticle. The industry uses countercurrent and multistage scalding so that birds are always moving toward cleaner water and so that temperatures can be adjusted to obtain the most optimum scald. Feather removal primarily involves in-line picking machines. There are generally multistage pickers, where each picker concentrates on a certain area of the carcass. Feathers are removed by rotating rubber fingers.

9.2.4 EVISCERATION

After feather removal, birds are eviscerated where their viscera, both edible and inedible, is removed. Inedible viscera (e.g., gastrointestinal tract, lungs) is generally rendered, whereas the edible viscera (e.g., hearts, livers, gizzards) is chilled and packaged for consumption. After evisceration, birds are rinsed using an inside–outside bird washer. This helps in reducing fecal contamination prior to the chiller. The process from hanging birds onto shackles through evisceration and rinsing generally takes 10 to 15 minutes.

9.2.5 CHILLING

The primary goal of chilling carcasses is to reduce carcass temperature. Carcass temperatures should reach <4°C within 4 hours for broilers and 8 hours for turkeys according to United States Department of Agriculture regulations. There are a few methods that are commonly used by the poultry industry worldwide. In the United States, water immersion chilling is the primary chilling regimen used. In addition to temperature reduction, an increase in carcass yield can be achieved due to water uptake in the chiller. However, most of the water is held in and under the skin and is easily lost as drip in the first few hours after chilling, especially if carcasses are cut up. Immersion chilling also helps in reducing the bacterial loads on carcasses but can increase the frequency of bacterial contamination at the same time, the cross-contamination effect. Changes in the industry have resulted in changes in antimicrobials used. In immersion chilling, a countercurrent flow system is used where the coldest water is in the final stages of the chilling process. This countercurrent flow system allows for a gradual decrease in body temperature to prevent cold shortening, a toughening effect, and for maximum water uptake, or yield increase. The process time typically lasts approximately 1.5 hours depending on carcass size.

Air chilling is another method of chilling that is commonly used in Europe and also in Canada. Traditional air chilling uses air temperature to decrease carcass temperature. With an air chilling system, a soft scald should be used so that the cuticle remains intact. The cuticle provides protection to the epidermis of the carcass so that it does not dry out and become discolored. There are advantages and disadvantages to this type of system. One advantage is that cross-contamination (between carcasses) is reduced because the carcasses are not in a common bath; however, if a single carcass is contaminated, the bacterial load will likely remain the same, with little or no reduction in contrast to immersion chilling, where bacterial loads can be reduced on carcasses due to a dilution effect. There are also advantages in the areas of consumer preference. Processors state that air chilling leaves the carcass with a crispier outer skin and locks in the moisture and natural juices of the carcass when it cooks. Some disadvantages are that air chilling systems require a great deal of space and the chilling process is relatively slow, taking approximately 2 to 3 hours. Yield generally decreases as well. In efforts to improve air chilling and maintain the same chilling method, improved air chills using water sprays have been developed. These systems provide a faster chill compared to normal air chilling. In addition, these spray chill systems can be used with either a hard- or soft-scalded bird.

9.3   AGING: RIGOR MORTIS DEVELOPMENT

The conversion of muscle to meat results in dramatic postmortem changes in the physical and biochemical aspects of muscle. *Rigor mortis* is a Latin term meaning "stiffening of death." It is characterized by stiffening and loss of extensibility of the muscle due to the formation of permanent actomyosin bonds.

ATP is required for muscle contraction and relaxation; its stores deplete quickly, and it must be synthesized by the muscle. Aerobic metabolism is capable of producing high amounts of ATP in order to keep the muscle functioning. However, the loss of the circulatory system upon slaughter of an animal ceases oxygen delivery to the muscle. This results in a shift from aerobic metabolism to anaerobic metabolism, which solely depends on glycolysis; it is relatively inefficient in producing ATP. ATP can also be formed from creatine phosphate and adenosine diphosphate; however, this supply of energy is short term. Continued use of ATP for relaxation coupled with reduced ability to produce ATP results in insufficient ATP to prevent the formation of the actomyosin bonds, and thus, rigor mortis develops.

Not only is there a depletion of ATP but the pH of the muscle also declines as rigor develops. The end product of glycolysis is lactic acid. In the live animal, the circulatory system removes lactic acid from the muscle. With the loss of the circulatory system after death, lactic acid accumulates in the muscle and lowers its pH. The pH decreases from 7.4 in living muscle to 5.5–5.7 after rigor development. The rate of decline in pH is important because it can affect many meat quality attributes including color, water-holding capacity, and texture.

If muscle is stimulated through cutting and/or freezing (extreme reduction in temperature) prior to rigor mortis completion, the muscle can contract because of the sufficient energy supplies that remain in the muscle at that time. Deboning in this state results in a tougher product because the muscle is denser owing to the contracted state. Also, limiting the resolution phase of rigor mortis, through freezing, for example, can interfere with proteolysis, which can also result in tougher meat.

9.4   SECONDARY POULTRY PROCESSING

Secondary processing generally refers to the steps that are used to cut up the carcass into parts. It has been a basic way of adding value to poultry for many years. After the carcasses have been chilled and aged, if aging is used, carcasses can be cut up into a variety of parts. Basic cut-up can be achieved by using a series of different blades making specific cuts. Many of the equipment manufacturers have lines of equipment to make an 8- or 9-piece cut. However, the equipment can be adjusted so that only specific parts (e.g., wing) can be cut.

Deboning breast meat and thigh meat can also be placed into the secondary processing category. Because of the increased demand for parts and boneless meat, deboning meat is a standard process. Breast or leg meat can be deboned by hand or by automated equipment. The time of deboning, especially the breast meat, can significantly impact quality. It is recommended that breast meat be deboned at least 4 h to 6 h postmortem to allow for rigor completion. Deboning prior to rigor completion results in tougher meat. Deboned meat, breast or leg, can then be prepared for distribution (e.g., freeze then package, or vice versa), portioned, or further processed.

Portioning generally involves cutting fillets horizontally (slitter) or vertically or a combination of both. Portioning meat can maximize the use of the product in markets. For example, a 4-oz portion can be cut from the breast fillet and be used for a product such as a grilled chicken sandwich, and the remaining portion of the fillet can be used in other products (i.e., trim). The breast piece with the intact membrane, or epimysium, is considered the premium piece because it most resembles the native fillet and can be used for a large variety of products (nonbreaded or breaded), while the remaining piece has less resemblance to the native fillet and is therefore commonly used for batter/breading operations or in other further-processing operations. The portioning process can be either an automated process (more common) or manual and can use a variety of techniques for cutting, including band-saw blades and water jet technology. In automated systems, a mold or template, using vacuum or pressure, can be utilized for horizontal/vertical cut combinations. Flattening-type pressure can also be used for horizontal cuts, while water jet technology is often used for vertical cuts. Often, breast fillets are subjected to horizontal cutting followed by vertical cutting, unless these steps are combined using specialized equipment. Portioning can immediately follow the deboning process, or the meat can be further aged and then portioned.

9.5   FURTHER PROCESSING

Further-processed products can be classified into three basic categories: whole muscle, comminuted, and emulsions. Some examples of these types of products are a whole-muscle deli loaf, a chicken nugget or patty, and a frankfurter, respectively. All of these products are generally processed using the same basic steps; however, there are numerous variations that can be used (Figure 9.2). The goal of further processing is to add value to the commodity products and provide high-quality products in various forms to fit the needs of consumers. The poultry industry has not only been very responsive to consumer preferences, but they have also been very proactive in providing a variety of new poultry products in efforts to increase per-capita consumption of poultry meats.

9.5.1   Particle Size Reduction

The first step in further processing is to harvest the muscle. This may occur in the secondary processing stage. After harvesting the muscle, whole muscle or trim, the next step would be to reduce the particle size. The main purpose of reducing particle size is to increase the area for protein extraction. Protein extraction is critical

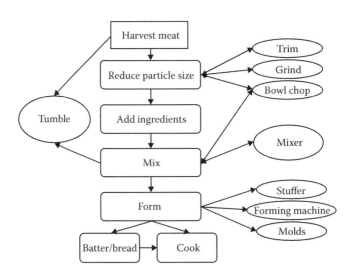

FIGURE 9.2   Process flow diagram of further processing of poultry.

for products that would require particle binding, such as a comminuted or emulsion products. Some whole-muscle products also go through some sort of particle size reduction in order to increase surface area. Whole muscles can be cut up into smaller pieces (e.g., chunks) or macerated, thereby puncturing the outer layer of connective tissue, the epimysium. For comminuted products, meat can be coarsely ground, flaked, or chopped. Comminuted products generally result in a texture with particle sizes greater than 0.25 inch. Emulsion products are finely chopped typically using a bowl chopper, or silent bowl cutter, to achieve a homogeneous product size less than 0.25 inch.

9.5.2 INGREDIENT ADDITION

The addition of ingredients generally occurs after or during particle size reduction. The main ingredients that are added in further-processed products are sodium chloride (salt) and sodium tripolyphosphate in the presence of water since these both aid in protein extraction and water retention. The muscle proteins, myosin and actin, are salt soluble and, therefore, require salt to solubilize them. If proteins are not adequately extracted, then the meat particles will not bind in the cooking process. Salt will also add to the flavor of the product.

9.5.3 PROCESSES FOR INGREDIENT ADDITION

Ingredients can be added in processes such as injection and/or tumbling (whole-muscle products) and mixing (comminuted) or during particle size reduction (emulsions, using bowl chopper). Injection of ingredients can be used in bone-in or boneless whole-muscle products. This process is sometimes followed by the process of tumbling depending on the importance of ingredient distribution and protein extraction for a particular product. The process of tumbling is used for whole-muscle products and can be used for those products that will be formed into a deli loaf (multiple pieces) or for products that will remain a single unit (e.g., breaded breast tender). For whole-muscle products, the addition of ingredients and tumbling generally occur simultaneously. Tumbling under vacuum is a process where the meat is tumbled in a barrel, much like a clothes dryer but on a larger scale. The marinade (containing ingredients) penetrates the fillet, and the fillet is agitated by the impact of the meat pieces against each other. Not only will this process distribute the marinade throughout the piece of meat, but the process will also aid in protein extraction, where proteins are brought to the surface of the meat. A variety of whole-muscle products can be tumbled; the length of the process depends on the amount of protein extraction desired, among other factors (e.g., thickness of meat portion, amount of product tumbled, etc.). For the production of deli loaves or rolls, where it is essential that whole muscles bind together upon cooking, it is critical that protein extraction occurs during the tumbling process. A lesser amount of protein extraction would be necessary for products remaining a single unit (e.g., marinated breast fillet, breaded fillet or tender, etc.). For comminuted products, mixers are used to incorporate the ingredients and extract proteins. Bowl choppers can also be used to accomplish these steps, especially for nugget and patty operations.

9.5.4 Forming

After tumbling or mixing the ingredients, the product is then formed. Whole muscles that will be used to make deli loaves or rolls are formed in the cook-in packages (e.g., premolded, casing, etc.). After forming, the packages are sealed and cooked, at which time the meat pieces are bound together resulting in a deli-style meat. Comminuted meat can be molded into the preformed packages or casing, much like the whole-muscle products, and then cooked. However, these products are generally lower-quality products compared to the whole-muscle products. Comminuted products are often used for nugget or patty formation using forming equipment. The forms have a variety of shapes: traditional nugget, cartoon shape, patty, etc. The possibilities are numerous and can potentially change, often depending on marketing. The meat is deep chilled (e.g., –3°C to –5°C), molded in the forming machine, and then generally conveyed directly to the batter and breading operation, followed by cooking (fry, oven, or combination). Emulsified products are stuffed into casings and cooked; these products can range from frankfurters to deli rolls. These products are cooked in ovens using steam and/or smoke applications. Products must then be cooled.

9.5.5 Batter and Breading

The comminuted products and some whole-muscle products (bone-in or boneless) are often battered and breaded as a final step prior to cooking. Nuggets, tenders, and wings are good examples of the various types of breaded products available in the market. The coating process generally includes a predust, batter, and breading layer on the product; however, the passes in these various coatings can vary depending on the targeted amount of coating pickup. There are several types of breadings available, which offer different sensory characteristics such as appearance, crispness, and color.

Although there are many further-processed products in the marketplace, many of these products are processed using the same basic steps. The goal of further processing is to add value to the commodity products and provide high-quality products in various forms to fit the needs of consumers. The poultry industry has not only been very responsive to consumer preferences, but they have also been very proactive in providing a variety of new poultry products in efforts to increase per-capita consumption of poultry meats.

9.6 PACKAGING AND DISTRIBUTION

After processing, products are ready for packaging and distribution. There are a variety of steps that poultry products can go through in preparation for distribution.

Packaging can be in a variety of forms, from bulk to retail packages. Many further-processed items are frozen prior to packaging. For example, after nuggets have been fully cooked, they are generally frozen in a spiral freezer and then packaged into appropriate packages. Another example would be marinated breast fillets that may be frozen using a spiral freezer and then packaged into bags. Some products are packaged first and then quick-chilled or frozen (e.g., retail breast fillets, parts, etc.). After packaging, products are ready for distribution. Refrigerated or frozen storage will be required prior to and during distribution.

ADDITIONAL SOURCES OF INFORMATION

Barbut, S. (ed.). 2002. *Poultry Products Processing, An Industry Guide*. CRC Press, Boca Raton, FL.
Owens, C.M., C.Z. Alvarado, and A.R. Sams. 2010. *Poultry Meat Processing*, 2nd ed. CRC Press, Boca Raton, FL.

10 Air and Osmotic Partial Dehydration, Infusion of Special Nutrients, and Concentration of Juices

Miriam E. Agnelli

CONTENTS

10.1 INTRODUCTION

Partial dehydration in high-water-content foodstuff such as fruits, vegetables, and juices has become of great interest in order to diminish water activity and thus prolong shelf life during refrigeration. In this chapter, we are going to describe some of the pretreatments that can be used before refrigeration to accomplish the preceding goals. In order to get a better understanding of the subject, we are going to approach it considering first, the possible pretreatments for fruits and vegetables, and then the available methods for juice concentration.

10.2   PRETREATMENTS FOR FRUITS AND VEGETABLES

Due to the rapid increase in consumption of minimally processed food (per capita) throughout the world, the industry is adapting itself to changing market forces, such as an increased demand for prepared food, a need of diversification, and enhanced control of quality and safety (Mallet 1994). This applies to fruits and vegetables in particular, whose market can be divided into two sections: whole refrigerated fruits and vegetables and fruit and vegetable pieces to be incorporated as ingredients into prepared foods.

Over the past years, many new fruit- and vegetable-based foods appeared on the market, for example, refrigerated pieces of vegetables ready to consume for salads, fruit pieces predipped in sugar, and predried refrigerated fruit pieces. They are widely used as basic materials or as additional components in many food formulations, for example, cooked dishes, pastry and confectionery products, ice cream, desserts and sweets, fruit salads, cheese, and yogurt. In most cases, fruits and vegetable pieces impart a fresh and healthy image to the food, giving it an additional value appreciated by the consumer and providing higher margins for the producer.

The fruit and vegetable pieces must preserve their natural flavor and color, retain a pleasant texture, and preferably be free of antioxidants or other additives. As a general rule, fresh like values are required for high-moisture foods, such as ingredients to be incorporated into fruit salads or ice cream, while some color and flavor changes, and relatively small increases in consistencies, may be still acceptable or wanted for lower-moisture foods such as cakes or various baked products. Besides sensory properties relating to acceptance and quality, well-defined functional properties are also required, with reference to the physicochemical environment and the shelf life of the food. The "compatibility" of ingredients with food, as a physicochemical system, is mainly dependent upon the equilibrium vapor pressures or water activities of the components. Water activity (aw) should be matched to avoid, or to allow, a controlled rate of diffusion of moisture between the fruit and the food and to properly adjust the shelf life and storage behavior according to the specific food formulation.

Chemical and physical actions of the operation units to obtain ready-to-use fresh-cut fruits and vegetable, such as peeling, cutting, size reduction, and washing, are highly detrimental to them since their texture is mainly ensured by turgor. Turgor is the ability to retain water inside the cells. Rupture of cell walls due to the operation units mentioned above will cause water to migrate out of the cell, and thus, texture losses occur. Also, interactions between previously separated substances (enzymes and subtracts) become possible and provide a medium for diffusion of all other molecules involved in deterioration reactions. Water can also participate directly in deterioration reactions, which include the following:

- Production of off-flavors
- Changes in color due to enzymatic or nonenzymatic reactions on pigments, especially browning

The use of pretreatments can help in avoiding or reducing these detrimental reactions by inactivating the deterioration reactions or by reducing the water content in the material. Conventional pretreatments include washing, blanching, soaking, and in some cases, presentation treatments such as comminuting, grinding, coating, and finally, packaging. These pretreatments present no major difficulties. Only blanching presents certain drawbacks. Enzymes responsible for food deterioration can be inactivated chemically or by using thermal treatments. Both of them affect cellular tissues similarly.

Currently, there is a renewed interest in implementing partial dehydration and formulation steps prior to refrigeration. The advantages over conventional refrigeration include (1) energy savings, since water load is reduced, and in transport and wraping costs as well, (2) a better quality and stability (color and flavor and reduced rate of microbial growth).

Depending on the food and on the final use of the product, many techniques can be used to partially dehydrate it. The most common are:

1. Partial air-drying.
2. Osmotic dehydration (OD). This technique can combine dewatering with impregnation soaking and leads to formulation.

Water elimination mechanisms are different in both methods, and the obtained products present completely dissimilar characteristics. These distinctive features of the intermediate product and its final use, will determine the best method to use.

10.2.1 PARTIAL AIR-DRYING

Partial dehydration is generally achieved by air-drying. When using partial air-drying, food ingredients with high water activity (aw > 0.96) are generally obtained, since water removal is limited to 50%–60% of the original content. To avoid browning during air-drying, blanching or other treatment such as dipping in antioxidant solutions (ascorbic or citric acid, sulfur dioxide) can be used.

The air dehydration step usually produces a weight loss of up to 50%, corresponding to 60% water reduction and a 10% relative increase in total solids in the fresh weight. These partially dehydrated fruits can be used as fresh substitutes in fruits salads, surface garnitures, or tart and pie fillings. The wetting effect of the fruits is reduced, and the shelf life is longer owing to lower water content. Weight losses of 75%–80% with solid gains of 20%–50% can also be obtained. Such products are suitable for pastry foods, where wetting has to be avoided, and for yogurt preparations, where they can absorb moisture, preventing the separation of whey (Giangiacomo et al. 1994).

During air-drying treatment, elimination of water involves phase changes; thus, although low temperatures are used during the process, a loss of cell functionality may occur and, consequently, considerable changes in sensory and nutritional quality (Spiess and Beshnilian 1998). Air dehydration treatments on strawberries promote changes in the composition of sugars and citric acid other than those expected from the simple concentration in the tissue. These changes reflect the cellular activity in the volume of viable cells in the tissue that is affected by process conditions. Drying treatments seem to accelerate the metabolic activity, which depends on the decrease in acidity and increase in the native sugar content (Moraga et al. 2006a,b). This effect depends on the ratio of viable cells present in the treated fruit. Slight cellular stress in the live cells caused by mild treatments might provoke the maturation processes. So, time processing for the air-drying dehydration step must be reduced as much as possible, as for certain fruits, such as kiwi or strawberry, important color damages appear due to heat treatment. Kiwi fruit shows a definite yellowing of the typical green color when air-dehydrated to 50% weight reduction even at 45°C. For these fruits, air-drying must be replaced by osmotic dehydration, which is effective even at room temperature and operates away from oxygen.

10.2.2 Heat and Mass Transfer Characteristics in Air Partial Dehydration

Dehydration involves the application of heat to evaporate water and some means of removing water vapor after its separation from the fruit/vegetable tissues. Hence, it is a combined/simultaneous (heat and mass) transfer operation for which energy must be supplied.

A current of air is the most common medium for transferring heat to a drying tissue, and convection is mainly involved.

The two important aspects of mass transfer are the transfer of water to the surface of material being dried and the removal of water vapor from the surface.

In order to assure products of high quality at a reasonable cost, dehydration must occur fairly rapidly. Four main factors affect the rate and total drying time: the properties of the products, especially particle size and geometry; the geometrical arrangement of the products in relation to heat transfer medium (drying air); the physical properties of drying medium/environment; and the characteristics of the drying equipment.

It is generally observed with many products that the initial rate of drying is constant and then decreases, sometimes with two stages of different rates. The drying curve is divided into the constant rate period and the falling rate period.

10.2.2.1 Surface Area

Generally, the fruit and vegetables to be dehydrated are cut into small pieces or thin layers to speed up heat and mass transfer. Subdivision speeds up drying for two reasons: large surface areas provide more surface in contact with the heating medium (air) and more surface from which moisture can escape, and smaller particles or thinner layers reduce the distance that the heat must travel to the center of the food and reduce the distance through which moisture in the center of the food must travel to reach the surface and escape.

10.2.2.2 Temperature

The greater the temperature difference between the heating medium and the food, the higher the rate of heat transfer into the food, which provides the driving force for moisture removal. When the heating medium is air, temperature plays a second important role. As water is driven from the food in the form of water vapor, it must be carried away, or else the moisture will create a saturated atmosphere at the food surface, which will slow down the rate of subsequent water removal. The hotter the air, the more moisture it will hold before becoming saturated.

Thus, high-temperature air in the vicinity of the dehydrating food will take up the moisture being driven from the food to a greater extent than will cooler air. Obviously, a greater volume of air also can take up more moisture than a lesser volume of air.

10.2.2.3 Air Velocity

Not only will heated air take up more moisture than cool air, but also air in motion will still be more effective. Air in motion, that is, high-velocity air, in addition to taking up moisture, will sweep it away from the drying food surface, preventing the moisture from creating a saturated atmosphere, which would slow down subsequent moisture removal. This is why clothes dry more rapidly on a windy day.

Some other phenomena influence the drying process, and a few elements are summarized below.

10.2.2.4 Dryness of Air

When air is the drying medium of food, the drier the air, the more rapid the rate of drying. Dry air is capable of absorbing and holding moisture. Moist air is closer to saturation and so can absorb and hold less additional moisture than if it were dry. But the dryness of the air also determines how low of a moisture content the food product can be dried to.

10.2.2.5 Atmospheric Pressure and Vacuum

If food is placed in a heated vacuum chamber, the moisture can be removed from the food at a lower temperature than without a vacuum. Alternatively, for a given temperature, with or without vacuum, the rate of water removal from the food will be greater in the vacuum. Lower drying temperatures and shorter drying times are especially important in the case of heat-sensitive foods.

10.2.2.6 Evaporation and Temperature

As water evaporates from a surface, it cools the surface. The cooling is largely the result of absorption by the water of the latent heat of phase change from liquid to gas.

In doing this, the heat is taken from the drying air or the heating surface and from the hot food, and so the food piece or droplet is cooled.

10.2.2.7 Time and Temperature

Since all important methods of food dehydration employ heat, and food constituents are sensitive to heat, compromises must be made between maximum possible drying rate and maintenance of food quality.

As is the case in the use of heat for pasteurization and sterilization, with few exceptions, drying processes that employ high temperatures for short times do less damage to food than drying processes employing lower temperatures for longer times.

Thus, vegetable pieces dried in a properly designed oven for four hours would retain greater quality than the same products sun-dried over two days.

Several drying processes will achieve dehydration in a matter of minutes, or even less, if the food is sufficiently subdivided.

10.3   OSMOTIC DEHYDRATION

Some of the quality defects of the final product encountered during partial dehydration, by means of air convection, can be sometimes avoided using osmotic dehydration instead. Osmotic treatments are based on placing foods in a hypertonic solution in order to obtain some effects on concentration, dehydration, and impregnation formulation of food itself by exchanging solutes and water with the solution. The process is osmotic because it relies on the ability of the cell membranes to selectively allow the passage of water while hindering the mobility of larger molecules. Because of their larger molecular size, most of the soluble solids inside the cells cannot permeate the membrane, but water is readily exchanged between the cells and the surroundings. Maintaining the integrity of the cell membrane is important for producing an osmotic flow because osmotic dehydration aims at preserving the tissue for further use. Osmotic dehydration is defined as a part of the natural water removal process that is based on immersion of foodstuff, such as fruits and vegetables, in a hypertonic solution of sucrose, glucose, fructose, sodium chloride, or other osmotic agents. Because the hypertonic solution has a higher osmotic pressure and lower water activity, a driving force for water removal arises between solution and foodstuff. In osmosis, the cellular surface structure acts as an effective semipermeable membrane. Research is carried out on the different aspects of osmotic dehydration, such as the choice of the solutes and the concentration in the osmotic solution, the operating temperature, the treatment time, the solution agitation, the solution/food ratio, and the combination of methods of osmosis with other stabilization technology.

The main advantages of osmotic dehydration of food in comparison with other dehydration processes include the following:

1. Minimizing the heat damage to food tissue, color, and flavor, and less discoloration of fruits by enzymatic oxidative browning (Contreras and Smyrl 1981; Ponting et al. 1966). Karel (1975) also reported that the immersion of the plant material in an osmotic solution acted as a barrier to the entrance of oxygen.
2. Increasing retention of natural volatile components during further processing. Chirife et al. (1973), Flink and Karel (1970), and Flink (1975) pointed out the importance of the solid concentration on the retention of volatiles when osmotic dehydration was used as a pretreatment for a traditional preserving process.
3. Improving textural quality. Shipman et al. (1972) noted that the immersion of celery in a glycerol solution prior to air-drying improved the textural quality of the rehydrated products compared to the freeze-dried or air-dried products.
4. Reducing energy consumption. Because water was removed without thorough phase change, osmosis dehydration resulted in low operating costs (Bolin et al. 1983).

Osmotic dehydration technology for food preservation is mostly used as a pretreatment step for other stabilizing steps, because osmotic dehydration generally will not reduce the moisture content of a food product to a low-enough level for the product to be shelf stable. Therefore, osmotic dehydration is an intermediate step that is used to improve product quality during further processing. Most previous studies focused attention on rapid and effective removal of desired amounts of water from foods. A high osmotic rate would make the process more efficient and practical (Ponting et al. 1966). Some methods have been employed to speed up water transfer, such as using high concentration of the osmotic agent, high solution temperature, or low pressures.

Solutions of osmotic agents present high osmotic pressure and low water activity. The most commonly used osmotic agents are sucrose and sodium chloride. Other osmotic agents such as lactose, maltodextrin, ethanol, glucose, glycerine, and corn syrups have been used.

Karel (1975) pointed out that intermediate-moisture foods (IMF) have received attention since the development of new products based on new technological principles such as lowering water activity and retarding microbial growth by adding antimicrobial agents. These IMF can be produced safely with osmotic dehydration treatment.

Maltini and Torregiani (1981) examined the possibility of obtaining a shelf-stable product with no need for further treatments and lower preservative use by means of an osmotic treatment in a sucrose solution. These fruit products tasted fresh and were shelf stable with a water activity between 0.94 and 0.97 and water content in the range of 65% to 75%. Combined with vacuum packaging and pasteurization, the fruits remain stable for months at room temperature. This process causes minimal changes in the sensory and physicochemical characteristics of fruits.

In the case of refrigeration, using OD as pretreatment, shelf life can be extended to some days, with the extra time depending on the product and other factors such as pH, packaging, or the addition of antimicrobials.

The simplicity of the osmotic dehydration process, without the use of expensive equipment, and fewer or no energy requirements, makes it suitable for large-scale fresh food preservation. The osmotic treatment often provides better-quality intermediate-moisture fruits than air-drying, especially in fruits and vegetables sensitive to heat.

Osmotic dehydration reduces the water activity of solid foods by soaking the fruits or vegetables in a concentrated solution of sugars or salt. This treatment produces a flow of water out of the tissue. The dehydration levels reached depend upon specific needs, but products between 80% and 50% moisture content can be obtained through this treatment. Simultaneously, the solutes in the solution diffuse into the material contributing to the water activity depression. Because no phase change is usually carried out at temperatures below 50°C, the product suffers less damage than when treated with conventional direct dehydration methods. Osmotically dehydrated products are of better quality after further drying or freezing.

The final quality of osmotically dehydrated food products is mainly dependent on the nature of the food material, the composition and concentration of the osmotic solution, the treatment temperature, and the immersion time of the food.

Pretreatments with chemicals or blanching can modify the fruit tissue permeability and result in higher gain of solids and water loss because permeability increases and selectivity decreases after the impregnation phenomena (Karel 1975; Ponting 1973).

As mentioned before, the main aspects to be taken into account to achieve IMF with a certain quality are discussed in the following.

Composition of osmotic solution. The kind of sugar used in osmotic solutes affects the kinetics of water removal, sugar gain, and the final equilibrium water content. Obviously, the selection of osmotic agents cannot be made only on the basis of its osmocity. The choice of solutes and the concentration used depends on several factors. The most important factors are the organoleptic evaluation on the final fruit and the cost of solutes. Osmotic solution must have a low water activity. Moreover, the solution must be nontoxic and have a high solubility in water. The capacity of the compound to lower the water activity will affect the driving force responsible for the mass transfer. It is possible to select the most suitable solute or combine some solutes in order to obtain the best levels of water removal and sugar gain as well as desired changes in sensory characteristics. The common solutes used in osmotic dehydration are sucrose, high-fructose corn syrup, glucose, and salt.

Most studies on osmotic dehydration have used sucrose as the osmotic agent because of its effectiveness, low cost, availability, and desirable flavor. Because dry sucrose is troublesome to mix with fruits, it is more practical to use syrup during the process. Concentrated sucrose solution was used successfully to remove up to 50% of water from apples (Ponting et al. 1966; Flink 1975; Farkas and Lazar 1969; Ponting 1973).

Bolin et al. (1983) pointed out that the rate of penetration into the fruit pieces was faster with corn syrup than with sucrose. Fruits concentrated with corn syrup had lower water activity and moisture content than those concentrated with sucrose. However, taste panel evaluation showed a preference toward fruits osmotically dehydrated with sucrose. Corn syrup contains a large proportion of high-molecular-weight carbohydrates; it is suspected that larger molecules are able to penetrate the fruit to a great extent during OD. Besides producing an acceptable product of low sweetness, the use of corn syrup may be advantageous with respect to the secondary drying step. Water diffusion coefficients are extremely sensitive to dissolved solid content. The diffusion coefficient decreases as the solid content decreases.

Heng et al. (1990) showed that the use of DE 20 glucose syrup (more sweet) induced a slightly higher water loss (increasing about 3% to 5%) than in sucrose solution but lower solid gain. Salt solution is a good osmotic solution and usually used for dehydration of vegetable and meat products.

The choice of the osmotic syrup plays a very important role, and the specific effect of the solution has to be taken in account. The choice depends mainly on taste, cost, and aw-lowering capacity. Fruit juice concentrates have similar osmotic properties to high-fructose syrups (Maltini et al. 1990), and resulting products are of total fruit origin. If a concentrated fruit juice is used as osmotic solution, an even softer product can be obtained because of the higher content of monosaccharide in the fruit juice compared to the amount contained in the syrup from starch hydrolysis and because of the higher relative water content at a determined water activity (Torregiani et al. 1988). If a fructose syrup contains sorbitol, softer osmodehydrated apricot, clingstone peach cubes, and sweet cherry halves can be obtained when compared with the same fruit osmodehydrated in fructose alone (Erba et al. 1994; Torregiani et al. 1997). The presence of sorbitol in hydrolyzed lactose syrup also leads to a lower texture in osmo-dehydrated red pepper cubes (Torregiani et al. 1995). Moreover, as reported previously, sorbitol has a specific protective effect on color in the following step.

Concentration of sugar solution. In general, it has been shown that water loss in osmotically dehydrated fruits is improved by increasing solute concentration in the osmotic solution. Ponting et al. (1966) observed that the rate of dehydration is not increased greatly with sucrose solution at high sugar concentration, for example, above 65°Brix. The higher the starting concentration of the solution, the higher the percentage of water loss from the fruits in a given period of time. Nevertheless, the preparation of highly concentrated solutions involves heating and consequent coloration, which sometimes is bothering. Results also have shown that sugar solution with a concentration below 40°Brix was not effective for osmotic dehydration. Concentrated sucrose solution with 50°Brix to 70°Brix has been the most common osmotic solution. Increasing sugar concentration of the solution enhances water loss rather than solid gain (Ponting et al. 1966; Lenart and Flink 1984a; Conway et al. 1983; Raoult-Wack et al. 1991).

pH of sugar solution. Moy et al. (1978) have investigated the relationships of acid-ity and sucrose concentrations on the rate of osmotic dehydration of some tropical fruits such as papayas and mangos. The results showed that combining organic acid and sucrose increases water removal from fruits by inhibiting gelation. Aurand et al. (1987) also pointed out that the apple tissue softens under conditions of low pH. In all cases, maximum weight reduction is observed at pH 3. Increasing or decreasing the pH of a sugar solution by the addition of hydrochloric acid or sodium hydroxide, respectively, results in lower weight reduction. In more acidic solutions (pH 2), the apple rings become very soft, and firmness was maintained at pH values of 3 to 6 (Contreras and Smyrl 1981).

Additives in osmotic solutions. Some additives mixed with the sugar solution con-tribute to improve the quality of the final product. Addition of calcium in sugar solu-tion resulted in decreasing sugar gain and strengthening fruit texture while slowly increasing water loss. A 0.001-M concentration of calcium seems to be enough to produce this effect without a bitter apparition aftertaste (Heng et al. 1990).

Temperature of osmotic dehydration treatments. Ponting et al. (1966) and Lenart and Flink (1984a) found that the rate of osmosis and diffusion is markedly affected by temperature. A temperature increase in the osmotic solution results in an increase in the rate of water removal and sugar uptake. Diffusion of flavors from the fruits to the solution is also enhanced at high temperature. Above 45°C, enzymatic browning and flavor deterioration of fruit tissue begin to take place. At temperatures of about 60°C, fruit tissue characteristics such as cell membranes and many nutrient compo-nents such as ascorbic acid are destroyed.

Many experiments have shown that an increase in solution temperature has a favorable effect on water loss without modification of sugar gain because of the dif-fusion differences between masses of water and sugar (Raoult-Wack 1994; Fito et al. 1995). The optimum temperature depends on the property of the fruit material.

Immersion time. When the concentration of sugar solution is kept constant, an increase in contact time results in an increase in water loss (Ponting et al. 1966).

Although water loss increases as a function of time, its rate decreases. On the other hand, sugar gain in fruit pieces increases with contact time. Many experiments illustrated that maximum mass transfer occurred mainly in the first one or two hours of treatment.

Pressure condition in osmotic treatments. Experiments showed that dehydration was enhanced under vacuum. Hawkes and Flink (1978) pointed out that apple slices immersed in osmotic solution under vacuum gave a higher increase in solid content due to the primary uptake of osmotic solutes. Some fruits such as pineapples, bananas, strawberries, and pears were dehydrated by direct osmosis carried out at low pressure (70 mm Hg), and results showed that higher drying rates were obtained by the low-pressure method, whereas lower water removal and sugar gain occurred in the fruit material with higher solid content (Dalla Rosa et al. 1982). Adambounou and Castaigne (1983) studied partial dehydration by osmosis of banana slices dried at 65°C under vacuum and pointed out that sugar gain and water loss were faster, especially during the first hours.

Geometry properties of food material. Under the same osmotic condition, different sizes and geometry of fresh food samples can give final products with very different characteristics. The specific surface area (A/V) of the fruit samples is closely related to mass diffusion properties: high water loss and solid gain correspond to a higher A/V value.

10.3.1 MASS TRANSFER CHARACTERISTICS DURING OSMOTIC DEHYDRATION

After immersing water-rich fresh food material in a hypertonic solution (sugar, salt, etc.), the driving force for water removal is the concentration gradient between the solution and the intracellular fluid. If the membrane is perfectly semipermeable, the solute is unable to transfer into the cells. However, it is difficult to obtain a perfect membrane in food material due to its complex internal structure and possible damage during processing. Osmotic dehydration is therefore a multicomponent transfer of two simultaneous, countercurrent solution flows. There are two major mass transfer phenomena involved in osmotic dehydration: the movement of solute into the material and the flow of water out of the tissue. There have been numerous studies to describe the kinetics of these two countercurrent flows. Water and some solutes such as organic acids, reducing sugars, minerals and some flavor compounds, flow out of food materials, affecting the organoleptic and nutritional characteristics of the final product (Le Maguer 1988). In the opposite sense, soluble solids present in the osmotic solution are taken up by the food material.

Variables such as temperature, immersion time, nature, concentration, and composition of solutes, influence the mass transfer kinetics, thus the variability in reported results in the literature. The kinetics of mass transfer is usually described using terms such as water loss, solid or solute gain, and weight reduction. From an osmotic point of view, the pathways of sugar and hypertonic sugar solution from the external into the intercellular space and cell wall or vice versa are readily accomplished by diffusion during the passage of the substance from the outside of the plant cell to its interior or vice versa. During the OD of fruits, the behavior of mass transfer inside the material tissue depends on both processing variables and fruit biological microstructure properties. There is naturally wide variation in the physical nature of the food material. When a porous fresh food material is immersed in hypertonic sugar solution, the eventual effect of cellular dehydration on water transport properties will strongly depend on the material and initial biological microstructural characteristics, especially the intercellular space present in the tissue (Lenart and Flink 1984b; Islam and Flink 1982). It is generally accepted that the most important organ controlling the osmotic phenomena is the plasmalemma membrane. The selective permeability of cell walls and membranes controls the quantity and the rate of mass transfer. Any destruction or disruption of the cell structure during osmotic dehydration treatment will result in poor dehydration and undesired mass transfer behavior.

10.3.2 FORMULATION OF FOOD

Soluble solid uptake due to osmotic dehydration, in addition to improving color, aroma, and vitamin stability during storage, can also play a very important role in the preparation of new types of ingredients at reduced water activity (Torregiani et al. 1988). Due to the soluble intake, the overall effect of osmotic dehydration is a decrease in water activity, with only a limited increase in consistency. Consistency is actually associated with the plasticizing and swelling effect of water on the pectic and the cellulosic matrix of the fruit tissues. Hence, it depends primarily on the insoluble matter and water content rather than on the soluble solids and water activity; in this way, low water activities may be achieved while maintaining an acceptable consistency.

According to Maltini et al. (1993), a relationship can be found among the phase composition (i.e., the relative amount of insoluble solids, soluble solids, and water), the texture index, and the water activity of the fruits after processing. At equal water activity, there is a difference between phase composition and texture as a result of the solid gain after osmotic treatment. The higher the solid uptake becomes, the higher the difference will be. Compared to simple air dehydration, osmotic dehydration can produce softer product, which is more pleasant to eat by hand or to incorporate into pastry, ice cream, cheese, or yogurt (Giangiacomo et al. 1994).

Also, impregnation of other components important from the physiological point of view for human health can be done if they are properly introduced in the osmotic solution. Thus, the food can be enriched with physiological active compounds using atmospheric or vacuum OD treatments. Mineral supplementation has become a very popular method to complement inadequate diets. Fortifying fruits and vegetables with iron, calcium, or zinc can prevent many diseases, and these minerals can be easily introduced in the vegetable matrix by means of osmotic treatments. For that purpose, the vegetable matrix, the physicochemical characteristic of the solution, and the concentration of the active compound must be adequate to achieve an established percentage of the recommended daily intake in a serving of the final product (Alzamora et al. 2005).

10.3.3 OSMOTIC SOLUTION RECYCLING

As stated before, when food is placed in the solution, a water flux comes from the food to the solution, and a flux of solutes occurs countercurrently. The main phenomenon is then the dilution of the solution with the increase in mass and the lowering of the dewatering potential. From the process, the economic, and the environmental points of view, it is necessary to find an effective method. Thus, the osmotic solution management is the bottleneck of the OD process, and it restrains industrial implementation (Warczok et al. 2007).

The most promising way to reduce environmental impact caused by the spent osmotic solution would be to reuse the concentrated solution for as long as possible. However, loss of solutes and particles from food into solution was reported by many authors, leading to chemical, chemical–physical, and sensory changes of the osmotic solution itself. Specific research on the influence of the repeated use of a sugar solution has been carried out on treatments of papaya, pineapple, and peach (Argaiz et al. 1996); apple (Valdez-Fragoso et al. 1998); sour cherries (Szymczak et al. 1998); and apple and stoned cherries (Giroux et al. 2001). It was shown that it was possible to reuse the solution between 5 and 20 times, depending on the treated fruit, without any impact on the main mass transfers in the products and with good microbial quality of the solution. The use of activated carbon or polyvinylpolypyrrolidone for decoloration of used syrup has been proposed by Szymczak et al. (1998) and has led, especially for the activated carbon, to no significant differences in dehydrated fruit color between fresh and recycled syrups.

10.3.4 SOLUTION CONCENTRATION RESTORING

When reusing the osmotic solution, the first main problem is to restore the solute concentration. The technological answers that can be suggested include both phase-changing and nonphase-changing processes:

- Evaporation (atmospheric at high temperature, under vacuum at moderate temperature)
- Solute addition (no phase change)
- Membrane concentration (no phase change)
- Cryoconcentration

Evaporative restoring is probably the most popular technique to be implemented industrially for medium/large production plants, as the cost of the evaporators is relatively low. However, it is necessary to study the engineering and energetic aspects of the applied method of water removal, mainly the knowledge of the energy consumption in the osmotic process and the comparison of a per-unit energy consumption in this technique with that of other methods of water removal. The only studies on this aspect related to fruit processing are still those of Lenart and Lewicki (1988) and Collignan et al. (1992a). Results indicated that per-unit energy consumption during convection drying of fruits and vegetables was two or three times higher than that of osmotic dehydration and syrup reconcentration through an evaporator.

Restoring the solution concentration by adding dry solute or mixing with concentrated solution can save energy costs, as it avoids heat of evaporation and the need for expensive plants. The method can be suggested successfully for small-scale production at a low-technological-level process, where the initial solution mass is small.

Indeed, the main hurdle of this technique is the increase in the solution mass, even if a constant loss in the volume of syrup (9%–14%) is due to adherence of the food pieces (Bolin et al. 1983).

Salt solution recycling is often a necessity in other food processing industries such as in the cheese and table olive industry, where the reuse of salting brine is very common and the brines are worked for several cycles without any fresh brine addition. For this reason, great attention has focused on the innovative technological solutions applied in these sectors, which could also be used for syrup recycling during osmotic treatments, for example, the use of membrane concentration, which can reach goals other than solution restoring. This technique could be useful because it combines filtration and reconcentration without any energy cost other than the energy required for pumping. Actually, membrane processing has to take into account the fouling phenomena at the membrane interface and the difficulty of working with relatively high-viscosity fluids such as the osmotic solution (40–60 Pa s in the case of 60% of solids). The use of membranes has been applied successfully for remediation and/or recycling of diary effluents (Horton 1997) and for brine recycling in the table olive industry (Garrido-Fernandez et al. 2001).

10.3.5 MICROBIAL CONTAMINATION OF THE SOLUTION

Different sources of contamination can affect the microbial stability of the used solutions, although the water activity values, ranging around aw = 0.9–0.95, should be able to limit the growth of nontolerant bacteria and yeasts (Valdez-Fragoso et al. 1998).

During the processing of fruit and vegetables with a pH < 4.5, yeast, molds, and lactic bacteria are the most frequent microorganisms released from the products to the solution. In this situation, pathogenic bacteria are not able to grow. Depending on the environmental process conditions, the microbial load after several osmotic treatment cycles can range from 2×10^2 CFU ml^{-1} (Valdez-Fragoso et al. 1998) to high levels of yeasts and fungi only after the 15th cycle and 10^5 CFU ml^{-1} after 8 continuous treatments (Dalla Rosa et al. 1995).

10.4 JUICE CONCENTRATION

Industrial production of fruit juices is an important market activity. Nowadays, as health issues acquire more and more importance in the choice of food products, fruit juice consumption is on the rise. In view of the high water content of fruit juices (75%–90%), a concentration step is normally introduced in their industrial processing, with the aim of reducing packaging, transportation, and storage costs. The most common methods for juice concentration are discussed in the following (cryoconcentration is addressed in Chapter 13).

10.4.1 EVAPORATION

Heat treatment of juices is an area where the design of process requires careful consideration in order to avoid any detrimental effects on flavor and appearance of the product. Early evaporators had demonstrated that high-vacuum-temperature processing produced concentrate of good flavor quality, but it was soon discovered that there was a drawback in that the heat treatment was insufficient to deactivate pectin methylesterase, which gave rise to gelation in the final product. The effect was not immediately apparent but was seen to occur after a few weeks of storage, when the contents of a filled drum, of, say orange concentrate, might be found to have gelled. The introduction of short-term high-temperature pasteurization into the concentration process, bringing the juice temperature to around 95°C with sufficient "holding" time to eliminate microorganism and to denature the enzyme, was used to offset the effect. The use of multieffect multistage evaporators combines high efficiency in heat utilization and easy control when °Brix variations are present in the feed. Thermally accelerated short-time evaporators are generally used.

Although a combination of product quality and cost considerations will dictate the methods used for bulk processing of fruit juices, there are instances where the flavor components present in the juice are vulnerable to any form of heating during concentration. Strawberry juice is perhaps the best example of this, being one of the most sensitive of fruits, and it works well with alternative processes for concentration such as freeze concentration and hyperfiltration.

10.4.2 HYPERFILTRATION AND ULTRAFILTRATION

By use of a selective membrane, water can be removed by filtration from the juice in order to effect its concentration. Depending upon the molecular size of the compounds and the cutoff value of the membranes used, there is likely to be some loss of flavor components. These may be recovered from the permeate by distillation and returned to the juice concentrate. Concentration by these methods is less effective in terms of folding than other methods but can provide advantages in specific cases: for example, capital costs associated with hyperfiltration are around 10%–30% less than for evaporative systems with aroma recovery equipment.

Evaporative concentration of fluid foods presents major drawbacks. First is the heat-induced deterioration of sensory (color, taste, aroma) and nutritional value (vitamins, etc.) of the finished product (concentrate). It is well known that in the first few minutes of evaporative concentration, most of the aroma compounds contained in the raw juice are lost, and the aroma profile undergoes an irreversible change (Lazarides et al. 1990).

An additional drawback in the use of vacuum evaporation is the high energy demand, despite the use of energy-saving systems (thermocompression, mechanical compression, etc.). The results of several experimental studies were published in the area of applying reverse osmosis to concentrate liquid foods. Most of the research activity was in the area of fruit juices, tomato juice, and vegetable juices (Merson et al. 1980; Sheu and Wiley 1983; Merlo et al. 1986). Despite the extensive research efforts, it appears that reverse osmosis cannot be useful by itself, as the potential of the method is only to preconcentrate and not to achieve final concentration. For example, in tomato juice, which is one of the most popular concentrated food commodities, it has been shown that only a concentration of approximately 9°Brix can be achieved (economically speaking) by reverse osmosis. Efforts to achieve higher concentration values encounter serious membrane fouling and severe reduction of permeation fluxes.

An additional limitation is imposed by the limited resistance of reverse osmosis membranes to hydraulic pressure. Most of the new-generation membranes are capable of resisting a maximum of 60–80 atm, which indicates that the obtained concentrate has to have an osmotic pressure less than this limit in order for a driving force to exist.

These inherent problems of concentration by reverse osmosis have forced some researchers to design combined concentration processes, which are beyond the scope of this chapter.

REFERENCES

Adambounou, T. L., Castaigne F. 1983. Partial dehydration of bananas by osmosis and determination of isotherm curves. *Lebensm. Wiss. Technol. (Food Sci. Technol.)*, 16, 230–234.

Alzamora, S., Salvatori, D., Tapia M., Lopez-Malo, A., Welti-Chanes, J., Fito, P. 2005. Novel functional foods from vegetable matrices impregnated with biologically active compounds, *J. Food Eng.*, 67, 205–214.

Argaiz, A., Welti, J., Lopez-Malo, A. 1996. Effect of syrup reuse on the quality of fruits preserved by combined methods. In: *IFT Annual Meeting: Book of Abstracts*, p. 101, Chicago: Institute of Food Technologists.

Aurand, L., Woods, A. E., Wells, M. R. 1987. *Food Composition and Analysis*. Chapters 2 and 5, pp. 19–34, 178–230. An AVI Book. New York: Van Nostrand Reinhold.

Bolin, H. R., Huxsoll, C., Jackson, R., Ng, K. 1983. Effect of osmotic agents and concentration on fruit quality. *J. Food Sci.*, 48, 202–205.

Chirife, J., Karel, M., Flink, J. M. 1973. Studies on mechanism of retention of volatile in freeze-dried food models: system PVP-n-propanol. *J. Food Sci.*, 38, 671–678.

Collignan, A., Raoult-Wack, A. L., Themelin, A. 1992. Energy study of food processing by osmotic dehydration and air dehydration. *Agric. Eng. J.*, 1, 125–135.

Contreras, J. M., Smyrl, T. G. 1981. An evaluation of osmotic concentration of apple rings using corn syrups solids solutions. *Can. Inst. Food Sci. Technol.*, 14, 301–314.

Conway, J., Castaigne, F., Picard, G., Voxan, X. 1983. Mass transfer considerations in the osmotic dehydration of apples. *Can. Inst. Food Sci. Technol. J.*, 16, 25–29.

Dalla Rosa, M., Pinnavaia, G., Lerici, C. R. 1982. La Disidratazione della fruta mediante osmosi diretta. *Ind. Conserve*, 57(1), 3.

Dalla Rosa, M., Bressa, F., Mastrocola, D., Pittia, P. 1995. Use of osmotic treatments to improve the quality of high moisture-minimally processed fruits. In: *Osmotic Dehydration of Fruits and Vegetables*, Lenart, A., Lewicki, P. P. (eds.), pp. 69–87, Warsaw, Poland: Warsaw Agriculture University Press.

Erba, M., Forni, E., Colonello, A., Giangiacomo, R. 1994. Influence of sugar composition and air dehydration levels on the chemical–physical characteristics of osmodehydrofrozen fruit. *Food Chem.*, 50, 60–73.

Farkas, F., Lazar, M. 1969. Osmotic dehydration of apple pieces: effect of temperature and syrup concentration. *Food Technol.*, 23, 688–690.

Fito, P., Chiralt, A., Quan, X. 1995. *Influence of Vacuum Treatment on Mass Transfer during Osmotics of Fruits*. Valencia, España: Universidad Politécnica de Valencia. Departamento de Tecnología de alimentos.

Flink, J. M. 1975. Process conditions for improved flavor quality of freeze dried foods. *J. Agr. Food Chem.*, 23, 1019–1023.

Flink, J. M., Karel, M. 1970. Effects of process variables on retention of volatiles in freeze drying. *J. Food Sci.*, 35, 444–449.

Garrido-Fernandez, A., Brenes-Balbuena, M., Garcia-Garcia, P., Romero-Barranco, C. 2001. Problems related to fermentation brines in the table olive sector. In: *Osmotic Dehydration and Vacuum Impregnation: Application in Food Industries*, Fito, P., Chiralt, A., Barat, J., Spiess, W. and Beshnilian, D. (eds.), pp. 123–132, Lancaster, PA: Technomic Publishing.

Giangiacomo, R., Torregiani, D., Erba, M., Messina, G. 1994. Use of osmodehydrofrozen fruit cubes in yoghurt. *Ital. J. Food Sci.*, 3, 345–350.

Giroux, E., Castillo, M., Valdez-Fragoso, A. 2001. Recycling of concentrated solutions in osmotic dehydration processes and application to their automatic control. In: *Proceedings of the International Congress on Engineering and Food, ICEF 8*, Welti-Chanes, J., Barbosa-Canovas, G. and Aguilera, J. (eds.), Vol. 2, pp. 1347–1350, Lancaster, PA: Technomic Publisher.

Hawkes, J., Flink, J. M. 1978. Osmotic Concentration on fruit slice prior to freeze dehydration. *J. Food Process. Preserv.*, 2, 265–284.

Heng, K., Guilbert, S., Cuq, J. L. 1990. Osmotic dehydration of papaya: influence of process variables on the product quality. *Sci. Aliments*, 10, 831–848.

Horton, B. S., 1997. Water, chemical and brine recycle or reuse applying membrane processes. *Aust. J. Dairy Technol.*, 52(1), 68–70.

Islam, M. N., Flink, J. N. 1982. Dehydration of potato II. Osmotic concentration and its effect on air drying behaviour. *J. Food Technol.*, 17, 387–403.

Karel, M. 1975. Osmotic drying. In: *Principles of Food Science: Part II*, Fennema, O. R. (ed.), New York: Marcel Dekker.

Lazarides, H. N., Iakovidis A., Schwartzberg, H. 1990. Aroma loss and recovery during falling film evaporation. In: *Engineering and Food, Advanced Processes*, Vol. 3, Spiess, W. and Schubert, H. (eds.), pp. 96–105, Barking, U.K.: Elsevier Applied Science Publishers.

Le Maguer, M. 1988. Osmotic dehydration: review and future directions. In: *Proc. Int. Symp. Progress Food Preserv. Process*, Vol. 1. Brussels, Belgium.

Lenart, A., Flink, J. M. 1984a. Osmotic concentration of potatoes. II. Spatial distribution of the osmotic effect. *J. Food Technol.*, 19, 65–89.

Lenart, A., Flink, J. M. 1984b. Osmotic concentration of potatoes. I. Criteria for the end point of the osmotic process. *J. Food Technol.*, 19, 45–63.

Lenart, A., Lewicki, P. P. 1988. Osmotic preconcentration of carrot tissue followed by convection drying. In: *Preconcentration and Drying of Food Materials*, Bruin, S. (ed.), pp. 307–308, Amsterdam: Elsevier Science.

Mallet, C. P. (ed.). 1994. *Editorial Introduction in Frozen Food Technology*, pp. 19–25, London: Blackie Academic & Professional.

Maltini, E., Torregiani, D. 1981. The production of shelf-stable fruits by osmosis. In: *Proc. of 246th Event European Federation of Chemical Engineering*, pp. 471–476, Milan, Italy.

Maltini, E., Torregiani, D., Forni, E., Lattuada, R. 1990. Osmotic properties of fruit juice concentrates. In: *Engineering and Food, Physical Properties and Process Control*, Vol. 1, Spiess, W. L. E. and Schubert, H. (eds.), pp. 567–573, London: Elsevier Science Publishers.

Maltini, E., Torregiani, D., Rondo Brovetto, B., Bertolo, G. 1993. Functional properties of reduced moisture fruits as ingredients in food systems. *Food Res. Int.*, 26(6), 413–419.

Merlo, C., Rose, W., Pedersen L., White, M. 1986. Hyperfiltration of tomato juice during long term high temperature testing. *J. Food Sci.*, 51(2), 395–398.

Merson, R., Paredes, G., Hosaka, D. 1980. Concentrating fruit juices by reverse osmosis. In: *Ultrafiltration Membranes and Applications*, p. 405, New York: Plenum Press.

Moraga, G., Martínez-Navarrete, N., Chiralt, A. 2006a. Compositional changes of strawberry due to dehydration, cold storage and freezing-thawing processes. *J. Food Process. Preserv.*, 30, 458–474.

Moraga, G., Martinez Navarrete, N., Chiralt, A. 2006b. Water sorption isotherms and phase transitions in kiwifruit. *J. Food Eng.*, 72(2), 147–156.

Moy, J., Lau, N., Dollar, A. 1978. Effects of sucrose and acids on osmovac—dehydration of tropical fruits. *J. Food Proc. Preserv.*, 2, 131–135.

Ponting, J. D. 1973. Osmotic dehydration of fruits: recent modifications and application. *Process Biochem.*, 8, 18–23.

Ponting, J. D., Watters, G. G., Forrey, R. R., Jackson, R., Stanley, W. L. 1966. Osmotic dehydration of fruits. *Food Technol.*, 20, 125–131.

Raoult-Wack, A. L. 1994. Recent advances in the osmotic dehydration of foods. *Trends Food Sci. Technol.*, 5, 255–260.

Raoult-Wack, A. L., Guillbert, S., Le Maguer, M., Ríos, G. 1991. Simultaneous water and solute transport in shrinking media. Part 1. Application to dewatering and impregnation soaking process analysis (osmotic dehydration). *Drying Technol.*, 9, 580–612.

Sheu, M., Wiley, R. 1983. Preconcentration of apple juice by reverse osmosis. *J. Food Sci.*, 48, 422–429.

Shipman, J. W., Rahman, A. R., Segars, R. A., Kapsalis, J. G., Westcott, O. E. 1972. Improvement of the texture of dehydrated celery by glycerol treatment. *J. Food Sci.*, 37, 568–576.

Spiess, W. E. L., Beshnilian, D. 1998. Osmotic treatments in food processing, current state and future needs. In: *Proceedings of the 11th International Drying Symposium, IDS'98*, Akritidis, C. B., Marinos-Kouris, D. and Saravacos, G. D. (eds.), Vol. A, pp. 47–56, Thessaloniki, Greece: Ziti Editions.

Szymczak, J., Plocharski, W., Konopacka, D. 1998. The influence of repeated use of sucrose syrup on the quality of osmo-convectively dried sour cherries. In: *Proceedings of the 11th International Drying Symposium, IDS'98*, Akritis, C., Marinos-Kouris, D. and Saravacos G. D. (eds.), Vol. A, pp. 895–902, Thessaloniki, Greece: Ziti Editions.

Torregiani, D., Maltini, E., Bertolo, G., Mingardo, F. 1988. Frozen intermediate moisture fruits: studies on techniques and products properties. In: *Proc. Int. Symp. Progress in Food Preservation*, Vol. I, pp. 71–78, Bruxelles: CERIA.

Torregiani, D., Erba, M., Longoni, F. 1995. Functional properties of pepper osmodehydrated in hydrolysed cheese whey permeate with or without sorbitol. *Food Res. Int.*, 28, 161–166.

Torregiani, D., Forni, E., Longoni, F. 1997. Chemical–physical characteristics of osmodehydrofrozen sweet cherry halves: influence of the osmodehydrofrozen methods and sugar syrup composition. In: *1st Int. Cong. Food Ingredients: New Technologies. Fruits and Vegetables*, pp. 101–109, Tarantano (CN), Italy: Allione Ricerca Agroalimentare S.P.A.

Valdez-Fragoso, A., Welti-Chanes, J., Giroux, F. 1998. Properties of a sucrose solution reused in osmotic dehydration of apples. *Drying Technol.*, 16, 1429–1445.

Warczok, J., Ferrando, M., López, F., Pihlajamäki, A., Güell, C. 2007. Reconcentration of spent solutions from osmotic dehydration using direct osmosis in two configurations. *J. Food Eng.*, 80, 317–326.

11 Chilling and Freezing by Air

(Static and Continuous Equipment)

Judith Evans

CONTENTS

11.1   INTRODUCTION

Chilling and freezing are widely used to preserve foods. Most systems use air because it is flexible, hygienic, and relatively noncorrosive to equipment. Air chilling or freezing produces much lower rates of heat transfer than contact or immersion technologies. However, this may not be a major issue since conduction within the product is often the rate-controlling factor.

Heat transfer rates can be increased by raising the velocity or reducing the temperature of the air moving over the product. However, many chilled products are damaged at low temperatures, and reducing the air temperature is therefore not an option for all products. Where slower rates of chilling or freezing are acceptable, evaporation from the surface of unwrapped products is often a problem. Excessive dehydration caused by slow chilling or freezing is an economic loss and reduces product quality.

The most common air chilling or freezing systems use a fan to blow refrigerated air around an insulated room. Food products are either manually loaded or pass through the room/tunnel on conveyors. Batch systems are the simplest chilling systems but are often characterized by poor air flow and uneven cooling times. Continuous systems overcome the problem of uneven air distribution since each item is subjected to the same velocity/time profile. In the simplest of these systems, food is suspended from an overhead conveyer and moved through a refrigerated room. This process is often used in air chilling of poultry or in the prechilling of pork carcasses. Some small cooked products are continuously chilled on racks of trays, which are pulled or pushed through a chilling tunnel using a simple mechanical system. In more sophisticated plants, the racks are conveyed through a chilling tunnel in which the refrigeration capacity and air conditions can be varied throughout the length of the tunnel. In larger operations, products are conveyed through a linear tunnel or a spiral chiller/freezer.

11.2   DESIGN

Processes to cool food need to be designed to extract heat at the required rate while minimizing energy consumption and ensuring that the equipment fits into the process line and is simple to operate and maintain. Often, purchasing decisions are a compromise between initial cost and operation costs, and compromises are made to reduce the initial outlay. In addition, equipment is often not fully specified before purchase and therefore fails to fully meet demands during operation. Therefore, it is vital to provide a detailed and comprehensive specification of the required process prior to purchasing equipment.

11.2.1   KEY FACTORS IN DEVELOPING SPECIFICATION

A minimum process specification must include the types of products to be handled, whether products will be packaged, and the temperatures at which the products will enter and leave the chiller/freezer. The quantity of product to be handled and the required cooling times should be specified. If these are not already available, they should be obtained through trials possibly combined with mathematical modeling to extrapolate data.

Ideally, the plant should be designed for maximum efficiency at the most common conditions. This requires knowledge of local seasonal ambient conditions (temperature and humidity) to define condensing temperatures (the refrigeration system condenser is generally located outside the building) and of heat loads for the chilling or freezing process. In addition, future production plans should be taken into account to ensure that if higher throughputs are required, the refrigeration plant is capable of removing the additional heat.

Whenever possible, flexibility should be incorporated into the design as products may change in the future, resulting in alterations to cooling requirements. If the designer can incorporate flexibility into the system while still maintaining performance, the chiller or freezer is likely to be far more useful to the end user. Flexibility can be incorporated through use of variable-speed fans or variable belt speeds, baffles to change the direction of air flow over products or ability to adjust the product loading profiles. If product throughput is likely to change, it may be viable to incorporate larger high-pressure receivers (refrigerant storage vessels) and to keep pipe sizes as large as practical in the initial design. However, unless future increases in throughput are well defined, it may not be viable to incorporate larger condensers, spare evaporators, or spare compressors.

The specification should include a plan showing the location and space available for equipment and for maneuvering product into and out of the chiller/freezer. If the central refrigeration plant can be located close to the actual cooling facility, the capital and energy cost of installation will be reduced as the pipe runs will be as short as possible. It is also valuable to set a budget for both the initial capital expenditure and the running costs (operation and maintenance) of the equipment so that the operational costs can be balanced against the initial costs.

11.2.2 ENERGY

It is clear that the efficiency of most refrigeration plants could be improved. Currently, the food industry is responsible for 12% of the United Kingdom's industrial energy consumption and uses over 4500 GW h/year of electrical energy consumption (Market Transformation Programme, http://www.mtprog.com/). Overall figures would indicate that approximately 50% of the energy is associated with retail and commercial refrigeration and 50% with chilling, freezing, and storage (Swain 2006).

System efficiency can be targeted in two ways: either by ensuring that the current plant is operated as efficiently as possible or by replacing parts or all of a plant with a more efficient system. With a new plant, the efficiency of the plant should be an integral part of the initial specification. However, in an older plant, it may only be possible to maintain the plant well and to ensure that it operates at its initial design efficiency. In a chilling or freezing process, the measure of the overall efficiency is the amount of heat extracted from the food divided by the total energy consumed.

In all cases, the efficiency at which heat is removed from the product is extremely important. For example, if sufficient space is available when cooling cooked products, ambient cooling will substantially reduce the amount of heat that needs to be extracted by the refrigeration system. Predictions carried out using a mathematical model similar to that described by Evans et al. (1996) showed that a 1-hour initial

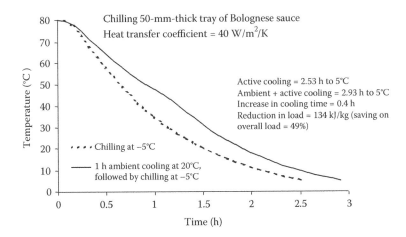

FIGURE 11.1   Effect of ambient cooling on 50-mm-thick tray of Bolognese sauce.

ambient cooling system could extract 49% of the heat required to cool a 50-mm-thick Bolognese-sauce ready meal from 80°C to 5°C (Figure 11.1). Ambient cooling extended the overall cooling time by only 24 minutes.

The main heat loads in food chilling or freezing are from the food and fans used to distribute air in the cooling chambers. However, infiltration loads can also be high in chilling or freezing rooms where food is loaded over periods of time (e.g., meat chillers). Infiltration not only adds heat directly to the room but also a latent load on the room and evaporator as moisture in the air is condensed and frozen. This is not only a heat load but also affects performance of the evaporator and can be a safety issue if ice is present on floors or ceilings where it can fall on operators. Therefore, door discipline is an important factor in reducing energy consumption and ensuring that air within the room is maintained at the correct temperature.

11.3   REFRIGERATION SYSTEMS

Refrigeration system choice is a vital part of selecting an efficient freezing or chilling system that has long-term use. The size of the refrigeration system will vary according to the amount of heat that needs to be removed from the product. The rate of heat removal from products will determine the size of the plant. A smaller, thinner product will release heat more quickly than a larger, thicker product. In small products, the heat release is mainly convection driven, and heat loads are more constant. In larger products, internal conduction has a much larger effect, and cooling load varies during the cooling period with the maximum heat load being at the beginning of the process (Figure 11.2).

Unless products are thin, cooling in the center of foods is controlled by conduction. Increasing the air flow (or heat transfer coefficient) over the product has minimal benefits above low heat transfer coefficients. It is essential in all refrigerated rooms that food is loaded correctly and does not impede air movement around the room and

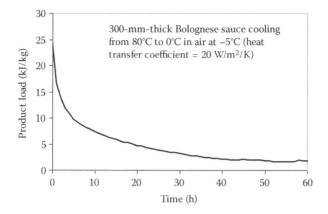

FIGURE 11.2   Product load during cooling of 300-mm-thick product.

that air does not bypass the food. Figure 11.3 shows the effect of increasing the heat transfer coefficient on the surface of varying thicknesses of a Bolognese sauce product. Increasing the heat transfer coefficient at the surface of a thick product reduces the cooling time substantially less than increasing the heat transfer coefficient over a thin product (Figure 11.4). Reducing the temperature of the process fluid also has limited gains in reducing cooling times in a thicker product, owing to poor conduction within the product (Figure 11.5). Reducing the temperature of the cooling fluid or increasing the heat transfer coefficient will increase energy input (through increased compressor power to reduce evaporating temperature or additional fan

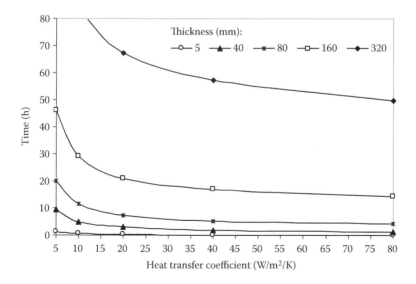

FIGURE 11.3   Effect of heat transfer coefficient and product thickness on cooling times for Bolognese sauce (cooling to 5°C from 80°C in air at –5°C).

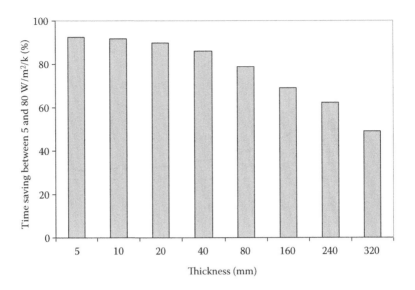

FIGURE 11.4 Difference in time required to cool varied thicknesses of Bolognese sauce between 80°C and 5°C in air at –5°C using heat transfer coefficient of 5 and 80 W/m²/K.

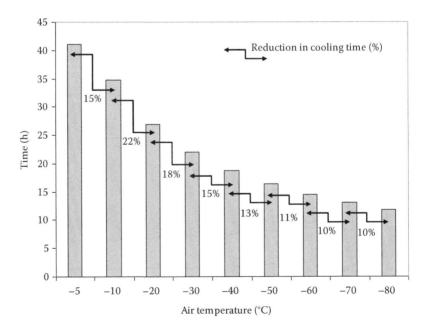

FIGURE 11.5 Effect of reducing the air temperature over 240-mm-thick Bolognese sauce (heat transfer coefficient = 20 W/m²/K) cooled from 80°C to 5°C (mean).

power to blow air around the product) into any cooling system, with limited gains in the product throughput. Once the food surface reaches a temperature close to the air temperature, it is possible to reduce the fan power and energy to the fans without unduly compromising the cooling times. Therefore, optimization of the cooling loads and throughputs is a vital part of refrigeration system efficiency.

11.3.1 Direct Expansion Refrigeration

11.3.1.1 Single-Stage Systems

Within the food industry, the majority of chillers/freezers are operated using a direct expansion refrigerant (DX system). Systems consist of two heat exchangers (a condenser and an evaporator), a means to pump and raise the pressure of the refrigerant (compressor), and an expansion device plus associated control devices, storage vessels, and safety devices. A basic refrigeration system is shown in Figure 11.6. The refrigerant is a volatile fluid that boils (evaporates) at a low-enough temperature to be useful (i.e., at a lower temperature than the food that is being refrigerated). The temperature at which the refrigerant boils is a function of pressure and the properties of the refrigerant. As the refrigerant boils in the evaporator, it gains heat from the environment (usually air but can be a liquid or a solid) and therefore gradually changes from mainly liquid at the entry of the evaporator to gas at the exit. The compressor draws the vapor away from the evaporator and controls the pressure (and therefore the temperature) in the evaporator. It raises the pressure of the gas to a value where condensation to liquid can take place. The high-pressure gas is then condensed at constant temperature to a high-pressure liquid by removing heat from the gas in a heat exchanger called the condenser. The temperature of condensation

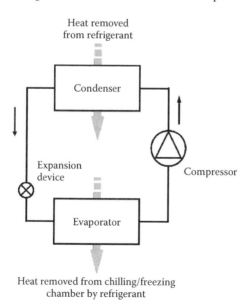

FIGURE 11.6   Basic direct expansion refrigeration system.

must be higher than the medium used for cooling for condensation to occur. The condensed liquid then enters the evaporator through a throttle valve or expansion valve that maintains the pressure between the condenser and evaporator. As the liquid passes through the expansion valve, the pressure of the liquid is reduced to the pressure in the evaporator.

11.3.1.2 Two-Stage Systems

In most cases, the refrigeration system used to cool food contains a hydrochlorofluorocarbon (HCFC) or hydrofluorocarbon (HFC) refrigerant. The working fluid is usually operated in a single stage but may on occasion be used in a two-stage system. In such systems, a primary stage is used to cool a secondary stage that may contain another DX refrigerant or may exchange heat with a secondary brine or liquid (usually water or glycol) that is pumped to a heat exchanger (Figure 11.7).

11.3.1.3 Pumped Recirculation Systems

In larger plants, pumped recirculation systems are common, often operating using ammonia (Figure 11.8). In this system, the refrigerant is contained in a large vessel termed a "surge drum" and is pumped or fed by gravity to the evaporators. The refrigerant boils in the evaporator, but unlike most DX systems, the evaporator is

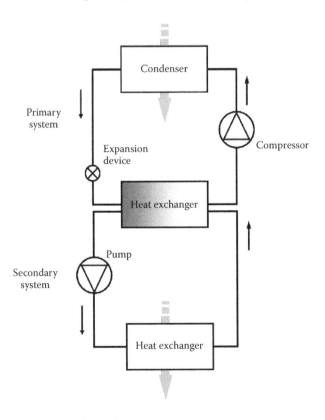

FIGURE 11.7   Two-stage refrigeration system.

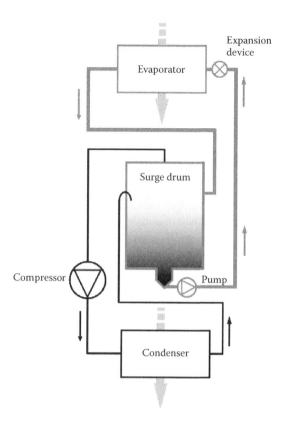

FIGURE 11.8   Pumped recirculation system.

fully flooded (i.e., the refrigerant does not fully boil and turn to gas). This allows the refrigerant to boil throughout the evaporator, enabling the evaporator to be more fully utilized than in a conventional DX system.

11.3.1.4   Absorption Systems

Very occasionally, absorption systems can be used for food chilling if excess heat is available to drive the system. Absorption systems vary the boiling point of a refrigerant by combining it with another fluid. For freezing, this is usually ammonia–water. The system comprises a condenser, an expansion device, an evaporator, and an absorber and regenerator. Refrigerant gas from the evaporator is absorbed in the carrier fluid, creating a weak solution. This is then pumped to condensing pressure and heated, and the refrigerant is driven off (regenerated) before being condensed to liquid for expansion in the evaporator. The regeneration process requires heat, so this type of system is really economic only if large quantities of excess heat are available.

11.3.1.5   Refrigerant Selection

Refrigerant selection is an important issue in terms of environment, safety, and suitability. However, if the system design is optimized for a particular refrigerant, there is unlikely to be greater than a 5% variation in efficiency between most common DX refrigerants. It should be noted that many previously popular refrigerants such as R22 are due to be phased out from 2010 onward as part of the Kyoto protocol. Therefore, a refrigerant with long-term sustainability that has low ozone depletion potential and low global warming potential should be selected. Current alternatives to HCFC and HFC refrigerants for a larger industrial plant include ammonia (R717), CO_2 (R744), and air (R729).

11.3.2 ALTERNATIVE REFRIGERATION SYSTEMS FOR AIR CHILLING/FREEZING

11.3.2.1 Air Cycle

Air cycle is one of the oldest refrigeration technologies. Air cycle machinery was used on board ships in the 1800s to maintain food temperature. However, the large reciprocating machinery was rapidly replaced at the beginning of the 1900s by smaller lighter systems using other refrigerants as new technology developed. Today, high-speed turbo machinery that is compact and lightweight is available, and therefore, the use of air as a refrigerant is once again a commercial possibility.

The principle of the air cycle is that when air is compressed, its temperature and pressure increase (1–2; Figure 11.9). Heat is removed from the compressed air at constant pressure, and its temperature is reduced, ideally while providing useful heat to high-temperature processes (2–3). The air is then expanded, and its temperature reduces as work is taken from it (3–4). The air then absorbs heat (gaining tempera-

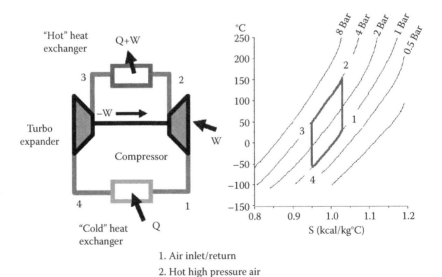

1. Air inlet/return
2. Hot high pressure air
3. Warm high pressure air
4. Cold high pressure air

FIGURE 11.9 Air cycle on a T–S (temperature–entropy) diagram and shown diagrammatically.

ture) from low-temperature processes at constant pressure (4–1), where it starts the cycle again.

The application of air cycle to food processing has many advantages, the most obvious being that air is safe. Any leakage of air from the system is not a risk to the workers, the food, or the environment. It is not flammable, neither does it suffocate, and it is food safe. Air is also very green; it does not deplete the ozone layer and, in specific applications, will decrease the energy used and therefore the CO_2 production.

The primary reason for using an air cycle for food processing is that the range of operating conditions available is greatly increased. A number of theoretical studies have indicated the potential for air cycle in food processing operations (Gigiel et al. 1992; Russell et al. 2000, 2001). Integrated heating and refrigeration is one of the applications with the highest theoretical potential. Theoretically, air temperatures up to 300°C can also be obtained, suitable for direct cooking or the production of steam. Alternatively, fast freezing of small products using the low temperatures available from air cycle systems will result in faster freezing, improving food quality, and either a smaller freezer or a larger throughput through an existing freezer.

11.3.2.2 Cryogenic Systems

Air-based cryogenic systems are primarily used for freezing food. In these systems, a cryogen (usually liquid nitrogen) is either sprayed directly onto a product in a tunnel or is expanded in a heat exchanger, which is used to cool air blown around a tunnel. Cryogenic systems are covered in more detail in Chapter 12.

11.4   CHILLING AND FREEZING EQUIPMENT

11.4.1   Batch Systems

Batch chilling and freezing systems are generally best suited to a larger, irregularly shaped product or a product that has been wrapped or palletized. Large individual items such as meat carcasses are hung from overhead rails; smaller products are placed either unwrapped or in cartons on racks, pallets, or large bins. Generally, products are placed in the chiller or freezer for relatively long periods of time and are handled in batches that fit into the working pattern of the factory (e.g., a 24- or 48-hour working cycle for meat chilling).

Most batch systems are based around an insulated room where air is distributed from an evaporator or evaporators. On occasion, this room may be a store room where the air flow is often uneven and slow. Although not recommended, chamber freezing is often carried out due to availability of space to freeze a product. When chamber freezing or chilling, care needs to be taken that the temperatures of food at the center of large pallets does not remain at a level where microbial growth can occur. For example, Wanous et al. (1989) found that sausages at the center of a pallet required 6 to 7 days to achieve –15°C from a starting temperature of 7°C.

The main advantage of batch freezers and chillers is the flexibility to be able to chill or freeze a range of products. In practice, air distribution is often a major problem, often overlooked by the system designer and the operator. As the freezing time of the product is reduced as the air speed is increased, an optimum value exists between the decrease in freezing time and the increasing power required to drive the fans to produce higher air speeds. This optimum value can be as low as 1.0 m s⁻¹ air speed when freezing beef quarters to at least 15 m s⁻¹ for thin products.

Bulk products are often split up to enable faster freezing. For example, pallets often have spacers ("egg crates") inserted throughout the load to enable air to be distributed throughout the load. Although this results in thinner product layers, it is vital to ensure that there is good air flow between each layer. This can be achieved only by blocking the free area around the pallet and forcing the air through the spacers (Figure 11.10).

Most chilling or freezing processes are single-stage systems where the air temperature and velocity remain constant throughout the process. An alternative is a two-stage system, where the air temperature and/or air velocity are changed at some point in the process. This can be especially advantageous when chilling at a low initial temperature can be used to rapidly reduce the surface temperature to a value just above its freezing point. The air temperatures can then be raised to prevent surface freezing. When freezing products, the air speed can be high in the initial stage to rapidly reduce surface product temperatures to close to the air temperature and can then be reduced in a second stage. Once the surface temperature of a large product is close to the air temperature, conduction will be the major heat transfer mechanism, and therefore, high air velocities are no longer necessary. In addition, lowering the air velocity will reduce the fan heat load on the room and reduce energy consumption. Data presented by James and Bailey (1990) for a range of two-stage chilling systems for beef showed that chilling times to below 7°C could be achieved in under 18 hours and that weight loss was reduced by up to 1.37%. Likewise, for pork, James et al. (1983) and Gigiel and James (1984) have shown that all the initial peak heat load can be extracted from a carcass by a rapid initial chilling procedure followed by a slowed second stage. Weight loss was reduced by half compared to controls.

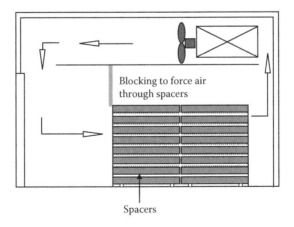

FIGURE 11.10   Batch pallet freezer.

Weight loss of products is not only a quality issue but also an economic loss. In a survey of beef slaughter houses, Gigiel and Collett (1989) found that the cost of weight loss was 20 times that of the energy used in the chilling process. Apart from two-stage processes, another method in reducing weight loss is the use of ice bank chillers. These provide a high-humidity air stream created from a bank of ice built up in a tank. The air from the room is cooled and humidified by direct contact with the cold water from the ice bank. Ice banks are commonly used for cooling fruits and vegetables but have also been used experimentally for meat cooling. In the work carried out by Gigiel and Badran (1988), pork carcasses lost 0.4% less weight when chilled, and an additional 0.9% when stored, in an ice bank when compared to conventionally chilled controls.

11.4.2 Continuous Systems

Continuous chilling and freezing systems are generally suited to smaller products with shorter chilling or freezing times. Generally, products tend to be uniform in shape, and the system is designed specifically for certain product types. Therefore, such systems have a limited flexibility and cannot often chill or freeze a large range of product sizes. In a continuous system, products are often conveyed through a tunnel or room by an overhead conveyor or on a belt. This overcomes the problem of uneven air distribution since each item is subjected to the same velocity/time profile. The time spent in the chiller or freezer (residence time) is generally low to reduce the amount of floor space required for the equipment.

11.4.2.1 Carton/Box Freezer

Carton freezers are used for freezing small cartons such as those containing ice cream. Box freezers are used for freezing boxed products such as meat, fish, and poultry. The cartons or boxes are loaded automatically onto an upper track and are hydraulically pushed through a freezer. Tracks are placed at many levels, and the product is lowered between each level to the exit (Figure 11.11). Freezing is usually accomplished in 3 to 24 hours.

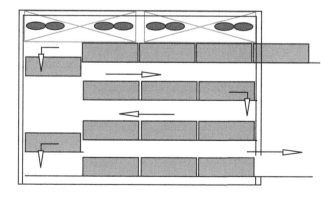

FIGURE 11.11 Carton/box freezer.

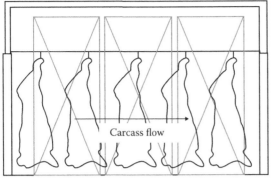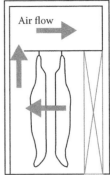

Section through length of room   Cross section

FIGURE 11.12 Tunnel freezer.

11.4.2.2 Tunnel/Belt

Products are conveyed through a tunnel, usually by an overhead conveyor or on a belt. Most commonly, air is directed across the products but may also be directed at the products from above when a belt is used to convey them (Figure 11.12). The products are evenly spaced, and so uniform air flows around the products can be achieved. Generally, most chillers/freezers are restricted to one product size and shape so that product loading can be optimized. However, often, racks or trolleys are used to enable a range of products to be processed.

11.4.2.3 Fluidized Bed

Fluidized beds are commonly used to freeze small individually quick frozen products such as small fruits or vegetables, meat mince, or prawns. Products travel through a tunnel on a mesh belt, and air is blown from underneath the belt onto the products. Air velocities are sufficient to partially fluidize the products, and therefore, the products do not clump together, and each item is frozen individually. The products are fed in at one end of the freezer and overflow from the exit (Figure 11.13). Often, the

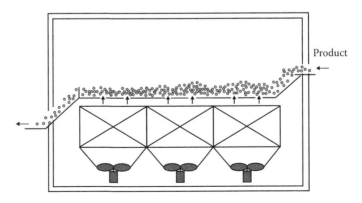

FIGURE 11.13 Fluidized bed freezer.

bed is angled or shaken to assist the product flow. High heat transfer coefficients are achieved owing to localized high air velocities over the surface, and therefore, freezing times are short. Due to small products and short freezing time, these systems can be compact.

11.4.2.4   Spiral Freezer/Chiller

Spiral chillers/freezers generally require a smaller footprint than tunnel chillers/freezers but tend to be taller. They are constructed on a belt that is stacked in a spiral of up to 50 layers (Figure 11.14). Therefore, they allow a long belt to be used in a small area of the production plant. Air can flow either horizontally through the stacks or vertically through the belt.

11.4.2.5   Impingement

Impingement technology increases the surface heat transfer in air and other freezing systems (Newman 2001; Sundsten et al. 2001; Everington 2001). Impingement is the process of directing a jet or jets of fluid at a solid surface to effect a change. The very-high-velocity (20–30 m s^{-1}) impingement gas jets "break up" the static surface boundary layer of gas that surrounds a food product. The resulting medium around the product is more turbulent, and the heat exchange through this zone becomes much more effective (Figure 11.15). Impingement freezing is best suited for products with high ratios of surface area to weight (e.g., burgers) or for products requiring crust freezing. Testing has shown that products with thickness less than 20 mm freeze most effectively in an impingement heat transfer environment. When freezing products thicker than 20 mm, the benefits of impingement freezing can still be achieved; however, the surface heat transfer coefficients later in the freezing process should be reduced to balance the overall process efficiency.

Impingement freezing has substantial advantages in terms of freezing times. In trials carried out by Sundsten et al. (2001), the time required to freeze a 10-mm-thick 80-g hamburger from +4°C to −18°C in a spiral freezer was 22 minutes, whereas in an impingement freezer, the time was 2 minutes 40 seconds. In addition, dehydration was significantly higher for hamburgers frozen in the spiral freezer (1.2%) compared to the impingement freezer (0.4%).

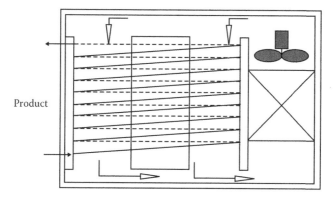

FIGURE 11.14   Spiral freezer/chiller.

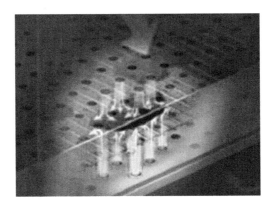

FIGURE 11.15   Impingement system. (Courtesy of Air Products, Allentown, Pennsylvania.)

11.5   FUTURE TRENDS

A number of novel freezing technologies are currently being developed and will be covered in Chapter 16. Those related to static and continuous air-based systems include magnetic resonance and ultrasonic freezing and solar, thermionic, magneto-caloric, electrocaloric, and thermoacoustic refrigeration.

In the shorter term, the food industry is becoming more aware of environmental issues. Due to consumer pressure and rising energy costs, energy is becoming increasingly important, and companies are seeking ways to reduce energy consumption by improving the efficiency of the process and optimizing the efficiency of equipment used for chilling or freezing. Environmental pressures are also encouraging the use of new refrigerants such as CO_2 and renewed interest in older refrigerants such as ammonia and air. The increasing need to produce new and novel products is fuelling the need for flexible equipment that can futureproof manufacturers against unexpected changes in the market. Owing to these trends, it is unlikely that air-based systems will disappear in favor of more specialized new technologies. Instead, new technologies are more likely to be added to air-based systems to retain the flexibility inherent in air-based freezing systems.

REFERENCES

Evans, J., Russell, S., James, S.J. (1996). Chilling of recipe dish meals to meet cook-chill guidelines. *Int. J. Refrig.*, 19, 79–86.

Everington, D.W. (2001). Development of equipment for rapid freezing. In *Rapid Cooling of Food*, Meeting of IIR Commission C2, Bristol, U.K. Section 2, pp. 173–180. Paris, France: International Institute of Refrigeration.

Gigiel, A.J., Badran, R.J. (1988). Chilling and storage of pig carcasses using high humidity air as produced by an ice bank cooler. *Int. J. Refrig.*, 11, 100–104.

Gigiel, A., Collett, P. (1989). Energy consumption, rate of cooling and weight loss in beef chilling in U.K. slaughter houses. *J. Food Eng.*, 10, 255–273.

Gigiel, A.J., James, S.J. (1984). Electrical stimulation and ultra-rapid chilling of pork. *Meat Sci.*, 11, 1–12.

Gigiel, A.J., Chauveron, S., Fitt, P. (1992). *Air as a Replacement for CFC Refrigerants*. IIR Congress 'Cold 92,' Buenos Aires, Argentina, 7–9 Sept., pp. 83–92. Paris, France: International Institute of Refrigeration.

James, S.J., Bailey, C. (1990). Chilling of beef carcasses. In *Chilled Foods—The State of the Art*, Elsevier Science, pp. 159–181.

James, S.J., Gigiel, A.J., Hudson, W.R. (1983). The ultra rapid chilling of pork. *Meat Sci.*, 8, 63–78.

Newman, M. (2001). Cryogenic impingement freezing utilizing atomized liquid nitrogen for the rapid freezing of food products. In *Rapid Cooling of Food*, Meeting of IIR Commission C2, Bristol, U.K., pp. 145–151. Paris, France: International Institute of Refrigeration.

Russell, S.L., Gigiel, A.J., James, S.J. (2000). Development of a fluidised bed food freezing system that can use air cycle technology. *Proceedings of IChemE Food and Drink* 2000. London: Institution of Chemical Engineers. ISBN 0 85295 438.

Russell, S.L., Gigiel, A.J., James, S.J. (2001). Progress in the use of air cycle technology in food refrigeration and especially retail display. *AIRAH J.*, 55, 11, 20–23.

Sundsten, S., Andersson, A., Tornberg, E. (2001). The effect of the freezing rate on the quality of hamburgers. In *Rapid Cooling of Food*, Meeting of IIR Commission C2, Bristol, U.K., pp. 181–186. Paris, France: International Institute of Refrigeration.

Swain, M.J. (2006). Improving the energy efficiency of food refrigeration operations. IChemE Food and Drink Newsletter, September 4, 2006.

Wanous, M.P., Olson, D.G., Kraft, A.A. (1989). Pallet location and freezing rate effect on the oxidation of lipids and myoglobin in commercial fresh pork sausage. *J. Food Sci.*, 54, 549–552.

12 Chilling and Freezing by Cryogenic Gases and Liquids
(Static and Continuous Equipment)

Silvia Estrada-Flores

CONTENTS

12.1 INTRODUCTION

In 2007, frozen and chilled processed foods represented a global retail value of U.S. $88,653 million and U.S. $177,709 million, respectively (Euromonitor International 2007). Examples of frozen and chilled processed foods include products such as the following:

- Chilled processed meats
- Chilled fish/seafood
- Chilled smoked fish
- Chilled pizza
- Chilled soup and pasta
- Chilled processed food
- Drinking milk
- Other dairy (cheese, butter)
- Frozen processed red meat
- Frozen processed poultry
- Frozen processed fish/seafood
- Frozen processed vegetables
- Frozen meat substitutes
- Frozen processed potatoes
- Frozen bakery and dessert products (including ice cream)
- Frozen ready-to-eat and pizza meals

In the current economic climate, it is likely that frozen foods will increase their market share as cost-effective alternatives to dining out. Additionally, continuing debate on the nutritional value of chilled and frozen foods is highlighting the advantages of a well-managed frozen cold chain in preserving the characteristics associated with "fresh" products (Fordred 2008).

Cryogenic freezing and chilling are well suited for many of the aforementioned chilled and frozen product categories. Benefits of this technology are fast cooling rates, high productivity, reduction of food safety risks, low setup costs, and low energy consumption during freezing/chilling. Cryogenic equipment is easy to operate, is compact, and has low maintenance requirements. However, there are also some disadvantages: the operative costs of cryogenic systems can be much greater than the costs associated with running mechanical refrigeration systems. Additionally, cryogenic systems may lead to higher carbon footprints for frozen/chilled foods due to the manufacturing and distribution of cryogenic refrigerants (Dalzell 2000).

Cryogenic refrigeration refers to the use of expandable gaseous refrigerants such as argon, oxygen, hydrogen, nitrogen, carbon dioxide, and other gases that evaporate or sublime at very low temperatures at atmospheric pressure. The most commonly

TABLE 12.1

Physical and Chemical Properties of Nitrogen and Carbon Dioxide

Chemical Formula	N_2	CO_2
Molecular weight (kg kg mol^{-1})	28	44
Latent heat (kJ kg^{-1})	199.1 (vaporization)	572.3 (sublimation)
Density of vapor at 0°C (kg m^{-3})	1.26	1.97
Processing temperature (°C)	−195.5 (boiling point)	−78.3 (solid)

used cryogenic substances in food manufacturing are nitrogen (N_2) and carbon dioxide (CO_2). The physical and chemical properties of both substances are presented in Table 12.1.

Not all products are suited to cryogenic processing: the American Society of Heating, Refrigerating and Air-Conditioning Engineers (ASHRAE) recommends the use of cryogenic cooling for small-scale production, new or seasonal products that require fast chilling/freezing, and situations where the installed freezing/chilling capacity cannot cope with the production needs (ASHRAE 2006). Further, cryogenic cooling is better suited to small- or medium-sized products. In large products with limited surface area, the rate of chilling/freezing is limited by the internal heat transfer (Cleland and Valentas 1997).

During cryogenic processing, the surface of the product can be exposed to (1) a spray of liquid N_2; (2) a mixture of solid ("snow") and gaseous CO_2; or (3) a direct immersion into the liquid cryogen. The handling of the refrigerant is much simpler than in mechanical refrigeration systems: the cryogenic substance is usually delivered to the food manufacturing plant as a high-pressure liquid. It is then stored in a tank, which must be insulated or refrigerated (Cleland and Valentas 1997). Piping is used to transport the cryogenic substance to the freezer, where an arrangement of nozzles and valves sprays the substance into an insulated tunnel transporting the product to be treated (North and Lovatt 2006). As indicated in Table 12.1, there are significant differences in the thermal properties of N_2 and CO_2. Therefore, although the equipment used in handling both cryogens are similar, the design parameters for CO_2 and N_2 are different (Kennedy 2008).

12.1.1 State of the Art

A patent survey covering the period of 1966 to 2002 was conducted using information from 40 patent-issuing authorities in the United States, Japan, PCT (World Patent Office), and major European countries (Estrada-Flores 2002). Figure 12.1 illustrates the rate of increase in the number of patented cryogenic inventions in this period.

About 6600 patents for cryogenic technologies in the most important industrial fields (e.g., polymer engineering, chemical engineering, and semiconductors) were

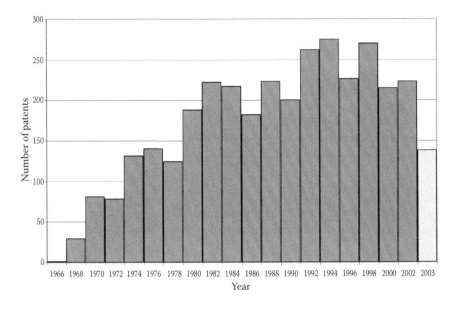

FIGURE 12.1 Patented inventions using cryogenics from 1966 to 2003, all fields of science.

granted worldwide. Less than 4% of these patents were related to cryogenic developments or improvements with direct application to food processing* (Figure 12.2a).
There is an emerging field of application that uses the physical characteristics of
frozen materials to aid in mixing, milling, and others. This field accounted for 7%
of the total patents granted during 1991–2003 (Figure 12.2b). Therefore, there is still
room for the development of new cryogenic equipment and processes of benefit to
the food industry.

Interestingly, many published patents have been developed by food manufacturing companies in need of specific freezing/chilling processes using cryogenics, as
distinct from manufacturers of food equipment. For example, the equipment and
process to prepare free-flowing ice cream in the form of deep frozen beads were
developed by Dippin' Dots founder Curt Jones (Jones 1992).[†] Therefore, the development of new cryogenic technologies in food manufacturing seems to follow a "user-
led" innovation framework, whereby customers modify or adapt existing products
according to their own needs of their own accord (Grunert et al. 2008).

---

* The patent search used the following criteria: titles containing the words cryo* OR ("low temperature"
AND nitrogen) OR ("low temperature" AND "carbon dioxide") were assumed to identify authentic
new patents. Patents that made use of established cryogenic technologies as secondary processes were
excluded. * Stands for wildcard.

† The patent was invalidated in 2007 due to undisclosed sales made one year prior to the patent's priority
date (United States Court of Appeals for the Federal Circuit, 2007. Dippin' Dots Inc. and Curt D. Jones
v T R Mosey, Dots of Fun International Laser Expressions Inc. Decided February 9, 2007).

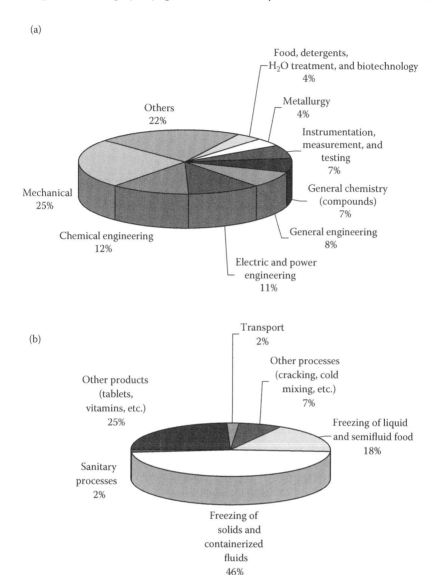

FIGURE 12.2   Patents granted in cryogenic technologies from 1991 to 2003: (a) per field of application and (b) in the food industry.

12.2   SPECIAL DESIGN AND OPERATION CONSIDERATIONS

12.2.1   Capital and Operating Costs

Although the capital cost of cryogenic freezers is only about one-fourth of the cost of an equivalent mechanical freezer, their operating costs are roughly eight times higher (Shaikh and Prabhu 2007). When designing a cryogenic process, the engineer must ensure that the size of the freezer/chiller matches the capacity of the production line. This is a key factor in achieving the lowest freezing costs (Lang 2006).

The consumption of cryogenic substances is the major cost component of cryogenic freezing operations; this consumption ranges between 0.4 and 1.6 kg and between 0.7 and 1.3 per kilogram foodstuff for N_2 and CO_2, respectively (James and James 2003). In addition to the refrigerant costs, the availability of cryogenic substances and the distance from the supplier to the manufacturing plant are other aspects that need consideration.

12.2.2   HEAT TRANSFER CONSIDERATIONS IN CRYOGENIC PROCESSING

In Part I of this book, several methods and equations for predicting freezing times are discussed. Additionally, Cleland and Valentas (1997) provide an excellent overview of the equations required to predict freezing times in a variety of equipment. The following paragraphs highlight some important heat transfer and design aspects specific to cryogenic freezers.

12.2.2.1   Heat Transfer Mechanisms

The rate of heat transfer during cooling of any kind is influenced by several factors. Some of these are the thermal properties (such as thermal conductivity and specific heat) of the food, the surface area of the product available for heat transfer, the size and shape of the product, the temperature difference between the food and the freezing medium, the insulating effect of air surrounding the food, and the presence of packaging materials. These factors are significant for mechanical and cryogenic cooling alike. However, the heat transfer phenomena occurring in cryogenic freezing differs from mechanical freezing in several aspects, as illustrated in Figure 12.3.

The most common process in cryogenic freezing consists of spraying the surface of the product with either N_2 or CO_2. When liquid CO_2 is fed into a spray nozzle (Figure 12.2a), the CO_2 expands and changes to approximately equal parts (by weight) of solid and vapor. As solid CO_2 particles contact the food surface, sublimation occurs almost instantly, drawing latent heat out of the product. This heat transfer mechanism provides approximately 85% of the total cooling capacity. The remaining cooling is a result of the currents created by the resulting CO_2 "snow," the flow of CO_2 through the distribution piping, and the internal convective mechanisms created inside the freezer. Effective freezing in these systems is better achieved through constant spraying of liquid CO_2 throughout the length of the freezer (Kennedy 1998).

In cryogenic systems that use liquid N_2, the liquid is sprayed into the freezer and separates into liquid droplets and vapor. As droplets touch the product surface, the liquid N_2 becomes gaseous N_2, extracting latent heat from the food surface in this process. Cryogenic freezing with N_2 draws 50% of the refrigeration effect through the N_2 phase change. The rest of the product cooling and freezing occurs via convective heat transfer between the gaseous N_2 flowing along the freezer and in countercurrent with the product flow (Awonorin 1997).

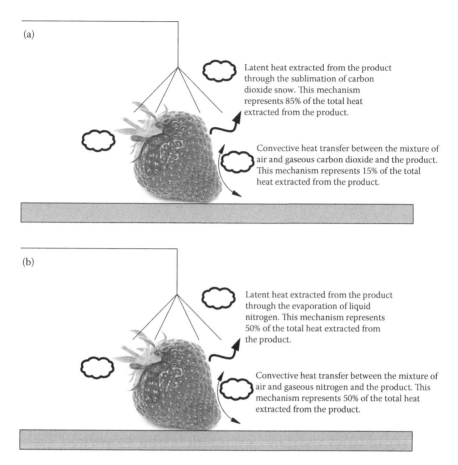

FIGURE 12.3   Representation of heat transfer phenomena during cryogenic freezing using (a) liquid carbon dioxide and (b) liquid nitrogen.

12.2.2.2   Design Aspects

In a well-designed cryogenic freezer, the product heat load represents 85% to 95% of the total heat load. Other minor load components include heat from fans, heat ingress through the insulation, storage vessel losses, and air interchange (Cleland and Valentas 1997). Therefore, the product heat load is the main determining factor of the rate of usage of the cryogen, which is also the most significant operating cost component. It is therefore critical to ensure that the cryogen is optimally used during the freezing process. In freezers using liquid N_2, the maximum freezing effect is achieved through the impingement of the liquid nitrogen droplets on the food surface since the phase change of liquid to gas will lead to the highest cooling capacity (Potter and Hotchkins 1998). Therefore, piping and spraying systems for the cryogen's delivery must be carefully designed to achieve the maximum cooling potential possible.

The impingement of liquid N_2 droplets on the surface of the product also promotes high heat transfer coefficients. There is a scarcity of information on measured heat transfer coefficients for cryogenic CO_2 systems. One study describes the heat transfer phenomena during the injection of pressurized liquid CO_2 into a permeable snow bag (da Veiga and Meyer 2002). The release of the liquid into atmospheric pressure converts the liquid cryogen into roughly equal parts of solid snow particles and a gaseous phase. The bag is used to separate the snow from the cold vapor, which escapes from the bag to precool an insulated shipping container. The measured heat transfer coefficients inside the bag ranged from 2.7 to 5.2 W m^{-2} K^{-1}. It is reasonable to expect low heat transfer coefficients between CO_2 snow and the product's surface due to poorer contact between these two solids, as distinct from spraying liquid N_2 onto the product's surface (Cleland and Valentas 1997).

12.2.2.3   Safety Considerations

The permissible exposure limits established by the Occupational Safety and Health Administration of the U.S. Department of Labor for CO_2 is 5000 ppm (OSHA 2008a,b). Although N_2 gas has no significant physiological effects on itself, N_2 leakages can displace oxygen from enclosed spaces, thus creating a risk of asphyxia for working personnel if the oxygen levels in the working area drop below 19.5% (OSHA 2008a,b). Therefore, the atmosphere of the processing room needs to be constantly monitored. Oxygen and CO_2 sensors are used for N_2 and CO_2 equipment, respectively (Lang 2006). The use of fans and air-conditioning systems in the cryogenic processing area is highly recommended to avoid gas buildup in corners or similar enclosures in the room. Exhaust systems particularly targeting the vaporized cryogenic substance need to be considered in the design of the freezer.

12.3   EQUIPMENT USED IN CRYOGENIC REFRIGERATION

Some examples of representative cryogenic equipment are described below.

12.3.1   FREEZERS

12.3.1.1   Immersion Freezers

The principle of these systems is the immersion of the product in a liquid N_2 bath, freezing the product superficially in 5 to 50 seconds, depending on the product surface available. The main objective is to achieve the formation of a superficial "crust" in the product, reducing dehydration and product clumping. The system leads to a significant thermal shock in the product, and hence, it is not suitable for delicate food items that may crack under these circumstances. Immersion systems are typically used in the meat and poultry industry.

A patented hybrid freezing system uses a two-stage process using one cryogenic immersion freezer followed by a spiral mechanical freezer (Kiczek 1993). The advantage of this combination over previous technologies is the creation of the superficial crust as a conditioning step prior to the traditional blast freezing process, thus avoiding moisture loss from the product. The blast freezing process is further enhanced by the recirculation of the cold, gaseous N_2 produced during the immersion bath to the freezing section. This reuse of N_2 significantly increases the refrigeration capacity of the mechanical system.

12.3.1.2 Tunnel and Spiral Freezers

A typical cryogenic tunnel freezer has a conveyor belt that transports the product throughout the freezer, while a liquid cryogenic substance is sprayed onto the food (Figure 12.4). The tunnel can have a straight, continuous belt, or a serpentine belt that moves the product through a number of flights and landings. The spiral freezers have the same principle as the tunnel freezers, although spiral equipment needs less floor area since a vertical-axis spiral belt replaces the straight belt. The design of both types of freezers can comprise one or more spray zones and one or more transfer fans to move the cold vapor along the tunnel.

Traditional tunnel designs use only one cryogenic substance. However, a patented tunnel freezer uses a mixture of liquid oxygen (O_2) and liquid N_2 in a composition similar to that found in normal air (Foss et al. 1999). This design avoids the safety hazards created by dangerous buildup of gaseous N_2 and the resulting lack of oxygen in the surroundings of the freezer. The tunnel has an immersion bath and oxygen sensors to monitor the levels and control the system accordingly.

12.3.1.3 Cryogenic Impingement Freezers

These systems use a combination of high-velocity air jets (i.e., air impingement) and atomized N_2 applied vertically downward on the surface of the product. The use of air jets drastically decreases the superficial resistance to heat transfer, and thus, the freezing rate relies mostly on the temperature gradient between the product and N_2, which is typically of 190°C, and the thermal properties of the product. Impingement

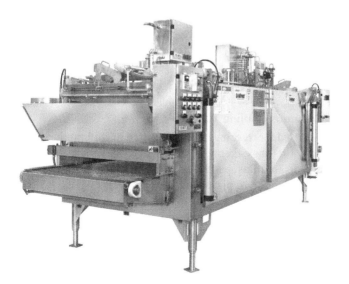

FIGURE 12.4 Example of a cryogenic tunnel freezer. ColdFront Ultra Performance Tunnel Freezer. (Courtesy of Praxair Inc., Danbury, Connecticut.)

freezers are generally better suited to products with high surface-to-weight ratios (Newman and Mc Cormick 2001). Although impingement freezers can raise the production rates to 20%–50% above the capacities obtained with traditional cryogenic tunnels, the latter are 15%–40% less costly than impingement freezers (Lang 2006).

12.3.1.4   Free-Flowing Freezers for Liquid Products

The case of a cryogenic freezer that delivers frozen beads of flavored liquid dairy was discussed previously (Jones 1992; Jones et al. 2001). In this system, the liquid product is fed into a perforated tray. Droplets form and fall into the freezing chamber filled with a mixture of gaseous and liquid cryogenic refrigerants. As the droplets fall, they solidify, forming solid beads of flavored ice cream, yogurt, or flavored ice. The frozen beads are removed from the freezing chamber and packed for distribution and later consumption.

12.3.1.5   Cryomechanical Freezer

Cryomechanical freezers normally encompass a cryogenic immersion stage followed by a blast freezing process. The aim is to achieve a balance between operating costs and product moisture and quality loss: while most of the heat is removed by the mechanical refrigeration system, the initial temperature pull-down is achieved quickly through the cryogenic stage, with the quality improvements normally expected from cryogenic systems (Cleland and Valentas 1997; Agnelli and Mascheroni 2001).

12.3.2   Chillers

12.3.2.1   Rotary Chilling System

In this equipment, the product is chilled in a rotating chamber through the use of a CO_2 injection system. The rotation of the product in the chamber allows homogeneous temperatures at the exit of the tunnel and separation of the product during the quick chilling, thus avoiding the formation of lumps (Figure 12.5). These systems specifically address the need for a homogeneous cooling as a step prior to loading these products into bulk shipping containers or as a conditioning stage prior to further processing (Praxair Inc. 2008).

12.3.2.2   Continuous Cooling Tunnels

Tunnels normally used for freezing can also be used to chill products through a fine-tuning of the control settings. Examples of products that can be chilled in continuous tunnels are meat cuts, muffins, pasta, fruits and vegetables, poultry parts, meat patties, foods, convenience foods, pizza, and seafood.

A particular example of continuous tunnels is the design used to chill bird carcasses. A well-designed tunnel can reduce the broiler's core temperature from an average of 38°C to 0°C in 4 to 6 minutes. This cooling is achieved through the use of CO_2 or N_2 sprayed in the surface of the carcass immediately after the washing stage.

12.3.2.3 CO_2 Snowing

Systems that use CO_2 snow for cooling place the product in direct contact with snow at –79°C. Therefore, it is expected that these systems would create an initial crust of

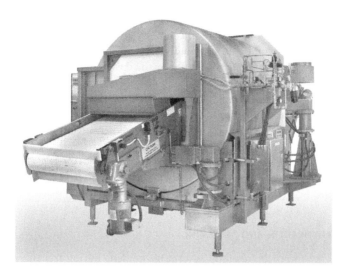

FIGURE 12.5   Example of a cryogenic rotary chilling system. ColdFront Continuous Rotary Chilling System. (Courtesy of Praxair Inc., Danbury, Connecticut.)

ice on the surface of the product. The application of snow varies depending on the equipment: some will apply the snow in a rather coarse manner, but others produce very fine snow particles at flow rates and directions that can be controlled precisely.

12.4   QUALITY OF CRYOGENICALLY FROZEN PRODUCTS

Freezing food materials is more complex than freezing pure water. All food materials contain solutes such as carbohydrates, salts, colorants, and other compounds, which affect their freezing behavior.

Most food products contain animal and/or vegetable cells, forming biological tissues. The water content of these tissues is either inside the cells (intracellular fluid) or surrounding these (extracellular fluid). Since the lowest concentration of solutes is found in the extracellular fluids, the first ice crystals are formed there. During a slow freezing, there will be time for the cell to lose water by diffusion, and the water will freeze on the surface of the crystals already formed. As the cells keep losing water, the cell shrinks more and more until it collapses (Reid 1993). The large ice crystals will exert pressure on the cellular walls, contributing to drip loss during thawing. A rapid freezing promotes a large number of small ice crystals distributed uniformly throughout the tissue, both inside and outside the cells (Goff 2002). Hence, products frozen with cryogenic technologies show a matrix of small ice crystals and a better texture than products frozen using slower heat transfer processes.

A published work (Jul 1984) has shown that fast freezing rates do not necessarily result in better product quality in frozen beef and lamb meat. Detrimental effects of quick freezing can occur in some products, owing to considerable internal strain, surface ruptures, and cell structural changes. However, frozen fish and shellfish, dairy, and horticultural products do benefit with cryogenic freezing.

Quality losses are also related to breakages of the cold chain, such as those commonly encountered in retail and domestic storage. Most frozen products will end up being stored in commercial and, ultimately, domestic freezers. In these situations, temperature abuse is not uncommon, owing to the design of defrost and temperature control systems and the practices followed by the users of domestic and commercial appliances. Even the best freezing technology will not prevent the loss of quality in frozen products that are subjected to temperature abuse before and after the freezing process. Cryogenically frozen products can be more sensitive to these temperature abuses owing to the delicate matrix of ice crystals formed initially and the higher temperature difference between freezing and storage.

The effect of cryogenic freezing on food quality is further explored in the following paragraphs.

12.4.1 DEHYDRATION AND SHRINKING

Product dehydration is the result of several phenomena during freezing, such as an increase in solute concentrations in the unfrozen medium as a result of the formation of ice, the water vapor pressure difference between the product surface and air, and structural damage to cells leading to rupture and loss of intracellular water (Reid 1993; Goff 2002).

Moisture loss in unpacked foods leads to surface desiccation. Dehydration also causes a decrease in the product volume (shrinking), which may range between 3% and 6% in mechanical systems, depending on factors such as the type of product, freezing practices, and the freezer's relative humidity (Zaritzky 2000; Dean 1994).

One of the most commonly used reasons for selecting cryogenic freezing over other methods is the low dehydration and shrinking rates in the former, primarily due to the rapid cooling and the fast decrease in water vapor pressure at the surface. In cryogenic systems, weight losses are reported to be less than 1% (Dean 1994). In some countries, food legislative bodies are already accepting the practice of compensating the dehydration loss by misting the surface of the product, such as in frozen meat patties in the United States (Kennedy 1998). This would prevent the issue of superficial dehydration, thereby making the product more tolerant to slow freezing. However, this practice raises food safety concerns related to the microbial quality of the water added.

As discussed previously, temperature fluctuations during retail and domestic storage can prompt partial melting of ice crystals and moisture migration within the product. This can lead to the presence of superficial liquid in the product. When, after defrosting, the air temperature returns to its initial control set point, superficial ice will be formed; if the heat transfer scenario is such that sublimation of superficial ice crystals occurs, then another quality issue can appear: "freezer burn" (Kennedy 1998; Goff 2002). This quality problem can appear in both cryogenically and mechanically frozen unwrapped products.

12.4.2 MICROBIOLOGICAL ACTIVITY

Freezing prolongs the shelf life of products by decreasing the rate of microbial growth. The negative effect of freezing on microorganisms relates to temperature shock, concentration of extracellular solutes, toxicity of intracellular solutes, dehydration, and ice formation (Zaritzky 2000). Nevertheless, early studies showed that some bacteria can grow and reproduce in 3 to 5 log units per gram at $-7.5°C$ (Schmidt-Lorenz 1963; Schmidt-Lorenz and Gutschmidt 1968) and that yeast can reproduce at temperatures above $-10°C$ (Schmidt-Lorenz and Gutschmidt 1969). Freezing maintains viral infectivity, and a number of viral outbreaks have been associated with consumption of frozen berries (Ramsay and Upton 1989). In general, quick freezing and thawing increase the survival of bacteria as compared with slow freezing and thawing (Davies and Obafemi 1985). Further, the use of cryogenic substances to freeze microbiological samples is the typical procedure used to keep these viable under long-term storage. Hence, consumers and manufacturers should not take microbiological safety for granted in cryogenically frozen products.

Having said this, cryogenic freezing has been successfully used to reduce the risks of *Vibrio vulnificus* in oysters. *V. vulnificus* became a concern worldwide given that, although few cases occur each year, *Vibrio* bloodstream infections in persons with liver disease are fatal in about 50% of the cases (Centers for Disease Control and Prevention 2000). After 35 days of cryogenic freezing, *Vibrio* presence in oysters was shown to decrease to nondetectable levels (Mestey and Rodrick 2006). However, there is evidence that mechanical freezing works equally well in decreasing the risks of *Vibrio*: both mechanical and cryogenic freezing of oysters were shown to inactivate 95% to 99% of *V. vulnificus* in samples of inoculated oyster homogenates, as long as the oyster temperature fell below $-20°C$ (Dombroski et al. 1999).

Freezing promotes yeast destruction in fermented bakery products, thus decreasing the raising potential of the dough (Phimolsiripola et al. 2008). In these products, cryogenic freezing to reach core temperatures ranging from $-40°C$ to $-60°C$ (Neyreneuf and Delpuech 1993) or a combination of cryogenic and mechanical freezing may help in reducing yeast death. The effect of cryogenic freezing/chilling on products with benign microflora (e.g., dairy fermented products, beer, and wine) should be further investigated.

It is recommended that the investigation on the effect of freezing operations on the microbial quality of foods is performed considering the entire cold chain: the physical and chemical characteristics of the product, the freshness of the foods prefreezing, and the handling during and after freezing (e.g., transport, storage and display at retail, domestic storage, and thawing) significantly affect the microbiological quality of frozen foods.

12.4.3 ADHESIVE FORCES OF PRODUCTS DURING FREEZING

In freezing tunnels refrigerated by mechanical means, quality problems may arise due to product adhesion onto the metallic belt transporting the product. The adhesive force between superficial ice and the metallic surface follows a temperature-dependent, nonlinear behavior: initially, the adhesive force increases as the temperature of the metal decreases until this force becomes larger than the strength of the ice. However, as the metal temperature approaches $-80°C$, the adhesive force is reduced dramatically. When the metallic surface reaches $-80°C$, the product can be removed with a minimum effort. This temperature can be easily achieved in cryogenic equipment.

Zero Adhesive Technology has been implemented in ice cream manufacturing machinery to produce novel ice cream presentations (Kennedy 1998).

12.4.4 RECRYSTALLIZATION

Recrystallization, also known as Ostwald ripening, refers to the changes of the ice crystals following the completion of freezing and during storage. These changes include growth of larger ice crystals at the expense of smaller ones, changes in shape,* and changes in the orientation of ice crystals. Maintaining low constant storage temperatures can minimize these phenomena. Cryogenically frozen products might present more recrystallization problems due to the very low temperatures applied during freezing and the radical change to higher storage temperatures (Reid 1993; Zaritzky 2000).

12.4.5 MECHANICAL DAMAGE (FREEZE-CRACKING)

Potential mechanisms that have been proposed to explain the freeze-cracking phenomenon in frozen foods include the following:

- Mechanical damage from ice crystals occurs when flexible cell components are stressed in areas where ice is present. Ice crystals continue to grow in size, exerting additional stress on fragile cellular structures. As flexing of cellular tissues occurs, ice can grow into this newly created volume and prevent the structure from relaxing back into its original shape (Reid 1993; Zaritzky 2000).
- During cryogenic freezing, the formation of a superficial crust in the product may prevent the natural expansion of the internal mass of the product (Hung and Kim 1996). If the internal stress is higher than the frozen material strength, the product will crack during freezing (James and James 2003).
- In spherical foods undergoing freezing, there are opposite mechanical forces that result from phase change expansion and thermal contraction. While expansion causes tensile (positive) radial stress and compressive tangential stress in the frozen shell, contraction causes the opposite effect. The unfrozen core is at uniform and isotropic stress (uniform pressure). This stress is normally tensile because the effect of expansion dominates (Pham et al. 2006).

Precooling can prevent freeze-cracking by reducing the temperature differences between the product and the freezing medium and by reducing the difference between freezing time at the center and at the surface. If the phase change of the core region occurs before the surface becomes brittle, food products can withstand the internal pressure, thus avoiding freeze-cracking (James and James 2003).

---

* Sharper surfaces are less stable than flatter ones and will show a tendency to become smoother over time.

12.5 QUALITY OF CRYOGENICALLY CHILLED PRODUCTS

A well-known objective of the application of rapid chilling to foods is minimizing the residence time in the temperature range of $-10°C$ to $50°C$, in which rapid growth of pathogenic and spoilage microorganisms can take place (Light and Walker 1990). However, it has been hypothesized that other benefits of the use of cryogenic substances for rapid chilling of meats could include the following:

1. The bactericide action of CO_2, which may decrease microbial load on the surface of the unwrapped, chilled product.
2. The depletion of oxygen, which may avoid the oxidation of the meat's myoglobin and thus may contribute to the retention of the bright red color associated with fresh meat.
3. The rapid cooling achieved with cryogenic chilling would lead to lower dehydration rates than those observed in mechanical chilling systems.

A recent study (Kennedy 2008) reviewed the validity of these hypotheses using published research on the effect of CO_2 and N_2 on meat quality and correlating these effects with the operating conditions during cryogenic chilling. The author concluded that, although there is scientific evidence of the inhibition of growth of spoilage microorganisms under CO_2-rich atmospheres, the conditions during cryogenic chilling would not allow the required levels of absorption of CO_2 into the chilled meat. Given that the determining factor of the bactericide effect of CO_2 in modified atmosphere systems is the absorption of CO_2 into the product (Devlieghere et al. 1998), cryogenic chilling would not lead to an increase in the shelf life of fresh meat on its own.

Kennedy (2008) also found that, although CO_2 and N_2 could both lead to the depletion of oxygen during chilling, on returning to a normal atmosphere, any changes in color would be restored. Chilling in a CO_2-rich atmosphere would lower the pH of the meat and would lead to some lightening of the color. However, this is unlikely to be significant in comparison to the color changes observed in modified atmosphere packaging techniques, and the meat's color would again revert to the prechilling condition after the cryogenic cooling.

In regards to dehydration during freezing, Kennedy (2008) concluded that freezing with N_2 should lead to the lowest dehydration rates, when compared to CO_2 and mechanical freezing. This conclusion was reached solely on the basis of the rate of cooling on the surface temperature of the product. Similarly designed chilling equipment will produce less dehydration with nitrogen as a coolant because of the larger temperature differential between the coolant and the product and the faster reduction of the surface temperature, as compared with CO_2 and air chilling.

12.6 ALTERNATIVE USES OF CRYOGENIC TECHNOLOGIES IN THE FOOD INDUSTRY

Cryogenic freezing can be used in separation processes of delicate food. For example, a patented freezing method for separating juice sacs from citrus fruits involves freezing the fruits with a cryogen, crushing the frozen fruits, and finally, separating the juice sacs from the crushed fruits (Ando et al. 1988). The method has been optimized recently by the Florida Department of Citrus (Davis et al. 2004).

In 2005, scientists from the Massachusetts Institute of Technology developed a carbon dioxide flash-freezing process that produces a free-flowing powdered dessert with a carbonation level similar to that found in sodas. The solid flakes, which were made with an ice cream base, produced a tangy fizz and did not cause a burning sensation (or frostbite) in the mouth, possibly owing to the insulation of the ice cream mix ingredients. Potential advantages of the flash-freezing process include energy savings, novel texture and consistency, and opportunities to produce ice cream for convenience markets (Brisson et al. 2007).

Cryogenic milling of spices has replaced traditional grinding techniques in some countries, owing to the higher retention of flavor and the extension of shelf life obtained after low-temperature grinding treatments (Duxbury 1991; Pesek and Wilson 1986).

A heat/cool technology to shuck oysters uses a low-pressure steam treatment followed by a cooling period in either a cryogenic tunnel or an ice bath. Among the advantages of the system are simplicity, relatively low cost, effectiveness, and maintenance of a near-raw oyster with modest but measurable reduction in microbial content (Martin et al. 2007).

Superchilling as a preservation process for fish and other seafood can also be accomplished through the use of cryogenic technology. It is a process where a minor part of the most external water content of the product is frozen. During superchilling, the temperature of the foodstuff is lowered about 1°C or 2°C below the initial freezing point of the product. After an initial surface freezing, the ice distribution levels out to a predefined value, and a uniform temperature is obtained within the product, which must be maintained during storage and distribution (Magnussen et al. 2008). The advantages of superchilling over conventional chilling techniques are improved quality, extended shelf life, and potential energy savings during distribution (Magnussen et al. 2008; Magnussen 1993).

In regards to freezing, technologies such as high-pressure shift freezing and ultrasonic freezing (also reviewed in this book) have been investigated as alternative treatments to cryogenic freezing. The development of new freezing treatments could avoid quality defects such as freeze-cracking and drip loss, which are often associated with cryogenic freezing (Zhu et al. 2005; Martino et al. 1998). However, new freezer designs should include considerations such as capital and operating costs, productivity, and safety. Moreover, an understanding of the entire supply chain of frozen foods is required for the successful introduction of new products and processes that make use of new freezing technologies (Billiard et al. 1999). For example, the benefits of fast freezing at cryogenic temperatures are lost if the products are later subjected to high-temperature fluctuations during the cold chain (Cleland and Valentas 1997). Such fluctuations are common during current transport and domestic refrigeration operations.

12.7   CONCLUSIONS

The analysis of recent patents worldwide revealed that there are significant opportunities to further develop cryogenic technologies for food freezing and chilling applications, possibly borrowing principles from the most advanced applications such as superconductor and spatial technologies or biotechnology. The development of cryogenic solutions appears to follow a user-led innovation process, whereby several novel equipment and processes have been driven by frozen food manufacturers, as distinct from equipment manufacturers.

New cryogenic equipment for the food industry should be developed considering the capital and operating costs of the freezing/chilling operation. Products with relatively high value addition may benefit from cryogenic cooling. The effect of processing variables on the physicochemical properties and quality attributes of the products should also be considered: cryogenic freezing/chilling can be advantageously applied in small- to medium-size products with a high surface-area-to-volume ratio.

Management practices during the food cold chain also need to be addressed. There is no benefit in fast freezing/chilling a product using very low temperatures if the same product is later subjected to high-temperature fluctuations during the cold chain (i.e., during transport or storage). In these cases, the quality of the product at the end of the cold chain might be just as poor as the quality of a product that underwent slow freezing. The development of new product formulations and new cooling technologies should be assessed taking into account potential temperature abuse occurring in commercial and domestic chains.

REFERENCES

Agnelli, M.E. and R.H. Mascheroni. 2001. Cryomechanical freezing. A model for the heat transfer process. *Journal of Food Engineering* (47):263–270.

Ando, T., T. Suzuki, K. Isii, H. Omura and J. Yamazaki. 1988. Apparatus for separating juice sacs of citrus fruits. Nippon Sanso Kabushiki Kaisha.

ASHRAE. 2006. Industrial food freezing systems. In *ASHRAE Handbook: Refrigeration.* Atlanta, Georgia: The American Society of Heating, Refrigerating and Air-Conditioning Engineers.

Awonorin, S.O. 1997. An appraisal of the freezing capabilities of tunnel and spiral belt freezers using liquid nitrogen sprays. *Journal of Food Engineering* 34(2):179–192.

Billiard, F., J. Deforges, E. Derens, J. Gros and M. Serrand. 1999. Control of the cold chain for quick-frozen foods handbook. In *IIR Technical Guide.* Paris, France: International Institute of Refrigeration.

Brisson, J.G., Jr., J.L. Smith and T. Baker. 2007. Carbon dioxide flash-freezing process applied to ice cream production. In *International Cryocooler Conference Inc.*, S. D. Miller and R. G. J. Ross (eds.). Boulder, CO: International Cryocooler Inc.

Centers for Disease Control and Prevention. 2008. *Vibrio vulnificus: Technical Information.* Centers for Disease Control and Prevention 2000 [cited 18 November 2008]. Available from http://www.cdc.gov/nczved/dfbmd/disease_listing/vibriov_ti.html.

Cleland, D.J. and K.J. Valentas. 1997. Prediction of freezing time and design of food freezers. In *Handbook of Food Engineering Practice*, K. J. Valentas, E. Rotstein and R. P. Singh (eds.). Boca Raton, FL: CRC Press.

da Veiga, W.R. and J.P. Meyer. 2002. Heat transfer coefficient of a snow bag. *International Journal of Refrigeration* (25):1043–1046.

Dalzell, J.M. 2000. *Food Industry and the Environment in the European Union: Practical Issues and Cost Implications.* London: Kluwer Academic/Plenum Publishers.

Davies, R. and A. Obafemi. 1985. Response of microorganisms to freeze–thaw stress. In *Microbiology of Frozen Foods.* London: Elsevier Applied Science Publishers.

Davis, C.L., M. Thomas and S. Pao. 2004. *Citrus Juice Vesicle Separation Method and System*. United States: Florida Department of Citrus (Lakeland, FL).

Dean, R. 2002. *Cold, Hard Facts: Food Product Design*. Available from http://www.foodprod uctdesign.com/.

Devlieghere, F., J. Debevere and J. Van Impe. 1998. Concentration of carbon dioxide in the water-phase as a parameter to model the effect of a modified atmosphere on microorganisms. *International Journal of Food Microbiology* 43:105–113.

Dinglinger, G. 1969. Problems of heat transfer when spray-freezing with liquid nitrogen. In *Frozen Foods*. Paris, France: International Institute of Refrigeration.

Dombroski, C.S., L.A. Jaykus, D.P. Green and B.E. Farkas. 1999. Use of a mutant strain for evaluating processing strategies to inactivate *Vibrio vulnificus* in oysters. *Journal of Food Protection* 62(6):592–600.

Duxbury, D.D. 1991. Cryogenically milled natural spices extend stability. *Food Processing* (52):70.

Estrada-Flores, S. 2002. Cryogenic technologies for the freezing and transport of food product. *Ecolibrium: The Official Journal of AIRAH* 6(1):16–21.

Euromonitor International. 2007. State of the Market: Global Market for Meal Solution Products. Retrieved October 12, 2008.

Fordred, C. 2008. Is 'fresh' always best? *Ecolibrium: The Official Journal of AIRAH* (5):10–14.

Foss, J., M. Mitcheltree, P. Schvester, K. Renz, J. Paganessi, L. Hunter, R. Patel and D. Baumunk. 1999. *Liquid Air Food Freezer and Method*. Alexandra, Virginia: United States Patent and Trademark Office.

Goff, G.H. 2002. *Theoretical Aspects of the Freezing Process*. University of Guelph 2002 [cited October 16, 2002]. Available from http://www.foodsci.uoguelph.ca/dairyedu/ freeztheor.html.

Grunert, K.G., B.B. Jensen, A. Sonne, K. Brunsø, D.V. Byrne, C. Clausen, A. Friis, L. Holm, G. Hyldig, N.H. Kristensen, C. Lettl and J. Scholderer. 2008. User-oriented innovation in the food sector: Relevant streams of research and an agenda for future work. *Trends in Food Science and Technology* 19:1–13.

Hung, Y.C. and N.K. Kim. 1996. Fundamental aspects of freeze-cracking. *Food Technology* 50(12):59–61.

James, C. and S. James. 2003. Cryogenic Freezing. In *Encyclopedia of Food Sciences and Nutrition*, B. Caballero and L. Trugo (eds.). London: Elsevier Science Ltd.

Jones, C. 1992. *Method of Preparing and Storing a Free Flowing, Frozen Alimentary Dairy Product*. Alexandra, Virginia: United States Patent and Trademark Office.

Jones, M., C. Jones and S. Jones. 2001. *Cryogenic Processor for Liquid Feed Preparation of a Free-Flowing Frozen Product and Method for Freezing Liquid Composition*. Alexandra, Virginia: United States Patent and Trademark Office.

Jul, M. 1984. *The Quality of Frozen Foods*. London: Academic Press Inc.

Kennedy, C.J. 1998. The future of frozen foods. *Food Science and Technology* 14(4):7–14.

Kennedy, C.J. 2008. Comparison of Chilling Meats by Carbon Dioxide and Nitrogen. Available from http://www.airproducts.co.uk/food/pdf/Kennedy-study_LINreport.pdf.

Kiczek, E. 1993. Cryo-mechanical system for reducing dehydration during freezing of foodstuffs. In *US Patent Trade Office*, U.S.A.

Lang, G. 2006. Cryogenic freezing. In *Industrial Refrigeration Consortium Research and Technology Forum*. Madison, WI: University of Wisconsin-Madison.

Light, N.D. and A. Walker. 1990. *Cook-Chill Catering: Technology and Management*. Elsevier Applied Science.

Magnussen, O.M. 1993. Energy consumption in the cold chain. *Cold Chain Refrigeration Equipment by Design* 171–177.

Magnussen, O.M., A. Haugland, A.K. Torstveit Hemmingsen, S. Johansen and T.S. Nordtvedt. 2008. Advances in superchilling of food—process characteristics and product quality. *Trends in Food Science and Technology* 19(8):418–424.

Martin, D.E., J. Supan, J. Theriot and S.G. Hall. 2007. Development and testing of a heat–cool methodology to automate oyster shucking. *Aquacultural Engineering* 37(1):53–60.

Martino, M.N., L. Otero, P.D. Sanz and N.E. Zaritzky. 1998. Size and location of ice crystals in pork frozen by high-pressure-assisted freezing as compared to classical methods. *Meat Science* 50(3):303–313.

Mestey, D. and G.E. Rodrick. 2006. A comparison of cryogenic freezing techniques and their usefulness in reduction of *Vibrio vulnificus* in whole oysters. Paper read at International Proceedings for Molluscan Shellfish Safety.

Newman, M.D. and S. Mc Cormick. 2001. *Modular Apparatus for Cooling and Freezing of Food Product on a Moving Substrate*. Alexandra, Virginia: United States Patent and Trademark Office.

Neyreneuf, O. and B. Delpuech. 1993. Freezing experiments on yeasted dough slabs. Effects of cryogenic temperatures on the baking performance. *Cereal Chemistry* 70:109–111.

North, M.F. and S. Lovatt. 2006. Freezing methods and equipment. In *Handbook of Frozen Food Processing and Packaging*, S. Da-Wen (ed.). Boca Raton, FL: CRC Press.

Occupational Safety and Health Administration (OSHA). 2008a. *Carbon Dioxide in Workplace Atmospheres*. U.S. Department of Labor [cited 19 Nov 2008]. Available from http://www.osha.gov/dts/sltc/methods/inorganic/id172/id172.html.

Occupational Safety and Health Administration (OSHA). 2008b. *Occupational Health and Safety Guideline for Nitrogen*. U.S. Department of Labor [cited 19 Nov 2008]. Available from http://www.osha.gov/SLTC/healthguide lines/nitrogen/recognition.html.

Pesek, C.A. and L.A. Wilson. 1986. Spice quality: Effect of cryogenic and ambient grinding on color. *Journal of Food Science* 51(5):1386–1388.

Pham, Q.T., A. Le Bail and B. Tremeac. 2006. Analysis of stresses during the freezing of solid spherical foods. *International Journal of Refrigeration* (29):125–133.

Phimolsiripola, Y., U. Siripatrawana, V. Tulyathana and D. Cleland. 2008. Effects of freezing and temperature fluctuations during frozen storage on frozen dough and bread quality. *Journal of Food Engineering* 84(1):48–56.

Potter, H.N. and J.H. Hotchkins. 1998. *Food Science*. 5th ed. Berlin, Germany: Springer.

Praxair Inc. 2008. Available from http://www.praxair.com/.

Ramsay, C.N. and P.A. Upton. 1989. Hepatitis A and frozen raspberries. *Lancet* 1(8628):43–44.

Reid, D. 1993. Physical phenomena in the freezing and thawing of plant and animal tissues. In *Frozen Food Technology*, C. P. Mallet (ed.). London: Blackie Academic & Professional.

Schmidt-Lorenz, W. 1963. Microbieller Verderb gefrorener Lebensmittel während der Gefrielagerung. *Kältetechnik* 15:39.

Schmidt-Lorenz, W. and J. Gutschmidt. 1968. Mikrobielle und sensorische Veränderungen gefrorener Lebensmittel bei Lagerung im Temperaturbereich von –2.5°C bis –10°C. *Lebensmittel-Wissenschaft und-Technologie* 1:26.

Schmidt-Lorenz, W. and J. Gutschmidt. 1969. Mikrobielle und sensorische Veränderungen gefrorener Brathähnchen und Poularden bei Lagerung im Temperaturbereich von –2.5°C bis –10°C. *Fleishwirtschaft* 49:1033.

Shaikh, N.I. and V. Prabhu. 2007. Mathematical modeling and simulation of cryogenic tunnel freezers. *Journal of Food Engineering* 80(2):701–710.

Zaritzky, N.E. 2000. Factors affecting the stability of frozen foods. In *Managing Frozen Foods*, C. J. Kennedy (ed.). London: Woodhead Publishing Ltd.

Zhu, S., H.S. Ramaswamy and A. Le Bail. 2005. Ice-crystal formation in gelatin gel during pressure shift versus conventional freezing. *Journal of Food Engineering* (66):69–76.

13 Chilling and Freezing by Contact with Refrigerated Surfaces
(Plate Freezers, Surface Hardeners, and Scraped Surface Freezers)

Rodolfo H. Mascheroni

CONTENTS

13.1 INTRODUCTION

The different types of cooling options to be covered in this chapter have a common feature: the main path of heat transfer between the refrigerating medium and the food product is by conduction through a surface of the refrigerator, normally metallic—of very high thermal conductivity—to which the food or its package is in direct contact, assuring a very high heat transfer rate when this contact is sufficiently intimate.

As will be seen later when describing the diverse types of equipment used for this purpose, the refrigerating system and the product to be treated (its shape, consistency and packaging, if any) must be perfectly matched to take advantage of the possibilities of these high transfer rates.

13.2 PLATE FREEZERS

The bases of contact freezing by refrigerated plates (plate freezers) are the high thermal conductivity of the materials that make up the plates that act as flooded plane surface evaporators, (generally built of extruded aluminum, aluminum–zinc, or aluminum–magnesium alloys or stainless or galvanized steel) and the possibility of achieving a very good thermal contact with the products to be cooled (provided they have perfectly plane surfaces with no air gaps or frozen water attached to the plate). Both characteristics ensure a high heat transfer rate. Besides, the lack of circulating air drastically decreases weight losses by sublimation of surface ice. Other advantages include the following: highly efficient, promotes energy conservation, easy to clean, covers small space, and can be operated at room temperature.

With regard to this type of equipment, plate freezers are designed to freeze large blocks of product, so they are aimed at bulk storage and distribution rather than at individual product portions for retail sale. The blocks are strong and of consistent dimensions, so they are easy to handle and stack.

Almost all of them correspond to both horizontal and vertical double contact plate freezer types. There is some other equipment for specific use to be described later.

If the design conditions are complied with, then for a given product and refrigerating fluid temperature, the plate freezer will provide the shortest freezing time as well as the lower operating cost (Gruda and Postolki 1986; American Society of Heating, Refrigerating and Air-Conditioning Engineers [ASHRAE] 2010). Below are its characteristics.

Design of plates. In the most modern versions, they are 2.5 cm to 5.0 cm thick, in which in the inner part, there is a continuous channel of rectangular or cylindrical profile covering the plate area uniformly. The refrigerating fluid circulates through this channel, which is being fed in a liquid state and evaporates inside the plate (direct evaporation) leaving it as a mixture of liquid and vapor. Different brands provide alternatives for diverse refrigerants (freons, ammonia, CO_2). Also there is the possibility of using brines and secondary systems, in which case the plates behave as plate heat exchangers. The design guarantees a uniform temperature distribution on the entire surface and on both faces of the plate. Operating temperatures are of the order of –30°C to –40°C with very high heat transfer coefficients of about 550–600 W/(m^2 K) (Amarante and Lanoisellé 2005).

Contact with the products to refrigerate. The only ways to take advantage of those high transfer coefficients are to work with products whose surface is totally plane and to avoid the presence of any air gap that, existing between the plate and the product, hinders heat transfer. Another cause of a bad contact is the formation of a crust or drops of frozen water due to bad maintenance, condensation, or leaks from the containers.

The previous condition restricts the application to single products having a plane shape, such as fillets, or to those that can be reshaped to a plane after being pressed between two plates, such as anatomic cuts of meat. Because of that, the most frequently processed products are packed in cartons or in metallic trays or molds having perfectly plane bases to cover the whole area of plates and to occupy the space between them.

The condition of no air gap is accomplished when exerting a certain pressure between the plates to ensure removal of any airbag that would otherwise remain between plates and the product. In the case of trays and molds, their whole internal volume must be occupied by products such as minced beef or offal blocks. Air will fill any hole between portions or will remain in free spaces below the cover of boxes or molds acting as insulation that increases freezing times markedly.

The plate freezers appear in two spatial arrangements: those of horizontal plates and those of vertical plates. Figure 13.1 presents a scheme of each type of equipment. In both cases, there exist a series of plates (up to 24 according to brand and model) in between which the products to be frozen are placed. Each plate has its own inlet of coolant, which comes from a common feed, as well as its own outlet that joins to those of the other plates toward a common return line. Flexible hoses convey both the feed and the discharge of refrigerating fluid to allow the plate to have freedom of movement during loading and unloading.

There may be an individual compressor and condenser or the freezer can be fed of liquid refrigerant—or less frequently of a brine—coming from a central facility. To ensure good contact between the plate and the product, a pressing mechanism is always used.

Horizontal plate freezers. They are the most traditional and frequent type, and they are mostly used for boxes, trays, and molds. The products that are to be frozen under these conditions are blocks of fillets, surimi, minced meat, meat or poultry portions, sausages, offals, boxes with individual products (strawberries, peas, spinach, hamburgers, scallops, prawns, fish sticks and fillets, ice creams), bags of products (mashed spinach, egg paste, fruit pieces), product portions or products in molds, strawberries with sugar, fruit pulps, soups, ready-to-eat meals, etc.

The surface area of each plate can vary from 1 to 2.5 m^2, the number of plates is between 6 and 24, and the production per equipment can reach 3000 kg/h. Usually, the plates assembly is enclosed in an insulated cabinet with front—and sometimes also back doors. To obtain proper performance, the plate freezer must be piped correctly. The refrigerant overfeed rate of 16 to 20 to 1 requires engineered piping or there will be a major temperature difference between the top and bottom plates. It can vary by as much as 8.5°C at −40°C suction; if the refrigerant piping is overhead, the bottom plate would be at −31°C.

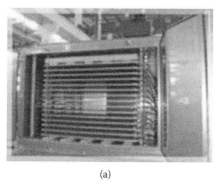

(a)

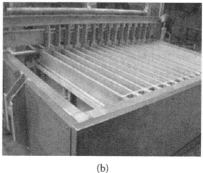

(b)

FIGURE 13.1   Examples of double contact plate freezers: (a) horizontal; (b) vertical.

Vertical plate freezers. They allow bulk products to be processed. They were first used to freeze whole fish and fillets on the high seas because this design suits applications where the headroom is tight, such as on board a ship.

At present, its use was also extended to blocks of minced meat or meat portions, meat and poultry byproducts, squids, octopus, fruit juice or pulp, spinach, liquid egg, concentrated coffee and tea, condensed milk, cream, food additives, etc. Most models have fixed plate spacing ranging between 5 cm and 10 cm, although other types have plates that can be slid horizontally to facilitate the unloading stage. They are very useful for freezing those foods that need no packing, providing higher transfer rate and lower packaging and processing costs.

The equipment is loaded from the top and can be bulk-filled with individual products. Once freezing is complete, the plates are loosened and separated from the blocks (sometimes by pumping hot refrigerant into the plates for a brief period), which are then generally unloaded from the bottom, although high-production equipment on board factory ships can be unloaded from above with moving cranes.

In general, foods frozen in vertical plate freezers are of superior quality, as the ice crystals that are formed are small due to the high transfer rate; thus, little cell damage occurs with lower drip production on thawing and best final texture.

The cost to freeze foods in plate freezers is the lowest possible as compared to alternative options (cold air and cryogenic). Elimination of expensive cartons normally used for blast freezing saves money in initial cost, avoids the need for storage space for the boxes, reduces shipping costs, obviates box makeup and handling costs, and reduces cold storage space requirements. The end user saves money by eliminating the carton handling and disposal problem (Briley 2001).

Specialized plate freezers. For particular applications (normally for certain foods in bulk or packaged in irregular shaped containers or of mixed sizes), certain horizontal plate designs lack the hydraulic mechanism and freeze from the bottom of the blocks only (like refrigerated shelves). Some variants of this type of freezer add a flow of cold air to freeze the top surfaces. These freezers resemble, to a certain degree, carton freezers, although they are suitable for a wider range of products because they can simultaneously accommodate any type of flat-based containers. Some models include the possibility of also being used as a normal double-contact horizontal plate freezer.

One important disadvantage of plate freezers is that loading/unloading operations are slow and labor-intensive, mainly the horizontal types; besides, being batch equipment, they are not suitable for high output continuous production lines. To partially overcome this limitation, some models are provided with doors on both sides (as option) for easy loading and unloading of freezing pans and quick cleaning of the freezer plates and with a systematized loading–unloading operation using forklifts. For very high outputs (several tons per hour), there are automated horizontal plate freezers that, by using a belt conveyor system fitted with an elevation and descent mechanism, automatically load and unload the plates placed at different levels.

In all cases (see subsequent calculations), the great advantage of the high heat transfer coefficient is lost if very thick materials are frozen. In those materials, the freezing time is mainly determined by the thickness and not by the transfer coefficient. Because of that, freezing of products that are more than 8 cm thick is very unusual; it is better to split the thickness whenever possible (e.g., blocks of meat, fillets, offals, byproducts, juices, or pulp).

13.3   SPECIALIZED CONTACT FREEZERS (SURFACE HARDENERS)

There are a series of soft, humid, sticky food materials that tend to lose their shape, to bend, to stick, and/or to dehydrate over the belts of continuous freezers. They include all kinds of pulps and purees, de-skinned chicken pieces, fish fillets, marinated pieces of fish, meat or poultry, bakery and pastry products, filled pasta, ice creams, peeled fruit pieces, etc. One adequate solution to avoid these disadvantages during the refrigeration process is to crust-freeze these items fast to avoid deformation, sticking, and dehydration.

One idea that has originated diverse designs from freezing equipment manufacturers is to use a refrigerated metal belt (fixed or moving) that slides the food to be crust-frozen. The high heat transfer rate that originated in contact with the cold metal surface (Lanoisellé et al. 1998) allows a 1-mm-thick crust that can be formed in less than a minute, which assures shape stability and nonstickiness. An additional advantage of crust freezing is the lower weight loss by water evaporation and sublimation—with respect to conventional cold-air freezing—owed to the rapid temperature descent of the food surface.

This kind of equipment is normally intended for freezing processes but can also be used as a quick and continuous surface hardener to assist the process of cutting soft meats and fishes, obtaining constant thickness slices, and maintaining product shape and integrity.

Some models have a fixed belt over which a very thin disposable plastic film carrying the food slides. The film comes in rolls that are unfolded in one end—the entrance—of the belt and are received and rerolled at the other end (see Figure 13.2b for a scheme of this system).

In another design, the belt is fixed and hollow and acts as a flooded plane surface evaporator—as in plate freezers—through which a refrigerant at about –40°C flows, reaching a similar temperature in the belt surface in contact with the sliding film. This design can incorporate an upper forced air freezing zone (with downward air flow generated by axial fans and independent mechanical refrigeration system), so

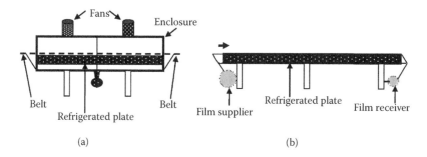

FIGURE 13.2   Scheme of continuous contact surface hardeners: (a) with continuous belt and upper and lower face refrigeration; (b) with lower face refrigeration and sliding plastic film.

as to refrigerate both faces of the product, in a design that resembles a horizontal belt freezer tunnel (see Figure 13.2a). A similar design uses a continuous belt that is refrigerated in its upper and lower surface by cold air. The evaporators are placed just below the belt. Heat transfer occurs mainly on the lower surface of the food.

Other equipment designs offer a system with the surface of the belt refrigerated at −196°C with vaporized liquid nitrogen. High-speed fans circulate the cold atmosphere generated by the plates in the freezing tunnel to quickly freeze the upper surface of the product and ensure efficient use of cold air.

The belts of these hardeners are linear, so to reach high throughputs of completely frozen foods of the order of several tons per hour, important belt lengths of up to 30 m are needed. This is a clear disadvantage with respect to spiral freezers, and normally their use is restrained to a prefreezer stage, which allows for shorter belt lengths.

An alternative design that is also refrigerated by liquid nitrogen claims to save valuable floor space and eliminate costly plastic film using a rotating device based on a rolling cylinder-shaped freezing zone, on which the product and nitrogen are spread. This continuous contact freezing system can be used to crust-freeze prior to entering a spiral freezer or other freezer that may leave belt marks on the product and is intended for freezing products without having individual pieces sticking to each other, such as products consisting of separate pieces (granules or crumbs) like risottos, pasta with sauces, peas, berries, strawberries, sliced or minced mushrooms, and French fries.

13.4 COMPARISON OF FREEZING TIMES IN PLATE FREEZERS, DOUBLE FACE REFRIGERATED SURFACE HARDENERS, AND BELT FREEZERS

Although each type of freezer is intended for processing diverse types of products, there is a frequent need to use one of these equipments for unusual foods or the possibility to choose between alternative designs and/or optimize process conditions—like product thickness and packaging—to reach higher processing rates.

The foods to be frozen are slabs of minced beef of seven different thicknesses (2L: 1.5, 2.25, 3.0, 3.75, 4.5, 5.25, and 6.0 cm). In the case of the surface hardener, they are unpackaged, and in the other two freezers, they are kept in metallic pans of the exact height of the slab (assuring very good contact and heat transfer coefficient). Their initial temperature Ti is 10°C, the refrigerant temperature Ta is −40°C, and the final temperature in the center of the slab Tc is −18°C.

The food characteristics are as follows: water content, 74%; thermal diffusivity $\alpha_o =$ 1.370×10^{-7} m^2/s; thermal conductivity $k_o = 0.500$ W/(m K); density = 1000 kg/m^3.

The freezing time (tf) predictive equation (Salvadori et al. 1997) is given by

$$tf = L^2/\alpha_o (-1.272Tc + 65.489) (1/Bi + 0.184) (1 + Ti)^{0.096} (-Ta - 1)^{-1.070} \quad (13.1)$$

where Bi = hL/k_o, and L is the half-thickness of the slab. In the case of nonsymmetric freezing due to different heat transfer coefficients on both faces, Equation 13.1

is modified (Salvadori and Mascheroni 1992) using the higher h and a corrected L given by the method of Uno and Hayakawa (1979).

The assumed heat transfer coefficient h is 560 W/(m^2 °C) for unpackaged food in the plate freezer (Amarante and Lanoisellé 2005), from which for the metallic tray covered by a plastic film, an effective heat transfer coefficient h of 200 W/(m^2 K) was considered; 28.1 and 39.6 W/(m^2 °C) were taken for the upper and lower faces of the food in the hardener (Amarante and Lanoisellé 2005) and 40 W/(m^2 °C) in the belt freezer (Tocci and Mascheroni 1995). No loading–unloading time is considered for continuous freezers, and 15 min is assumed for each batch in the plate freezer. The three equipments are calculated—in their combined belt width and length or in the total area of plates—to deliver 1 ton/h of frozen slabs with 1.5-cm thickness.

Table 13.1 presents the results of such predictions.

From the results of Table 13.1 for $L = 0.75$, the needed area of plates or belts for a 1 ton/h productivity can be deduced based on the number of freezing cycles per hour. The calculated values are 25.35 m^2 of plates and 40.96 and 31.06 m^2 of belts for the hardener and the belt freezer, respectively. In the case of plate freezers and belt freezers of the spiral type, these are usual values and in both cases they have a small footprint. In the case of hardeners, these are linear equipments, and this size implies a length of the order of 30 m, varying with the different brands, which is usually difficult to manage in existing food factories.

From the calculated values of productivity, conditioned by product thickness and "dead time" (for plate freezers), it can be deduced that for continuous freezers, an increase in thickness always leads to lower productivity. In the case of plate freezers, the relative weight of the loading–unloading time lowers as the thickness increases leading to higher ft. This balance between the two factors leads to a maximum in productivity, for the assumed h and loading–unloading period, for slabs of about 3.5-cm thickness.

All these factors—and the differences in final quality for diverse freezing methods—have to be balanced when deciding the best freezing method for a range of products to be frozen at the same factory.

TABLE 13.1
Freezing Times, Cycle Duration, and Productivity for the Three Types of Freezers and for Different Food Thicknesses

	Plate Freezer			Surface Hardener		Belt Freezer	
2L (cm)	ft (min)	Cycle (min)	Productivity (tons/h)	ft/Cycle (min)	Productivity (tons/h)	ft/Cycle (min)	Productivity (tons/h)
1.50	7.815	22.815	1.0000	36.867	1.0000	27.957	1.0000
2.25	13.807	28.807	1.1880	57.524	0.9613	44.020	0.9526
3.00	21.189	36.189	1.2609	77.827	0.9474	61.473	0.9096
3.75	29.961	44.961	1.2686	98.072	0.9398	80.316	0.8702
4.50	40.123	55.123	1.2417	120.600	0.9171	100.549	0.8341
5.25	51.674	66.674	1.1977	143.680	0.8981	122.171	0.8009
6.00	64.616	79.616	1.1463	169.157	0.8718	145.184	0.7703

Finally, these calculations can be easily extended to diverse freezing conditions (Ta, h, plate/belt area, dead time) and food characteristics (Ti, α_o, k_o).

13.5 SCRAPED SURFACE HEAT EXCHANGERS

Scraped surface heat exchangers (SSHEs) are a type of turbulent film heat exchangers. Their design is based on the concept of continuous heat transfer of the moving product with periodic removal of fluid from the wall in contact with the heating or cooling medium. Most commercial designs have a hollow cylindrical shape, with a rotating shaft in the center, with the product being pumped through the annular gap between the shaft and the outer cylindrical heat transfer tube (Figure 13.3a). As the product is pumped under pressure through the annulus, the central shaft with the scraper blades is rotated. The rotating scraper blades continually remove any deposition on the heat exchanger surface and also provide the means for mixing the fluid material, thereby maintaining the heat transfer rate and enabling extended runs without fouling. There exist other types that use plate heat exchangers, which are to be described in the next paragraphs.

SSHEs have a very high transfer rate per unit surface owed to the intimate contact between the food material and the metallic heat transfer surface. SSHEs are one of the most versatile pieces of processing equipment. They can handle products that are viscous, that contain particles, and that normally tend to clump or to deposit and form films on the heat transfer surfaces (Rao and Hartel 2006). SSHEs are intended for a wide range of applications in food manufacturing, such as processing of virtually any pumpable fluid or slurry involving slush freezing, crystallizing, cooling, heating, cooking, aseptic processing, mixing, plasticizing, gelling, polymerizing, etc. This versatility lies in the possibility of using different heat transfer media (water, brines, steam, diverse refrigerants, ammonia, etc.), according to the intended heat transfer process.

Different manufacturers offer diverse models with a centered or eccentric shaft (Figure 13.3b), a circular or elliptical process section, and a different annulus free section (always of a few millimeter thickness), so as to deal with fluids of different

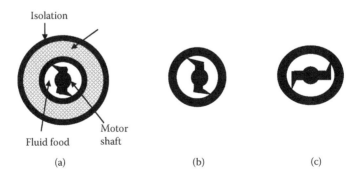

FIGURE 13.3 (a) Scheme of the cross-section of an SSHE with centered shaft; (b) scheme of an eccentric shaft; (c) scheme of an elliptical section.

viscosities, with or without suspended material, that may tend to collapse, crystallize or polymerize, etc. The design of these equipments must deal with products of different viscosities and textures (pure liquids, emulsions, suspensions with fine or coarse materials, sticky or agglomerating, etc.), with the turbulence originating in the forced fluid flow in the annulus, with the centrifugal forces generated by the rotating shaft and with the mixture induced by the blades, as well as with the possibility of crystallization on the interior wall of the cylinder. These facts have been the motive of numerous studies on the complicated fluid dynamics in these exchangers, most of them with the aim to find ways to improve mixture homogeneity and maximal heat transfer rate. In general, the presence of blades dominates the evolution of flow patterns and shear profiles deforming the flow patterns, making prediction of residence time distribution extremely difficult. Among the most recent studies, Sun et al. (2004), Rao and Hartel (2006), and Benkhelifa et al. (2008) can be cited.

In food refrigeration, it is common to use SSHEs in processes that include cooling of liquid foods and crystallization of solutes or solvents, such as dewaxing of oils and fats and preparation of various fatty acids, in margarine processing and ice cream production as crystallizers, and in freeze-concentration operations, typically of fruit juices, milk, coffee, whey, protein concentrates, etc.

In freeze concentration, SSHEs have advantages with respect to other concentration procedures like evaporation (ambient or under vacuum) because loss of aroma compounds is avoided as is the development of undesirable compounds due to high temperature reactions of components. Notwithstanding, freeze concentration is more expensive than evaporation. Here again the coupled heat and mass transfer during crystal nucleation and growing is a difficult subject to deal with. In different applications, diverse types of crystals are generated. Their type and size are regulated by contact time and agitation, both of which depend on the diameter of the annulus gap between the shaft and the surface of the inner cylinder, the type of blades, rotation speed, etc. Much attention has been paid by researchers to this subject. Recent discussion on crystallization in scraped surfaces may be found in the works of Pronk et al. (2005), Miyawaki et al. (2005), Bongers (2006), Drewett and Hartel (2007), and Aider and de Halleux (2009).

Sánchez et al. (2009) presented a very up-to-date review on the cryoconcentration of fruit juices and sugar solutions, remarking that the advantage of the freeze concentration technique is the quality of the product obtained due to the low temperatures used in the process, which makes it a very suitable technology for the processing of fruit juices. They remember that, at present, scraped surface crystallization is the only processing technique available at the industrial level. Among recent studies, Nonthanum and Tansakul (2008) dealt with freeze concentration of lime juice and Hernández et al. (2009) presented a study on apple and pear juice concentration in a multiplate freeze concentrator.

In a similar sense, Sánchez et al. (2011) reviewed cryoconcentration in dairy processing, remarking that this technology offers the advantage of minimizing the heat abuse of sensitive milk components such as proteins and flavors, providing an opportunity for producing dairy ingredients with enhanced functional and organoleptic qualities. By freeze concentration, skim milk has been concentrated up to 40 wt% total solids (TS) and whole milk up to 44 wt% TS.

REFERENCES

Aider, M., de Halleux, D. (2009). Cryoconcentration technology in the bio-food industry. Principles and applications. *LWT—Food Science and Technology*, 42:679–685.

Amarante, A., Lanoisellé, J.-L. (2005). Heat transfer coefficients measurement in industrial freezing equipment by using heat flux sensors. *Journal of Food Engineering*, 66:377–386.

ASHRAE (2010). *ASHRAE Handbook. Refrigeration.* Chapter 29. ASHRAE, Inc, U.S.A.

Benkhelifa, H., Haddad Amamou, A., Alvarez, G., Flick, D. (2008). Modelling fluid flow, heat transfer and crystallization in a scraped surface heat exchanger. *Acta Horticulturae*, 802:163–170.

Bongers, P.M.M. (2006). A heat transfer model of a scraped surface heat exchanger for ice cream. 16th European Symposium on Computer Aided Process Engineering, pp. 539–544. Frankfurt am Main, Germany.

Briley, G. (2001). A primer on plate freezers. *ASHRAE Journal*, 43(12):44–45.

Drewett, E.M., Hartel, R.W. (2007). Ice crystallization in a scraped surface freezer. *Journal of Food Engineering*, 78:1060–1066.

Gruda, Z., Postolki, J. (1986). *Tecnología de la congelación de alimentos.* Zaragoza: Ed. Acribia.

Hernández, E., Raventós, M., Auleda, J.M., Ibarz, A. (2009). Concentration of apple and pear juices in a multi-plate freeze concentrator. *Innovative Food Science and Emerging Technologies*, 10:348–355.

Lanoisellé, J.-L., Guyomard, P., Piar, G., Lanoisellé, P., Munoz, Y. (1998). Surgélation de produits alimentaires convoyés par film plastique mince sur sole refroidie. *Revue Générale du Froid*, 88(3):39–45.

Miyawaki, O., Liu, L., Shirai, Y., Sakashita, S., Kagitani, K. (2005). Tubular ice system for scale-up of progressive freeze concentration. *Journal of Food Engineering*, 69:107–113.

Nonthanum, P., Tansakul, A. (2008). Freeze concentration of lime juice. *Maejo International Journal on Science and Technology*, 1:27–37.

Pronk, P., Infante Ferreira, C.A., Rodriguez Pascual, M., Witkamp, G.J. (2005). Maximum temperature difference without ice-scaling in scraped surface crystallizers during eutectic freeze crystallization. 16th International Symposium on Industrial Crystallization, 11–14 September 2005, Dresden, Germany, 6 pp.

Rao, C.S., Hartel, R.W. (2006). Scraped surface heat exchangers. *Critical Reviews in Food Science and Nutrition*, 46:207–219.

Salvadori, V.O., Mascheroni, R.H. (1992). Método aproximado para la predicción de tiempos de proceso durante la congelación asimétrica de bloques de alimentos. *La Alimentación Latinoamericana* 26(193):75–79.

Salvadori, V.O., De Michelis, A., Mascheroni, R.H. (1997). Prediction of freezing times for regular multidimensional foods using a simple formulae. *Lebensmittel-Wissenschaft und-Technologie*, 30:30–35.

Sánchez, J., Ruiz, Y., Hernández, E., Auleda, J.M., Raventos, M. (2009). Review. Freeze concentration in the fruit juices industry. *Food Science and Technology International*, 15(4):303–315.

Sánchez, J., Hernández, E., Auleda, J.M., Raventos, M. (2011). Review: Freeze Concentration Technology Applied to Dairy Products *Food Science and Technology International*: doi 10.1177/1082013210382479.

Sun, K.-H., Pyle, D.L., Fitt, A.D., Please, C.P., Baines, M.J., Hall-Taylor, N. (2004). Numerical study of 2D heat transfer in a scraped surface heat exchanger. *Computers and Fluids*, 33:869–880.

Tocci, A.M., Mascheroni, R.H. (1995). Heat and mass transfer coefficients during the refrigeration, freezing and storage of meats, meat products and analogues. *Journal of Food Engineering*, 26:147–160.

Uno, J., Hayakawa, K.-I. (1979). Nonsymmetric heat conduction in an infinite slab. *Journal of Food Science*, 44:396–403.

14 Chilling and Freezing by Immersion in Water and Aqueous Solutions
(Hydrocooling, Brines, Ice Slurries, and Refrigerated Seawater)

Brian A. Fricke

CONTENTS

14.1 INTRODUCTION

Immersion chilling or freezing involves immersing food items in a refrigerated liquid or spraying food items with a refrigerated liquid. As compared to air chilling or freezing, direct immersion of food items into a refrigerated liquid results in more rapid cooling or freezing due to the considerably higher heat transfer coefficients that can be achieved during immersion chilling or freezing. For example, the core temperature of an 8-kg block of cooked meat will decrease from 74°C to 10°C in approximately 500 minutes during air-blast cooling with an air velocity of 3 m s^{-1}.

The same 8-kg block of cooked meat will cool from 74°C to 10°C in approximately 420 minutes during water-immersion cooling with a water velocity of 0.15 m s⁻¹ (Wang and Sun 2002). Immersion chilling or freezing also results in minimal moisture loss and dehydration as compared to air chilling or freezing. In fact, if the food item is immersed bare, it may gain weight (ASHRAE 2006b; U.S. Department of Agriculture [USDA] 2004). Further advantages of immersion chilling and freezing of food include greater throughput, low initial investment, and low operational costs as compared to other cooling and freezing techniques (Fikiin et al. 2003).

A variety of fluids can be used for immersion chilling or freezing of foods. Chilled water is used to cool fresh fruits and vegetables as well as meat and poultry products. Aqueous solutions of water and salt, such as sodium chloride or calcium chloride, as well as aqueous solutions of water and propylene glycol are used to freeze a wide variety of food items. Ice slurries are used to chill fresh fruits and vegetables, and refrigerated seawater is used aboard fishing vessels to preserve freshly caught seafood.

Immersion chilling or freezing rapidly cools food items since the refrigerated liquid that flows around the food causes the food surface temperature to be essentially equal to that of the refrigerated liquid (Ryall and Lipton 1979). Thus, the resistance to heat transfer at the food surface is negligible, and the rate of internal cooling of the food is limited by the rate of heat transfer from the interior to the surface. This internal cooling rate depends upon the volume of the food item in relation to its surface area as well as the thermal properties of the food item.

14.2   HYDROCOOLING

Hydrocooling typically refers to the process used to remove field heat from freshly harvested fruits and vegetables, whereby the fresh fruits and vegetables are sprayed with chilled water or immersed in an agitated bath of chilled water. Hydrocooling is an effective and economical method of precooling. However, it has a tendency to produce physiological and pathological effects on certain commodities, and therefore, its use is limited or prohibited with these commodities (Bennett 1970). In addition, proper sanitation of the hydrocooling water is necessary to prevent bacterial infection of commodities. Commodities that are often hydrocooled include asparagus, snap beans, carrots, sweet corn, cantaloupes, celery, snow peas, radishes, tart cherries, and peaches. Commodities that are sometimes hydrocooled include cucumbers, peppers, melons, and early-crop potatoes. Apples and citrus fruits are rarely hydrocooled. Hydrocooling is not popular for citrus fruits because of their long marketing season and good postharvest holding ability. In addition, citrus fruits are susceptible to increased peel injury and to decay and loss of quality and vitality after hydrocooling.

In addition to rapid cooling, hydrocooling has the advantage of causing no commodity moisture loss. In fact, hydrocooling may even rehydrate slightly wilted product (USDA 2004). Thus, from a consumer standpoint, the quality of hydrocooled commodities is high, while from the producer's standpoint, the salable weight is high. In contrast, other precooling methods such as vacuum cooling or air cooling may lead to significant commodity moisture loss and wilting, thus reducing product quality and salable weight.

Commodities may be hydrocooled either loose or in packaging. If commodities are to be cooled in packaging, the packaging must allow for adequate water flow within and must tolerate contact with water without losing strength. Plastic or wood containers are well suited for use in hydrocoolers. Corrugated fiberboard containers can be used in hydrocoolers, provided that they are wax-dipped to withstand water contact (USDA 2004).

14.2.1 Types of Hydrocoolers

Hydrocooler designs can generally be divided into two categories, namely, shower-type hydrocoolers and immersion hydrocoolers. In a shower hydrocooler, the commodities pass under a shower of chilled water, as shown in Figure 14.1. The shower is typically achieved by flooding a perforated pan with chilled water. Gravity forces the chilled water to pass through the perforated pan and shower over the commodities. Shower-type hydrocoolers may incorporate conveyors for continuous product flow or may be operated in batch mode. Water flow rates for shower-type hydrocoolers typically range from 6.8 to 13.6 L s^{-1} per square meter of cooling area (Bennett et al. 1965; Boyette et al. 1992; Ryall and Lipton 1979). Immersion hydrocoolers, shown in Figure 14.2, consist of large, shallow tanks that contain agitated, chilled water. Crates or boxes of commodities are loaded onto a conveyor at one end of the tank. The commodities then travel submerged along the length of the tank and are removed at the opposite end. For immersion hydrocooling, a water velocity of 75 to 100 mm s^{-1} is suggested (Bennett 1963; Bennett et al. 1965).

In large packing facilities, flooded ammonia refrigeration systems are often used to chill the hydrocooling water. Cooling coils are placed directly in a tank through which water is rapidly circulated. The refrigerant temperature inside the cooling coils is typically −2°C, producing a chilled water temperature of about 1°C. Because of the high cost of acquiring and operating mechanical refrigeration units, they are typically limited to providing chilled water for medium- to high-volume hydrocooling operations.

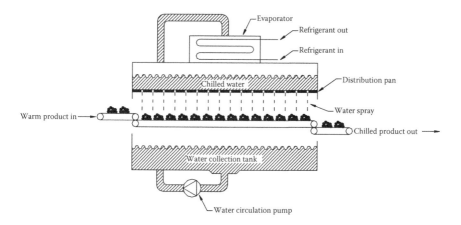

FIGURE 14.1   Schematic of a spray hydrocooler.

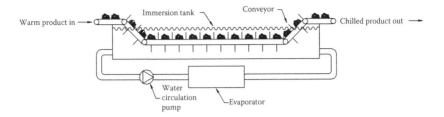

FIGURE 14.2 Schematic of an immersion hydrocooler.

Smaller hydrocooling operations may use crushed ice rather than mechanical refrigeration to produce chilled water. Typically, large blocks of ice are transported from an ice plant to the hydrocooler and then crushed and added to the water reservoir of the hydrocooler. The initial cost of an ice-cooled hydrocooler is much less than that of a hydrocooler utilizing mechanical refrigeration. However, in order for an ice-cooled hydrocooler to be economically viable, a reliable source of ice must be available at a reasonable cost (Boyette et al. 1992).

As an alternative to producing chilled water with mechanical refrigeration or ice, well water can be used to precool commodities. Well water temperature is about 10°C, and it is feasible to use this water for precooling of commodities if the temperature difference between the commodities and the water is about 5°C. However, the well water must not contain chemicals and biological pollutants, which could render the product unsuitable for human consumption (Gast and Flores 1991).

14.2.2 HYDROCOOLER EFFICIENCY

Hydrocooling efficiency is reduced by heat gain to the hydrocooling water from the surrounding air. Other sources of heat that reduce the effectiveness of hydrocoolers include solar loads, radiation from hot surfaces, and conduction from the surroundings. Protecting the hydrocooler from these sources of heat gain will enhance efficiency. Further energy losses occur if a hydrocooler is operated at less than full capacity, if it is operated intermittently, or if more water than necessary is used (Boyette et al. 1992).

In order to increase the energy efficiency of a hydrocooler, the following factors should be considered during design and operation (Boyette et al. 1992):

- Insulate all refrigerated surfaces and protect the hydrocooler from wind and direct sunlight.
- Use plastic strip curtains on both the inlet and outlet of conveyor hydrocoolers to reduce infiltration heat gain.
- Because both intermittent operation and operation at reduced capacity waste energy, operate the hydrocooler at maximum capacity.
- Consider the use of thermal storage in which chilled water or ice is produced and stored during periods of low energy demand. This chilled water or ice is subsequently used along with mechanical refrigeration to chill hydrocooling water during periods of peak energy demand. The use of

thermal storage will reduce the size of the required refrigeration equipment and may result in decreased energy costs.

• Use an appropriately sized water reservoir. Since energy will be wasted when the hydrocooling water is discarded after operation, this waste can be minimized by not using an oversized water reservoir. On the other hand, it may be difficult to maintain a consistent hydrocooling water temperature and flow rate if an undersized water reservoir is used.

14.2.3 HYDROCOOLING WATER TREATMENT

The surface of wet commodities provides an excellent site for diseases to thrive. In addition, since hydrocooling water is recirculated, decay-producing organisms can accumulate in the hydrocooling water, and thus, these organisms can be easily spread to other commodities that are being hydrocooled. Thus, to reduce the spread of disease, hydrocooling water must be treated with mild disinfectants.

Typically, hydrocooling water is treated with chlorine to minimize the levels of decay-producing organisms (USDA 2004). Chlorine in the form of hypochlorous acid from sodium hypochlorite or gaseous chlorine is added to the hydrocooling water, typically at the level of 50 to 100 parts per million. However, chlorination of hydrocooling water provides only a surface treatment of the commodities. Chlorine will not be effective at neutralizing an infection if it has developed below the surface of a commodity.

The chlorine level in the hydrocooling water must be checked at regular intervals to ensure that the proper concentration is maintained. Chlorine is volatile and will disperse into the air. The dispersion rate of chlorine into air increases with increasing temperature, and thus, as the temperature of the hydrocooling water increases, the rate of dispersion of chlorine increases (Boyette et al. 1992). Furthermore, if ice cooling is used, the melting of ice in the hydrocooling water dilutes the chlorine in solution.

The effectiveness of the chlorine in the hydrocooling water strongly depends upon the pH of the hydrocooling water. The pH of hydrocooling water should be maintained at 7.0 in order to achieve the maximum effectiveness from the chlorine (Boyette et al. 1992).

To minimize the accumulation of debris in the hydrocooling water, it may be necessary to prewash the commodities prior to hydrocooling. Nevertheless, hydrocooling water should be replaced daily or more often if necessary. Special care should be taken when disposing of hydrocooling water since this water often contains high concentrations of sediment, pesticides, and other suspended matter. Depending upon the municipality, hydrocooling water may be considered an industrial wastewater, and thus, a hydrocooler owner may be required to obtain a wastewater discharge permit (Boyette et al. 1992). In addition to the daily replacement of hydrocooling water, hydrocooler shower pans and/or debris screens should be cleaned daily, or more often if necessary, to provide maximum efficiency.

14.2.4 VARIATIONS ON HYDROCOOLING CONCEPT

Other concepts that are similar to hydrocooling include hydraircooling as well as chilling and freezing using aqueous solutions. Henry and Bennett (1973) and Henry et al. (1976) described a method of precooling called hydraircooling. In this technique, a combination of chilled water and chilled air is circulated over the commodities. Hydraircooling reduces the volume of water required for cooling as compared to conventional hydrocooling and also reduces the maintenance required to keep the cooling water clean.

Robertson et al. (1976) described a process in which vegetables are frozen by direct contact with an aqueous freezing medium. The aqueous freezing medium consists of a 23% NaCl solution. Freezing times of less than 1 minute were reported for peas, diced carrots, snow peas, and cut green beans, and a cost analysis indicated that freezing with an aqueous freezing medium was competitive with air-blast freezing.

Lucas and Raoult-Wack (1998) conducted a review of immersion chilling and freezing using an aqueous refrigerating medium and noted that immersion chilling and freezing have the advantage of shorter process times, energy savings, and better food quality as compared with air-blast chilling or freezing. They reported that immersion chilling or freezing with an aqueous refrigerating medium can be applied to a broad range of food items including pork, fish, poultry, peppers, beans, tomatoes, peas, and berries. A shortcoming of the immersion chilling or freezing process with aqueous solutions is the absorption of solutes from the aqueous solution by the food item.

14.3   SPRAY CHILLING AND IMMERSION CHILLING OF MEAT

Commercial beef slaughtering operations currently utilize chilled water sprays to chill beef carcasses during the initial hours of postmortem cooling. Spray chilling consists of intermittently spraying cold water on the carcasses. Typically, the carcasses are sprayed for about 1 minute at 10-minute intervals for a period of 3 to 8 hours. Compared to air chilling, spray chilling is faster and reduces evaporative losses, thereby minimizing carcass weight loss (Allen et al. 1987). Reduction in evaporative weight loss of the carcasses has been found to range from 0.5% to 1.5% during the first 24 hours postmortem (Hippe et al. 1991). Allen et al. (1987) found that spray-chilled beef sides lost 0.3% of their weight compared with 1.5% for nonspray-chilled sides. A properly designed spray-chilling operation can nearly eliminate carcass shrinkage through careful consideration of factors such as carcass spacing, length of the spray cycle, and carcass fatness (ASHRAE 2006a).

Spray chilling has been found to have little detrimental effect on the quality of the meat. Muscle color is not affected by spray chilling; however, fat color can be lighter in spray-chilled carcasses compared with nonspray-chilled carcasses. It may seem reasonable to believe that spray chilling could remove bacteria from carcass surfaces; however, some studies with beef and pork indicate that spray chilling of carcasses had no effect on bacterial growth, as compared with conventionally chilled carcasses (Hamby et al. 1987; Greer and Dilts 1988). It has been shown that the use of a sanitizer, such as chlorine (200 ppm), or an organic acid (1% to 3%) can significantly reduce carcass bacterial counts.

Poultry products are typically cooled by immersion in chilled water, since immersion chilling is more rapid than air chilling and dehydration is reduced. In fact, immersion chilling of poultry in water can result in a net absorption of 4% to 12% water (ASHRAE 2006b). Poultry can be chilled using continuous drag chillers, in which suspended carcasses are pulled through troughs containing agitated cool water and ice slush. Immersion of poultry carcasses in a common bath ensures that pathogens are distributed to all carcasses. However, immersion chilling of poultry carcasses can reduce contamination if antimicrobial chemicals are maintained at appropriate concentrations. Objections to the resulting mass gain from the external water, as well as the high cost of disposing the wastewater in an environmentally sound manner, have encouraged some operators to consider returning to air chillers.

Due to the popularity of immersion chilling and freezing of poultry in both the United States and the United Kingdom, Redmond et al. (2001) investigated the feasibility of using the immersion chilling process with other types of carcasses, specifically, whole lamb carcasses. Their objective was to compare the effects of immersion chilling and conventional air chilling on meat tenderness and evaporative weight loss. Immersion chilling was used in an attempt to achieve increased rates of chilling of whole carcasses. Lamb carcasses wrapped in plastic were immersed in a 50% polypropylene glycol solution. This solution was chosen since, although it is relatively viscous at low temperatures, it is nontoxic, noncorrosive, colorless, odorless, and nearly tasteless (Frazerhurst et al. 1972). Immersion chilling led to reductions in weight loss when compared with air chilling at the same temperature, mainly due to the fact that the immersion-chilled carcasses were wrapped in plastic. In addition, Redmond et al. (2001) found that immersion chilling had no effect on tenderness as compared with air chilling. In a commercial trial, it was found that customers had no reservations about accepting immersion-chilled lamb.

14.4 IMMERSION FREEZING

While chilled water is an excellent medium for immersion cooling of food products, aqueous solutions that have freezing points substantially lower than 0°C must be used to freeze food products by immersion. Aqueous solutions of water and salt, such as sodium chloride (NaCl) or calcium chloride ($CaCl_2$), as well as aqueous solutions of water and propylene glycol can be used to freeze a variety of food items. In addition, soluble carbohydrates such as sucrose, glucose, fructose, and other monosaccharides and disaccharides, along with ethanol, salts, and glycerol, have been used as immersion media when freezing fruits (Fikiin et al. 2003). For example, by freezing fruits in sugar-ethanol-based aqueous solutions, new dessert products can be developed that have beneficial color, flavor, and texture due to the enzyme-inhibiting action of the sugar (Fikiin et al. 2003).

Provided that the aqueous solution is nontoxic, the food products may be immersed bare in the aqueous solution. However, a shortcoming of the immersion chilling or freezing process with aqueous solutions is the absorption of solutes from the aqueous solution by the food item. Absorption of solutes by the food items may affect their quality and taste. In addition, the food items can dilute the aqueous solution, which can change its concentration as well as modify the freezing process parameters (Barbosa-Cánovas et al. 2005). Further side effects, including spoilage of the food items and cross-contamination of pathogenic microorganisms, may occur if the food items are immersed bare in the freezing medium (Berry et al. 1998). Thus, in order to avoid the detrimental impact that product contact with the aqueous solution may cause, most food items must be packaged in flexible plastic films or bags prior to immersion (George 1993).

During immersion freezing, food items experience rapid temperature reduction through the direct heat exchange that results from immersing the food items directly in an aqueous freezing medium (Hung and Kim 1996). Verboven et al. (2003) have found that for immersion freezing of spheres in an aqueous solution of 30% ethanol and 20% glucose with a temperature range from −20°C to 0°C, surface heat transfer coefficients can vary from 154 to 1548 W m^{-2} K^{-1}. With such high surface heat transfer coefficients, immersion freezing is effective in freezing food items, particularly for irregularly shaped foods.

Immersion freezing systems have been commonly used for shell freezing of large food items in an effort to reduce product dehydration by quickly freezing the outer layer of the food (Barbosa-Cánovas et al. 2005). Both crab and shrimp are commonly frozen by immersion in a brine solution. Whole cooked and eviscerated crabs in the shell can be dipped into a circulating brine at −15°C to −18°C for 45 minutes and then placed into fresh cold water to remove excess brine and provide an ice glaze (Hui et al. 2004). For immersion freezing of shrimp, a brine solution consisting of a mixture of salt and sugar is generally used (Hui et al. 2004). More recently, liquids such as glycerol, glycol, and calcium chloride solution have been used. Another common application of immersion freezing includes the freezing of turkeys and chickens wrapped in plastic bags by immersion in a propylene glycol solution (Fenton 1999; Smith 2002). During immersion freezing of tightly packaged poultry, surface freezing occurs, creating a desirable white surface due to ice crystal formation. Following surface freezing by immersion, final freezing of the poultry carcasses can be achieved by placing the carcasses in blast freezers (Smith 2002).

14.4.1 IMMERSION FREEZER CONFIGURATIONS

Liquid-immersion freezers are configured to operate in either batch mode or continuous mode (Sun 2006). For an immersion freezer designed to operate in batch mode, the food items are placed in baskets, which are in turn immersed in a tank containing the freezing medium. When the food items are frozen, the baskets are removed from the tank. To reduce the labor associated with loading and unloading batch-immersion freezers, continuous-immersion freezers may be used, in which a conveyor system is used to move the food items through the freezing medium. Current conveyor designs include auger systems, shown in Figure 14.3, and solid or mesh belts that incorporate baffles to ensure that all of the food items are continuously moved through the freezer, as shown in Figure 14.4. The choice of conveyor system depends upon the buoyancy of the food items in the freezing medium. For example, food items that float cannot be conveyed on a belt system where the belt is placed at the bottom of the immersion tank. Since it is often undesirable for the food items to remain wet following removal from a liquid-immersion freezer, most commercial immersion freezing systems either allow the cooling liquid to drain off or use fans to blow the liquid off the product (Sun 2006).

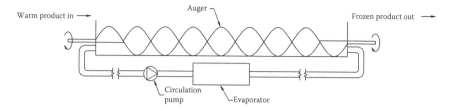

FIGURE 14.3 Schematic of an immersion Auger freezer.

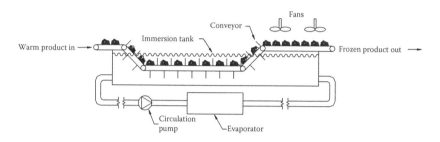

FIGURE 14.4 Schematic of an immersion freezer.

In a typical immersion freezing process, the liquid freezing medium is cooled by a central refrigeration plant and then pumped to the immersion freezer. After absorbing heat from the food, the liquid freezing medium returns to the refrigeration plant to be cooled. The resulting refrigeration load can be determined from the flow rate and temperature change of the liquid freezing medium.

14.4.2 PROPERTIES OF AQUEOUS SOLUTIONS

The design of immersion freezing equipment and the associated refrigeration system requires knowledge of the physical properties of the liquid freezing medium, including its specific heat, viscosity, initial freezing temperature, and thermal conductivity.

The refrigerating capacity, $\dot{Q}$, provided by the liquid freezing medium is a function of its specific heat and mass flow rate:

$$\dot{Q} = \dot{m}c_p\left(T_o - T_i\right) \tag{14.1}$$

where $\dot{m}$ is the mass flow rate of the freezing media, c_p is the specific heat of the freezing media, T_o is the temperature of the freezing media exiting the immersion freezer, and T_i is the temperature of the freezing media entering the immersion freezer. From Equation 14.1, it can be seen that in order to maximize the refrigerating capacity, a liquid freezing medium with a high specific heat is desirable. A high specific heat indicates that the fluid can remove a relatively large quantity of heat from the food to be frozen. In addition, Equation 14.1 shows that for a given refrigerating capacity, a fluid with a high specific heat requires a lower flow rate than a fluid with a low specific heat. Thus, lower pumping power is required for fluids with high specific heat. The mass flow rate and specific heat of the liquid freezing medium are also required to properly size the evaporator of the refrigeration system, which is used to cool the liquid freezing medium.

The pumping power required to circulate the liquid freezing medium also depends upon the viscosity of the liquid. Thus, to reduce the pumping power, a liquid freezing medium with a low viscosity should be used. In addition, the viscosity of the liquid freezing medium influences whether the flow in the system is laminar or turbulent. Turbulent flow is preferred in immersion freezers and evaporators since this type of flow results in higher heat transfer coefficients (Melinder 2000).

In addition, the lowest possible temperature that can be obtained in an immersion freezer is specified by the initial freezing temperature of the liquid freezing medium. The initial freezing-point temperature of the liquid freezing medium should be somewhat lower than the operating temperature of the freezing equipment. The liquid freezing medium used should provide an adequate ability to freeze the food product, but not more than is needed since a higher concentration of the freezing-point depressant will result in less water in the solution and thereby degrade the thermophysical properties of the solution.

Finally, the liquid freezing medium should not affect the quality of the end product and should be environmentally friendly (Fikiin et al. 2003). Thus, an ideal liquid freezing medium would possess a low viscosity, a high specific heat, and a low initial freezing temperature while being safe and environmentally friendly.

Water is an excellent fluid due to its relatively low viscosity and relatively high specific heat. Thus, water can absorb a relatively large quantity of heat, and the required pumping power will be relatively low. Water is also safe and environmentally friendly. However, the freezing temperature of water is relatively high, and thus, it is not suitable for freezing food products.

Aqueous solutions consisting of water with salts or glycols can be used to achieve low initial freezing temperatures. For example, the initial freezing temperature of calcium chloride brines can range from nearly −6.5°C for a 10% calcium chloride brine solution to less than −40°C for a 28% calcium chloride brine. Also, the initial freezing temperature of ethylene glycol solutions can range from nearly −4°C for a 10% ethylene glycol solution to less than −45°C for a 70% ethylene glycol solution. However, the specific heats of the aqueous freezing medium are generally less than that of water, and the viscosities of these aqueous freezing media are considerably greater than that of water. Thus, when compared to water, aqueous freezing media require greater flow rates to obtain the same heat transfer rate, and greater pumping power is necessary to achieve the desired circulation.

The selection of an appropriate liquid freezing medium also depends upon the physical and chemical interaction of the medium with the various components of the refrigeration system, including pumps, piping, valves, gaskets, and seals. In the presence of air or oxygen, aqueous solutions containing salts or glycols will corrode the metallic components of the system. If preventive measures are not adopted, leaks will eventually develop in the refrigeration system, and if brine enters the refrigeration system, there will have to be costly and elaborate flushing, cleaning, and drying of the system. Thus, to combat corrosion, inhibiters must be added to the aqueous freezing solutions (Ananthanarayanan 2005).

For example, both calcium chloride and sodium chloride brines must be treated to inhibit corrosion and to control the production of deposits. The treatment for calcium chloride typically involves the routine testing for pH to maintain the appropriate alkalinity as well as the addition of corrosion inhibitors. The ideal pH range for sodium and calcium chloride brines is 7.5 to 8.5; a brine that is slightly alkaline is considered safer than one slightly acidic. To correct an acidic condition, caustic soda (an alkali) is dissolved in warm water and added to the brine, while to correct an alkaline condition, acetic, chromic, or hydrochloric acid is added to the brine (Ananthanarayanan 2005). Inhibiters such as sodium chromate or sodium dichromate have been found to be effective in reducing corrosion when using calcium chloride and sodium chloride brines.

Generally, ethylene glycol and propylene glycol solutions are inhibited by the manufacturer using proprietary chemicals for corrosion control. The inhibitor's basic function is to form a protective surface barrier between the metal surface and the solution (Fenton 1999).

To prolong the life of immersion food-freezing equipment, it is very important to monitor the concentration of the inhibitor, and replenish the inhibitor if necessary, to ensure that corrosion will be minimized (Ananthanarayanan 2005). Typically, pH measurements of the solution should be regularly recorded, and periodic chemical analysis of the solution should be made.

14.5 ICE SLURRIES

Ice–water mixtures have been used throughout history to cool food, and presently, ice slurries are being used to commercially cool or freeze foods.

The most basic commercial ice slurry consists of small ice crystals, typically 1 mm in diameter or smaller, suspended in water, resulting in a mixture with a temperature of 0°C. The amount of ice in the mixture is typically between 5% and 40%. In order to achieve ice slurry temperatures of less than 0°C, ice crystals can also be suspended in seawater or aqueous solutions. An ice slurry consisting of seawater will typically have a temperature of about −2°C. Fikiin et al. (2003) have produced ice slurries based on sugar-ethanol aqueous solutions with temperatures of −25°C. Other common ice slurry mixtures are based on salts (sodium chloride or calcium chloride), glycols (ethylene glycol or propylene glycol), alcohols (isobutyl or ethanol), and sugars (sucrose or glucose).

Very high heat transfer coefficients can be obtained by directly immersing food items in an ice slurry because the ice slurry can absorb both sensible heat and latent heat resulting from the melting of the ice. Thus, an ice slurry has even greater heat absorption capability than single-phase aqueous solutions. For example, Torres-de Maria et al. (2005) reported that surface heat transfer coefficients for small cylindrical products immersed in an ice slurry vary between 200 and 450 W m^{-2} K^{-1} for an ice slurry at rest and between 1600 and 2900 W m^{-2} K^{-1} for an agitated ice slurry.

In addition, since an ice slurry has more cooling capacity than an aqueous solution, pumping volumes and line sizes are minimized (ASHRAE 2006e).

The high heat transfer coefficients obtained by direct immersion in an ice slurry result in short cooling and freezing times. Because of this, naturally occurring ice slurries, such as snow–water mixtures or crushed ice–water mixtures, have been used throughout history to cool food (Egolf 2004). When using a seawater-based ice slurry, bulk chilling of herring to −1.8°C can be achieved within 30 minutes, while capelin can be chilled from 11°C to 0°C in 6 to 8 minutes (Venugopal 2006). In addition, Fikiin et al. (2003) reported that strawberries, apricots, and plums can be frozen from 25°C to an average final temperature of −18°C in 8 to 9 minutes using an ice slurry based on a sugar–ethanol aqueous solution.

In the fresh-produce industry, body icing of fresh produce can be accomplished by pumping an ice slurry through the containers of the product. The ice, which is suspended in the water, is carried throughout all the void spaces in the containers, and drain holes in the containers allow the water to be removed, leaving the ice behind. Thus, the product is completely surrounded with ice, thereby providing uniform cooling of the produce and eliminating any hot spots within the containers (ASHRAE 2006e).

Figure 14.5 shows a schematic of an ice slurry refrigeration system. The ice is produced in the evaporator of the refrigeration system, which is typically a scraped-surface heat exchanger. The evaporator is a double-walled cylinder in which refrigerant flows through the wall cavity, and ice crystals are formed on the inner wall of the evaporator. The ice crystals are then mechanically removed from the interior wall of the evaporator with scrapping devices, such as rotating knives, rotating brushes, or augers. The ice crystals then mix with the slurry solution in the ice slurry mixing tank. Pumps discharge the ice slurry through pipelines to the point of use. The pumps are of the centrifugal type, modified for pumping slurry (ASHRAE 2006e). The slurry should be kept moving at all times to minimize the possibility of ice blockage in the lines.

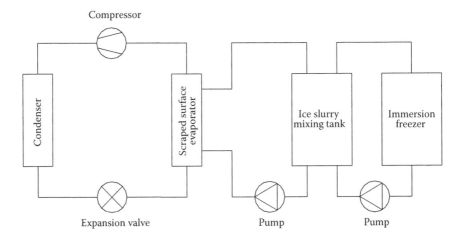

FIGURE 14.5   Schematic of an ice slurry refrigeration system.

14.6 REFRIGERATED SEAWATER

To preserve the quality of freshly caught fish and seafood during storage aboard fishing vessels, the fish and seafood are stored either in ice or refrigerated seawater. When using ice to preserve seafood aboard fishing vessels, it is necessary to divide the fishing vessel's hold into bins. Ice is then stored in alternate bins so that it is readily available for packing around the fish as the fish are loaded into adjacent bins. Each bin is generally divided horizontally by inserting boards across the bin while the bin is being filled with ice and fish. The horizontal boards support the weight of the ice and fish, thus ensuring that the fish at the bottom of the bin will not be crushed by the weight of the ice from above. The approximate amount of ice required to maintain proper storage conditions is 1 Mg of ice for each 2 Mg of fish during the summer and 1 Mg of ice for each 3 Mg of fish during the winter, based on a voyage of about 8 days (ASHRAE 2006c).

An alternative chilling method to icing is refrigerated seawater. In this method, seawater is continuously pumped through external chillers and then to fish storage tanks, where the chilled seawater passes around the fish (ASHRAE 2006c). Storing fresh seafood in refrigerated seawater has several advantages over storage in ice. The catch can be cooled more rapidly and with less effort using refrigerated seawater. Also, the fish are less likely to be crushed or lose weight when cooled by refrigerated seawater. On the other hand, it has been noted that some species of fish keep better in ice than in refrigerated seawater. Thus, refrigerated seawater is typically used for short-term storage of specific species of fish that are caught in large quantities within a short time, such as herring, mackerel, sprats, and blue whiting (Kelman 2001).

More limited applications of refrigerated seawater include the storage of halibut and shrimp aboard a vessel. Groundfish and shrimp can be stored in refrigerated seawater for short periods, typically two to four days; however, longer periods of storage can cause excessive salt uptake, accelerated rancidity, poorer texture, and increased bacterial spoilage (ASHRAE 2006c). These problems can be partially overcome by introducing carbon dioxide (CO_2) gas into the refrigerated seawater, which can increase the storage life of some species of fish by about one week (ASHRAE 2006c).

Two methods for cooling seawater include mechanical refrigeration or the addition of ice. Many times, the term *chilled seawater* (CSW) is used to refer to seawater that has been cooled by the addition of ice, while the term *refrigerated seawater* (RSW) is used to refer to seawater that has been cooled by mechanical refrigeration (Kelman 2001).

Figure 14.6 shows a schematic of a refrigerated seawater system. The seawater is filtered and pumped to the evaporator of the refrigeration system for chilling. The evaporator is typically either a shell and tube type or a plate type and constructed of either a copper-nickel alloy or titanium to minimize corrosion. The refrigerated seawater then flows to the fish holding tanks to chill the fish.

During operation of the refrigerated seawater system, one of the fish storage tanks is typically filled with seawater soon after leaving harbor, and the water is cooled by the refrigeration system on the outward passage to bring the temperature down to approximately 0°C before the first catch of fish comes aboard. Then, the other tanks are one-quarter to one-third filled with precooled seawater from the precooled tank

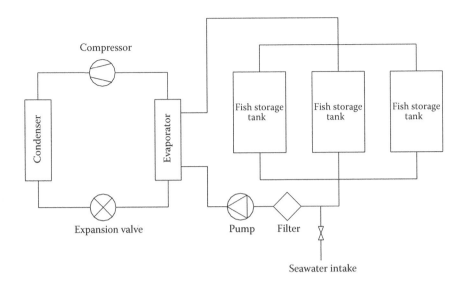

FIGURE 14.6   Schematic of a refrigerated seawater system.

before the fish are placed in the tanks. The water in each tank of fish is then circulated to the refrigeration system and returned to the tank, thus promoting rapid and uniform cooling of the fish. The water should be circulated continuously until the vessel is unloaded to prevent temperature stratification and the formation of pockets of warm water within the fish holding tanks (Kelman 2001).

14.7   CONCLUSION

Immersion chilling or freezing of foods has wide application, from hydrocooling of freshly harvested fruits and vegetables to chilling and freezing of meat products and packaged, processed foods. In addition, a wide variety of chilling and freezing media can be used to chill or freeze foods by immersion, including water, aqueous solutions of salts or glycols, ice slurries, and seawater.

As compared to air chilling or freezing, direct immersion of food items into a refrigerated liquid results in more rapid cooling or freezing due to the considerably higher heat transfer coefficients that can be achieved during immersion chilling or freezing. Immersion chilling or freezing also results in minimal moisture loss and dehydration as compared to air chilling or freezing. Further advantages of immersion chilling and freezing of food include greater throughput, low initial investment, and low operational costs as compared to other cooling and freezing techniques.

REFERENCES

Allen, D.M., M.C. Hunt, A. Luchiari Filho, R.J. Danler and S.J. Goll. 1987. Effects of spray chilling and carcass spacing on beef carcass cooler shrink and grade factors. *Journal of Animal Science* 64:165–170.

Ananthanarayanan, P.N. 2005. *Basic refrigeration and air conditioning.* New Delhi: Tata McGraw-Hill.

ASHRAE. 2006a. Chapter 17, Meat products. In *2006 ASHRAE Handbook Refrigeration.* Atlanta, GA: American Society of Heating, Refrigerating and Air-Conditioning Engineers.

ASHRAE. 2006b. Chapter 18, Poultry products. In *2006 ASHRAE Handbook Refrigeration.* Atlanta, GA: American Society of Heating, Refrigerating and Air-Conditioning Engineers.

ASHRAE. 2006c. Chapter 19, Fishery products. In *2006 ASHRAE Handbook Refrigeration.* Atlanta, GA: American Society of Heating, Refrigerating and Air-Conditioning Engineers.

ASHRAE. 2006d. Chapter 31, Marine refrigeration. In *2006 ASHRAE Handbook Refrigeration.* Atlanta, GA: American Society of Heating, Refrigerating and Air-Conditioning Engineers.

ASHRAE. 2006e. Chapter 34, Ice manufacture. In *2006 ASHRAE Handbook Refrigeration.* Atlanta, GA: American Society of Heating, Refrigerating and Air-Conditioning Engineers.

Barbosa-Cánovas, G.V., B. Altunakar and D.J. Mejía-Lorío. 2005. Freezing of fruits and vegetables: An agribusiness alternative for rural and semi-rural areas. In *FAO Agricultural Services Bulletin 158.* Rome: Food and Agriculture Organization of the United Nations.

Bennett, A.H. 1963. *Thermal Characteristics of Peaches as Related to Hydrocooling.* Technical Bulletin No. 1292. Washington, D.C.: Agricultural Marketing Service, United States Department of Agriculture.

Bennett, A.H. 1970. Principles and equipment for precooling fruits and vegetables. In *Symposium on Precooling of Fruits and Vegetables.* San Francisco, CA: American Society of Heating, Refrigerating and Air-Conditioning Engineers.

Bennett, A.H., R.E. Smith and J.C. Fortson. 1965. *Hydrocooling Peaches: A Practical Guide for Determining Cooling Requirements and Cooling Times.* Agricultural Information Bulletin No. 293. Washington, D.C.: Agricultural Research Service, United States Department of Agriculture.

Berry, E.D., W.J. Dorsa, G.R. Siragusa and M. Koohmaraie. 1998. Bacterial cross-contamination of meat during liquid nitrogen immersion freezing. *Journal of Food Protection* 61(9):1103–1108.

Boyette, M.D., E.A. Estes and A.R. Rubin. 1992. Hydrocooling. In *Postharvest Technology Series.* Raleigh, NC: North Carolina Cooperative Extension Service.

Egolf, P.W. 2004. Ice slurry: A promising technology. In *Technical Note on Refrigerating Technologies.* Paris: International Institute of Refrigeration.

Fenton, D.L. 1999. *Fundamentals of Refrigeration.* Atlanta, GA: American Society of Heating, Refrigerating and Air-Conditioning Engineers.

Fikiin, K., O. Tsvetkov, Y. Laptev, A. Fikiin and V. Kolodyaznaya. 2003. Thermophysical and engineering issues of the immersion freezing of fruits in ice slurries based on sugar-ethanol aqueous solution. *EcoLibrium* (August):10–15.

Frazerhurst, L.F., D.P. Haughey and L.G. Wyborn. 1972. Prototype development of an immersion freezer for fancy meats in monopropylene glycol. *Meat Industry Research Institute of New Zealand* 259:1–5.

Gast, K.L.B. and R.A. Flores. 1991. Precooling produce: Fruits and vegetables. In *Postharvest Management of Commercial Horticultural Crops.* Manhattan, KS: Kansas State University Cooperative Extension Service.

George, R.M. 1993. Freezing processes used in the food industry. *Trends in Food Science and Technology* 4(5):134–138.

Greer, G.G. and B.D. Dilts. 1988. Bacteriology and retail case life of spray-chilled pork. *Canadian Institute of Food Science and Technology Journal* 21:295–299.

Hamby, P.L., J.W. Savell, G.R. Acuff, C. Vanderzant and H.R. Cross. 1987. Spray-chilling and carcass decontamination systems using lactic and acetic acid. *Meat Science* 21:1–14.

Henry, F.E. and A.H. Bennett. 1973. "Hydroaircooling" vegetable products in unit loads. *Transactions of the ASAE* 16(4):731–733.

Henry, F.E., A.H. Bennett and R.H. Segall. 1976. Hydroaircooling: A new concept for precooling pallet loads of vegetables. *ASHRAE Transactions* 82(2):541–547.

Hippe, C.L., R.A. Field, B. Ray and W.C. Russell. 1991. Effect of spray-chilling on quality of beef from lean and fatter carcasses. *Journal of Animal Science* 69:178–183.

Hui, Y.H., P. Cornillon, I.G. Legaretta, M.H. Lim, K.D. Murrell and W.-K. Nip, eds. 2004. *Handbook of Frozen Foods*. Boca Raton, FL: CRC Press.

Hung, Y.C. and N.K. Kim. 1996. Fundamental aspects of freeze-cracking. *Food Technology* 50(12):59–61.

Kelman, J.H. 2001. Stowage of fish in chilled sea water. In *Torry Advisory Notes*. Rome: Food and Agriculture Organization of the United Nations.

Lucas, T. and A.L. Raoult-Wack. 1998. Immersion chilling and freezing in aqueous refrigerating media: Review and future trends. *International Journal of Refrigeration* 21(6):419–429.

Melinder, Å. 2000. Update on secondary refrigerants for indirect systems. IEA Annex 26 Workshop, Advanced Supermarket Refrigeration/Heat Recovery Systems, 2–3 October 2000, at Stockholm.

Redmond, G., B. McGeehin, M. Henchion, J.J. Sheridan, D.J. Troy, C. Cowan and F. Butler. 2001. Commercial systems for ultra-rapid chilling of lamb. Dublin: The Irish Agriculture and Food Development Authority (Teagasc).

Robertson, G.H., J.C. Cipolletti, D.F. Farkas and G.E. Secor. 1976. Methodology for direct contact freezing of vegetables in aqueous freezing media. *Journal of Food Science* 41(4):845–851.

Ryall, A.L. and W.J. Lipton. 1979. *Handling, Transportation and Storage of Fruits and Vegetables*, 2nd ed., Vol. 1. Westport, CT: AVI Publishing Company.

Smith, P.G. 2002. *Introduction to Food Process Engineering*. Berlin: Springer.

Sun, D.-W., ed. 2006. *Handbook of Frozen Food Processing and Packaging*. Boca Raton, FL.

Torres-De Maria, G., J. Abril and A. Casp. 2005. Surface heat transfer coefficients for refrigeration and freezing of foods immersed in an ice slurry. *International Journal of Refrigeration* 28(7):1040–1047.

U.S. Department of Agriculture. 2004. *The Commercial Storage of Fruits, Vegetables, and Florist and Nursery Stocks*. Washington, DC: Agricultural Research Service, U.S. Department of Agriculture.

Venugopal, V. 2006. *Seafood Processing: Adding Value through Quick Freezing, Retortable Packaging, and Cook-Chilling*. Boca Raton, FL: CRC Press.

Verboven, P., N. Scheerlinck and B.M. Nicolaï. 2003. Surface heat transfer coefficients to stationary spherical particles in an experimental unit for hydrofluidisation freezing of individual foods. *International Journal of Refrigeration* 26(3):328–336.

Wang, L. and D.-W. Sun. 2002. Evaluation of performance of slow air, air blast and water immersion cooling methods in the cooked meat industry by the finite element method. *Journal of Food Engineering* 51(4):329–340.

15 Special Precooling Techniques

Daniela F. Olivera and Sonia Z. Viña

CONTENTS

15.1 VACUUM COOLING TECHNOLOGY

15.1.1 INTRODUCTION

The overall objective of food refrigeration is to increase shelf life and thus increase the chances of preservation. It is also possible to define another set of specific objectives, which are characteristic of each different type of food. For foods that possess a definite and organized structure, such as plant and animal tissues, even more specific objectives can be defined (Casp and Abril 2003).

Many raw materials are cooled immediately after slaughter (beef, pork, and lamb meat) or harvest (fruits, vegetables) in order to slow down microbial growth and prevent metabolic processes that lead to deterioration.

Biochemical processes and structural changes that occur in muscle during the first 24 h *postmortem* play a great role in the ultimate quality and palatability of meat, and they are influenced by the chilling processes that are applied to carcasses after slaughter. When animals are slaughtered, muscles lose their blood supply and undergo a series of changes that will turn them into meat. The pH will start to drop from a value slightly higher than 7 to a final value (5.4–5.5) that regulates various meat characteristics (color, texture, juiciness, flavor, and microbial growth; Foegeding et al. 1996). When temperature diminishes, reaction rates are lowered, and the time required for the full range of *postmortem* changes to occur is increased.

If the pH drops very quickly due to rapid glycolysis while the carcass is still warm, denaturation of proteins does take place, and the meat loses its water-holding capacity, resulting in the so-called *PSE* (pale, soft and exudative) meat. This phenomenon is relevant in pork and poultry meat such as turkey (Barbut 1998; Solomon et al. 1998; Woelfel et al. 2002) and can be avoided by rapid cooling so as to reach low pH values only when the temperature is low enough that there is no denaturation of proteins (Van Laack et al. 2000).

On the other hand, when pre-rigor beef muscle is chilled, it undergoes a slow contracture from the coupling of mechanical and chemical events within the myofibrils. Cold-shortening occurs through a 30- to 40-fold increase in the concentration of ionic calcium in the myofibrillar region as the temperature of the pre-rigor muscle is reduced from 15°C to 0°C (Davey and Gilbert 1974; Thompson et al. 2006). Beef and lamb are particularly prone to this induced defect; toughness development in these species can be controlled, however, by cooling-rate adjustment during rigor onset (Marsh 1977).

These examples show that cooling processes must be strictly controlled since they influence many meat quality characteristics.

Chilling systems for fish have been in operation for well over a century (James 2001). Fresh fish and shellfish are highly perishable products due to their biological composition. Under normal refrigerated storage conditions, the shelf life of these products is limited by enzymatic and microbiological spoilage. However, with increasing consumer demands for fresh products with extended shelf life and increasing energy costs associated with freezing and frozen storage, the fish-processing industry is actively seeking alternative methods of shelf life preservation and marketability of fresh, refrigerated fish and, at the same time, economizing on energy costs (Ashie et al. 1996).

Fruits and vegetables (F&V) are living organisms that must be maintained as such during postharvest storage. The refrigeration of these products allows reducing dramatically the respiratory rate, weight loss caused by transpiration, ethylene production, and development of microorganisms (Casp and Abril 2003).

Immediate postharvest chilling of vegetables, salad crops, and fruits is a very common practice. As the maintenance of market quality is of vital importance to the success of the horticultural industry, it is necessary not only to cool the harvested product but also to cool it as quickly as possible. The process of precooling consists of the removal of "field heat" (respiration heat, sensible heat), which arrests the deteriorative and senescence processes so as to maintain a high level of quality that ensures customer satisfaction (Brosnan and Sun 2001). Precooling is applied before the products are transported to market or placed in a cold store. The temperature of the vegetables must be reduced quickly, in a few minutes or a few hours, so that they remain fresh. Thus, on-farm cooling facilities are a valuable asset for any produce operation. A grower who can cool and store produce has greater market flexibility because the need to market immediately after harvest is eliminated. The challenge, especially for small-scale producers, is the setup cost (Bachmann and Earles 2000).

It is well known that any subsequent handling operation in the food chain will increase the temperature of raw materials and processed products. Therefore, a secondary chilling process is often required for cooling products at temperatures close to the optimal ones.

According to James (2001), the need for rapid chilling systems lies in many noticeable advantages concerning product quality and processing efficiency (Table 15.1).

To capitalize these multiple advantages, the industry continually demands the development of rapid cooling systems and technologies that enable their implementation. Likewise, current trends are oriented to reduce to a minimum the time between obtaining raw materials and final product elaboration, distribution, and sale.

According to Sun and Wang (2001), vacuum cooling (VC) is a rapid cooling method for porous and moist foods that fulfills special cooling requirements. VC has been used as an effective precooling method for certain types of vegetables and fruits. This technique has also been applied to processing procedures for some foods such as liquid and baked foods to shorten the cooling time and thus to improve the general processing efficiency.

TABLE 15.1

Reasons Supporting the Implementation of Rapid Chilling Systems in Foods

Benefits about product quality	Reduced weight loss
	Better texture
	Improved appearance
	Better handling response
	Retention of nutritional quality
	Controlled microbial load
	Improved taste, reduction of off-flavors production
	Longer shelf life
	Distribution at near optimum temperature
Benefits about processing performance	Higher efficiency
	Lower costs
	Reduced energy input
	Shorter processing time
	Less space requirement
	Less variability
	Continuous processing
	Increased automation
	Increased flexibility

Source: Adapted from James, S. J., Rapid chilling of food—A wish or a fact? Bulletin 2001.6 IIR, 2001.

Sun and Wang (2001) have pointed out that the advantages of VC include shorter processing time, uniform cooling, reduced freezing damage, and sanitary and precise temperature control. Speed and efficiency are the main features of VC, which are unmatched by any other cooling alternative, especially when products are cooled in boxes or pallets.

Likewise, development in the refrigeration industry requires the exhaustive research of different cooling methods to meet the different needs of each product and process, and in that sense, the opportunities related to VC are remarkable.

15.1.2 PRINCIPLE

VC is a well-established technique for rapid cooling processing, which has been referred as one of the most efficient methods available (Sun and Hu 2003). VC was first introduced on a commercial scale in Salinas, CA, in 1948. The plant was used to cool iceberg lettuce, and from then on, VC has been traditionally applied in precooling leafy vegetables.

Zheng and Sun (2004) have pointed out that in the past decade, VC has started to be applied to other sectors of the food industry, for example, meat industry, bakery, fishery, and viscous food processing. In these areas, VC technology has provided a way to shorten processing time and improve product quality.

Liquid evaporation is the most widespread cooling source in the refrigeration industry. The characteristics of the liquid evaporation phenomenon are that high-kinetic energy molecules escape from the surface of liquid. Thus, whenever any portion of liquid evaporates, an amount of heat equal to the latent heat of evaporation is absorbed either from the mass of the liquid or from the surrounding objects, reducing their temperature as the mass supplies the latent heat of evaporation to the evaporating portion (Sun and Wang 2001).

VC is a specific application of evaporative cooling (Thompson et al. 1987). The absolute pressure of the atmosphere surrounding the product is reduced, which results in lowering the boiling temperature of water in the product. VC consists of placing products in a closed precinct in which the pressure is reduced to a value low enough for part of its water constitution to vaporize. Thus, the same product provides the vaporization heat necessary to the change state and to achieve a reduction in temperature.

The temperature at which a fluid changes from the liquid to the vapor phase or vice versa is called the saturation temperature. If the temperature of a liquid rises above its saturation point, part of the liquid will evaporate until reaching a new equilibrium state (Sun and Wang 2001).

The boiling temperature of water is a function of the pressure under which the process develops; the lower the pressure, the lower the boiling temperature (Figure 15.1). Thus, the diminution of pressure makes the product water constitution begin its vaporization at room temperature while the product temperature decreases to achieve the conservation one. In this case, the heat dissipation occurs by mass transfer, under a varying regime, from the product to the cooling medium (Casp and Abril 2003). The transfer is originated by the reduction in pressure that occurs in

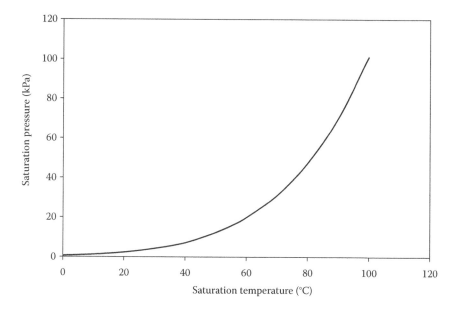

FIGURE 15.1 Relationship between saturation pressure and saturation temperature of water.

the environment surrounding the food until it reaches the saturation vapor pressure corresponding to the ambient temperature.

The saturation pressure for water at 100°C is 101.3 kPa. At 0°C, the saturation pressure is 0.610 kPa (Figure 15.1). Commercial vacuum coolers normally operate in this range. Although, for example, the cooling rate of lettuce could be increased without danger of freezing by reducing the pressure to 0.517 kPa corresponding to a saturation temperature of –2°C, most operators do not reduce the pressure below the freezing temperature of water because of the extra work involved and the freezing damage potential (American Society of Heating, Refrigerating and Air-Conditioning Engineers [ASHRAE] 2010).

Most of the water is vaporized on the surface of the product, although some of the vapor can be generated in the underlying layers of the structure or tissues, so that the decrease in surface temperature causes a temperature gradient from the center of food. This phenomenon produces the resulting flow of heat from the interior of the piece to the surface. For this reason, when working with plant fresh produce, VC works best with vegetables having a high ratio of surface area to volume.

VC is a batch process (McDonald and Sun 2000). The product to be cooled is loaded into the flash chamber, the system is put into operation, and the product is cooled by reducing the pressure to the corresponding desired saturation temperature. The system is then shut down, the product removed, and the process repeated. Since the product is normally at ambient temperature before it is cooled, VC can be thought of as a series of intermittent operations of a vacuum refrigeration system in which the water in the flash chamber is allowed to come to ambient temperature before each start.

15.1.3 EQUIPMENT

A vacuum cooler consists basically of a sealed enclosure connected to a vacuum system. The precinct may have cylindrical or parallelepiped shape and is constructed of steel plate. Thickness must be enough to withstand a depression near a bar inside. The internal dimensions are adapted to contain the standard pallet sizes. Vacuum coolers are currently manufactured for lifting capacities ranging between 1 and 12 pallets, which represents between 300 kg and 3500 kg per charge, in the case of leafy vegetables (Casp and Abril 2003).

The main components of a vacuum cooler are as follows (Figure 15.2):

1. The product chamber, where the produce is held and the VC process takes place. The product chamber must be large enough to hold the batch quantity for each cycle, along with additional room for working clearance and anticipated future requirements.
2. Vacuum pump, used for the initial evacuation of the chamber from atmospheric pressure to the flash point of water (23 mb at 20°C).
3. Vapor-condensing unit that acts to recondense the vapor back into water for removal. In fact, the vapor-condensing unit acts as an auxiliary vacuum pump to handle the large amount of refrigerant vapor practically and economically (Sun and Wang 2001).
4. Infeed system, necessary for assist rapid turnaround times, especially in larger vacuum coolers. There are a variety of manual and automatic options.

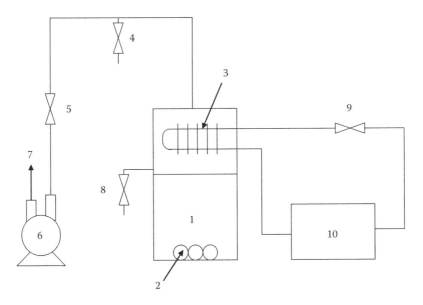

FIGURE 15.2 Schematic diagram of a vacuum cooler unit. (1) Vacuum chamber; (2) product; (3) condenser; (4) vent valve; (5) valve; (6) vacuum pump; (7) air exit; (8) draining valve; (9) expansion valve; (10) condensing unit.

5. Control systems. Careful attention must be paid to matching the pressure with the temperature of the product during the VC process. The maximum pressure of the vacuum system is the saturation pressure at the temperature of the product. Thus, the pressure must be reduced gradually with the decrease in temperature. Pressure should be controlled strictly during the VC process, since the effect of pressure difference on the evaporation rate differs from one product to another, its contribution to the evaporation rate is limited, and the low pressure will increase the condensing load of the vapor-condensing unit and the pumping load of the vacuum pump (Sun and Wang 2001).

The vacuum-pumping system in a vacuum cooler unit will discharge the refrigerant vapor to the atmosphere directly, without completing a cycle. Therefore, the VC process is an open cycle, and the heat is transferred from the mass of the cooled product to the evaporating water directly.

One issue that constrains the operational capacity is the pumping capacity, since it determines the time required for the VC operation. The vacuum equipment must be designed to remove the humid air efficiently in the enclosure at the beginning of the process and to continue removing the vapor that will be produced during the vaporization, until the desired final pressure is reached.

Since at the pressures to which the cooling processes conclude (6.6 mbar = 0.66 kPa) the specific volume of vapor is very high (193 m^3 kg^{-1}), a high pumping capacity is required (Casp and Abril 2003). For low- and medium-capacity VC installations, mechanical pumping systems are used, usually with volumetric vacuum pumps (positive displacement), whether rotary or alternatives. Both types are capable of achieving the pressure levels required for cooling, but they have the disadvantage of a small mass flow extracted at low pressure levels. Therefore, to achieve the required cooling times, pumps should have unacceptable dimensions.

The use of an additional refrigerator allows overcoming this drawback, so that virtually all the steam generated from boiling condenses in its evaporator. In this way, the quantity of water vapor that must be removed by the pump is minimized, so the pump should only extract the initial moist air and to obtain and maintain the desired pressure (Casp and Abril 2003).

For high-capacity cooling installations, the use of steam ejectors is more economical. In this case, the vacuum is generated by the expansion of water vapor at high pressure through the ejectors placed in series. The vapor produced is liquefied by means of multitubular vacuum condensers.

15.1.4 Technical Considerations

In a VC operation, the thermodynamic process is assumed to take place in two phases (ASHRAE 2010). In the first phase, the product is loaded into the chamber at room temperature, most of the air there is removed, and the temperature remains constant until saturation pressure is reached (Figure 15.3). At the onset of boiling, the remaining amount of air in the chamber is replaced by water vapor. At this point, the first phase ends while the second phase begins simultaneously. The second phase continues at saturation until the product has cooled to the desired temperature.

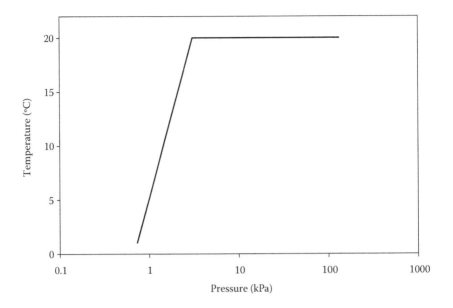

FIGURE 15.3   Temperature variation as a function of pressure during a VC process.

According to Sun and Wang (2001), there are two ways for reducing pressure after the flashpoint: (1) dropping to final pressure quickly and holding at that pressure to the final cooling temperature, and (2) dropping the pressure to the final pressure gradually to match the pressure of the vacuum chamber with the temperature of the cooled product.

If the ideal gas law is applied for an approximate solution in a commercial vacuum cooler, the pressure–volume relationships are

$$\text{Phase 1: } pv = 8.697 \text{ kN m/kg}$$

$$\text{Phase 2: } pv = 16.985 \text{ kN m/kg}$$

where p is the absolute pressure and v is the specific volume.

The pressure–temperature relationship is determined by the value of the room and the product temperature. Based on 20°C for this value, the temperature in the flash chamber theoretically remains constant at 20°C as the pressure is reduced from atmospheric to saturation after which it declines progressively along the saturation line. These relationships are illustrated in Figure 15.3. The product temperature would respond similarly to the temperature in the flash chamber but would vary depending on product physical characteristics and the amount of surface water available from it.

As it has been stated previously, most of the water is vaporized off the surface. However, some vaporization might occur within the intercellular spaces beneath the product surface. The heat required to vaporize this water is also taken off the product surface where it flows by conduction under the thermal gradient produced. Thus, the rate of cooling depends on the relation of the surface area to the volume of the product and the rate at which the vacuum is drawn in the flash chamber.

Since water is the refrigerant involved, the amount of heat removed from the product depends on the amount of water vaporized per time unit and its latent heat of vaporization, according to the following expression:

$$Q = \frac{dM}{dt} \lambda \qquad (15.1)$$

where Q is the total heat removed from the product (in watts), M is the product mass (in kilograms), t is the time (in seconds), and λ is the latent heat of vaporization (in joules per kilogram).

Assuming an ideal condition, with no heat gain by the surroundings, the heat dissipated on the surface of the product involves a reduction of its enthalpy that can be expressed by the following equation:

$$Q = Mc_p \frac{dT}{dt} \qquad (15.2)$$

where c_p is the specific heat (in joules per kilogram Kelvin) and T is the temperature.

Since the heat needed for vaporization of water coincides with the enthalpy change of the product, the equality between Expressions 15.1 and 15.2 can be established. Grouping variables and integrating between the initial and final conditions of the process, the following expression is obtained:

$$\frac{dM}{M} = \frac{c_p}{\lambda} dT. \qquad (15.3)$$

To simplify calculations, variations in the specific heat and latent heat in the range of temperatures at which the process occurs are neglected, so both can be considered constant. Therefore, the following expression arises:

$$[\ln M]_{M_0}^{M_f} = \frac{c_p}{\lambda} [T]_{T_0}^{T_f} \qquad (15.4)$$

Expressing the final temperature and mass as a function of their decreases ($M_f = M_0 - \Delta M$; $T_f = T_0 - \Delta T$) and substituting these values in Equation 15.4, an equation that links the temperature diminution of the product with the weight loss caused by the process is obtained:

$$\Delta T = -\frac{\lambda}{c_p} \left(1 - \frac{\Delta M}{M_0} \right). \qquad (15.5)$$

Weight loss is a result of water vaporization.

Then, the amount of moisture removed from the product during VC is directly related to the specific heat of the product and the temperature reduction accomplished. A product with a heat capacity of 4 kJ kg^{-1} K^{-1} would theoretically lose 1% moisture for each 6°C reduction in temperature (ASHRAE 2010). Thus, if produce must be cooled from 20°C to 0°C, the expectable weight loss will be close to 3.2%. In a study of VC of 16 different vegetables, Barger (1963) showed that cooling of all products was proportional to the amount of moisture evaporated from the product. Temperature reductions averaged 5 K to 5.5 K for each 1% of mass loss, regardless of the product cooled. This mass loss may reduce the amount of money the grower receives and cause a loss in product turgor and crispness. Some vegetables are sprayed with water before cooling to reduce this phenomenon (ASHRAE 2010).

As aforementioned, for cooling 1000 kg of produce from 20°C to 0°C, it is necessary to get the vaporization of ~30 kg of water. Therefore, at the pressure levels that are achieved, the volume of saturated vapor will be 5790 m^3. This volume should be removed from the site within about 15 min, so that the theoretical average flow for this example would be more than 23,000 m^3 h^{-1}. When vaporization is maximum, the instantaneous flow rate will actually be eight to ten times higher. Thus, the proper selection and sizing of the pumping equipment are relevant for VC efficiency.

VC is usually one of the systems that require more investment; therefore, its use is restricted to products for which this alternative is the quickest method or possibly the most convenient for any other reason.

VC allows cooling packaged products, provided that the packaging used has proper water vapor permeability. Carton boxes that were kept dry (and therefore have maintained all their mechanical properties) during the cooling process can be used (Casp and Abril 2003).

In general, the cooling of the load is homogeneous because the accessibility of the pressure at produce is uniform. Largest temperature differences are observed between the surface and the center of the product, especially when time is not enough to allow temperature to uniformly distribute.

As aforementioned, VC is a discontinuous process that requires high investment and consequently implies a high fixed cost. The use of VC is more appropriate to highly perishable products with an elevated commercial value. The product to be cooled has to have a high surface/volume ratio; the availability of water from the interstices and the external layers must be high, and the porosity of the product and the permeability of the skins should allow the easy transfer of the evaporation water to the surface.

15.1.5 APPLICATIONS

15.1.5.1 Plant Foods

According to Kader (2002), losses in quantity and quality affect horticultural crops between harvest and consumption. The magnitude of postharvest losses in fresh F&V is an estimated 5% to 25% in developed countries and 20% to 50% in developing countries, depending upon the commodity, cultivar, and handling conditions. To reduce these losses, the use of postharvest techniques that delay senescence and maintain high-quality levels must be implemented.

Fresh F&V are living tissues that are subject to continuous change after harvest. While some changes are desirable, most are not. Postharvest changes in fresh produce cannot be stopped, but they can be slowed within certain limits. Senescence is the final stage in the development of plant organs, during which a series of irreversible events leads to breakdown and death of the plant cells (Kader 2002).

Biological factors involved in deterioration of F&V are respiration, ethylene production, growth and development (e.g., sprouting of potatoes, onions, garlic, and root crops; elongation and curvature of asparagus spears that are accompanied by increased toughness and decreased palatability), transpiration or water loss, and physiological and pathological breakdown. All of them are influenced by temperature that is the environmental factor that most affects the deterioration rate of harvested commodities. For instance, the effect of temperature on biochemical reactions and decay is dramatic (Iqbal et al. 2009). Exhaustive research has shown the benefits of reducing temperature in products such as fresh-cut celery, shredded carrots, rocket leaves, minimally processed cabbage, and fresh chives, among others (Viña and Chaves 2006; Gómez-López et al. 2007; Koukounaras et al. 2007; Iqbal et al. 2009; Viña and Cerimele 2009).

VC can rapidly and conveniently reduce fresh produce temperature (McDonald and Sun 2000). According to Everington (1993), vacuum-cooled lettuce from ~25°C to 1°C in less than 30 min could be distributed by refrigerated vehicles into cold storage depots and retail outlets. Although Ozturk and Ozturk (2009) pointed out that VC of iceberg lettuce at 0.7 kPa was about 13 times faster than conventional cooling at 6°C, they also found that it was not possible to decrease the iceberg lettuce temperature below 10°C if vacuum pressure was set to 1.5 kPa.

He et al. (2004) analyzed whether the pressure reduction rate in a vacuum cooler would have an effect on physical and chemical quality characteristics as well as on the ultrastructure of iceberg lettuce after cooling and storage at 1°C and 85% RH for 2 weeks. Three different pressure reduction rates (from 10,000 to 600 Pa in 15, 30, and 60 min, respectively) were taken. The authors pointed out that the moderate pressure reduction rate allowed to achieve the maximum values of tissue firmness and ascorbic acid, and that membrane systems observed by transmission electron microscopy were kept intact compared to the other two pressure reduction rates.

Martínez and Artés (1999) evaluated several types of modified atmosphere packaging techniques, whether combined or not with VC, in improving the commercial quality of winter cultivated iceberg lettuce. The authors concluded that VC was useful in decreasing both pink rib and heart-leaf injury after the marketing period.

Sun and Wang (2001) have extensively reviewed the literature on the application of VC for plant products and have indicated that lettuce is most suited to vacuum precooling. According to the authors, VC is nowadays the standard commercial process used for lettuce not only in the United States but also in many European countries.

VC has also been applied to mushrooms. Mushroom quality is vitally important to the competitiveness of the production, and it contributes to price, sales, and customer loyalty. Mushroom quality can be lost as a result of biochemical degradative processes and by bruising caused by mishandling during harvest, packaging, and transport, among many other factors. Mushrooms start to deteriorate immediately after picking from the beds and have been one of the most difficult vegetables to be cooled properly (Sun and Wang 2001). They are over 90% water, and their porous structure allows water to escape readily, which makes them suitable for VC.

Tao et al. (2006) have investigated the effects of different storage conditions after VC on weight loss, respiration rate, soluble solid content, membrane permeability, and browning degree. They concluded that modified atmosphere packaging (5 ± 1% O_2 with 3 ± 1% CO_2) was the most suitable for mushrooms stored at 4 ± 1°C and 75% RH, following VC. In another study, Tao et al. (2007) pointed out that there were some effects of VC treatment on lipid oxidation, superoxide anion generation, and antioxidant enzyme system of mushrooms, and this technique had a positive influence on prolonging the shelf life of the mushrooms whenever they are stored under cold room or modified atmosphere packaging.

For some vacuum-cooled plant products with a complex internal structure that is tightly wrapped (e.g., cabbage), the cooling effect on the interior center is rather poor (Cheng and Hsueh 2007). To solve this inconvenience, multistage vacuum pressure reserving process can lower the surface and interior temperature of the cabbage effectively and uniformly. Cheng and Hsueh (2007) suggested that the air and aqueous moisture should be pumped by stage during the vacuum pressure reserving process in the vacuum chamber. The authors concluded that the handling process of the multistage vacuum pressure reserving could effectively reduce both the internal and external temperatures of the cabbage, make the two temperatures nearly identical, and effectively save energy required by the VC process.

Likewise, research on the application of VC on root and stem vegetables is rather scarce. This kind of products (e.g., bamboo shoots) has a small surface-to-volume ratio. Thus, they are less likely to yield a satisfactory temperature drop (Cheng 2006). When VC was combined with hydrocooling and vacuum-drying processes on bamboo shoots, cooling speed increased and the preservation period was longer. The temperature of cooling water could be reduced and controlled by means of multiple declines of vacuum chamber pressure, multistage vacuum pressure reserving process, and induction of external air into the chamber. The method was proven to shorten the cooling time and reduce the energy required for the cooling process (Cheng 2006).

15.1.5.2 Meat Products

Drummond and Sun (2008a) have mentioned that regulations and guidelines for the meat processing industry recommend cooling cooked meat as quickly as possible. Slow cooling of meat products implies the risk of pathogenic spore-forming microorganisms normally associated with meat products that could grow and/or produce toxins. The U.S. Department of Agriculture (USDA), through The Food Safety and Inspection Service, has recommended that noncured meat products should take no more than 1.5 h to cool from an internal temperature of 54.4°C to 26.6°C and should reach 4°C within an additional 5 h (USDA 1999). However, these requirements are difficult to fulfill for large joints of meat. Alternatives to solve this inconvenience are to portion meat before cooling or to ensure cooling to 10°C within 2.5 h after cooking.

Although VC has often been linked to substantial moisture loss, lower product yield, and adverse effects on quality properties, following water immersion cooking of meat joints, VC of cooked meat together with some of its cooking solution [immersion VC (IVC)] was employed for its potential use for rapid cooling of the cooked product (Cheng et al. 2005; Drummond and Sun 2006).

According to Drummond and Sun (2008a), during IVC, heat is transferred from the meat joint immersed in the liquid by a combination of evaporation, conduction, and convection. Thus, meat product size would have an effect on process parameters. The aforementioned authors compared mass losses and cooling times for different sample sizes when applying IVC, air blast cooling, or VC. IVC of water-cooked beef joints allowed to cooling beef samples of up to 3 kg from 72°C to 4°C in approximately 4 h, compared to almost 8 h for air blast and 1.5 to 2 h for VC. Drummond and Sun (2008b) have indicated that the introduction of mechanical agitation in the liquid will further improve the process and reduce cooling time by increasing convective heat transfer and maintaining a uniform low temperature around the joint during the last stage of cooling.

Taking into account that an increasing wide range of ready-to-eat meat products is available to the consumer in the present, rapid cooling of meat after cooking is imperative, in order to maintain sensory and nutritional quality, as well as microbiological safety. Both the relatively low thermal conductivity of meat and the need to maintain the temperature of the cooling medium above –2°C to avoid surface freezing are constraints to increase the cooling rate. Meat industry is developing value-added products made from sectioned or comminuted meat with added salt, phosphates, and other ingredients to supply the demands of a convenience-driven market (Drummond et al. 2009). When comparing IVC against air blast cooling (AB) and VC, to establish its applicability to cool water-cooked beef joints, Drummond et al. (2009) found that IVC could overcome some adverse effects normally associated with vacuum-cooled samples, such as loss of tenderness and juiciness, and it was capable of offering a product of comparable quality properties to the ones cooled by conventional methods. The authors reported that cooling loss was significantly lower for IVC and AB than for VC; Warner Bratzler shear average values for IVC and VC were similar, but both were lower than for AB samples; results from sensory analyses showed no differences between samples cooled by the three cooling methods in terms of color, overall flavor, binding, saltiness, and acceptability; meanwhile, IVC scores for tenderness and juiciness were significantly higher than for VC samples (Drummond et al. 2009).

Investigating moisture movement characteristics and their effect on the ultrastructure of vacuum-cooled cooked meat (2–5 kg boned-out pork meat, cooked in water), Jin et al. (2006) described that for the evaporation rate during VC, there were an accelerating period and a falling one. The authors pointed out that moisture movement of free water within cooked meat during VC consisted of two parts: one was free water migration within cooked meat caused by chemical potential, and the other was water vapor movement produced by evaporation or ebullition caused by pressure drop, with the latter being dominating in the process of moisture movement of free water. Likewise, VC allowed to reduce the surface temperature of cooked meat from 63°C to 10°C within the first 4–5 min; the average moisture content of the product decreased from 71% to 60.7% during VC, and the weight loss of cooked meat was 10.3% (Jin et al. 2006).

Schmidt et al. (2010) have developed and evaluated integrated processes of cooking and VC of chicken breast cuts in a same vessel with the aim of avoiding product manipulation and reducing the processing time. The vapor-cooking and the immersion-cooking both followed by VC were similar in terms of process time and global weight loss, in spite of using different cooking methods. On the other hand, the integrated process "immersion-cooking + IVC," which implied cooking and cooling the product immersed in the cooking water, reduced drastically the weight loss at the cost of increasing the cooling time (Schmidt et al. 2010). Thus, the choice of the best process would depend on the use of the cooked meat after cooling. When water loss is an important issue and the focus is not put on reducing the cooling time, the immersion cooking–IVC process is an interesting alternative, according to the authors.

In order to find a method for minimizing mass loss during VC, Houska et al. (2003) performed VC of meat together with salted water in which the meat was previously boiled (*in soup*). They reported that the mean mass change for 16 experiments was +0.48%, indicating weight gain during the cooling. For injected sirloin, the mass was found to increase by +7.7%. With this method, the soup penetrated into meat pores at the end of the VC. The authors suggested that this effect can be used in industrial practice for flavoring meat with the desired natural aroma of sauce (or gravy).

15.1.5.3   Seafood

On meat products, a great part of the investigations referring to VC have been made on the use of this technique for pieces of precooked meat and fish and also to cooling hams and joints (Desmond et al. 2000, 2002; Sun and Wang 2000; McDonald et al. 2000; McDonald and Sun 2001; Wang and Sun 2002a,b; Sun and Hu 2002; Landfeld et al. 2002; Houska et al. 2003).

Although the commercial use of VC requires that products have large superficial areas for mass transfer, results reported in the literature have also stated the viability of VC use for products with a low superficial area (e.g., cooked mussels, *Perna perna*).

Heat treatment is one of the more commonly used preservation techniques for mussels after their collection. The bivalve mollusks are submitted to steam for a time that varies according to their size and to the heating conditions (Huber et al. 2006). According to Huber et al. (2006), the temperature–time combination required to guarantee the destruction of pathogenic bacteria is 70°C for 4.5 min. After heat treatment, there is a manual extraction of the meat, which requires temperatures lower than 40°C. Mussel meat can be destined to be canned in brine and sterilized or be cooled (until an internal temperature of 7.2°C is achieved and afterward stored at this minimum temperature) or frozen. According to Huber et al. (2006), in small mussel processing units, water aspersion and immersion in cold water are used to cool the precooked product. Both offer risk of microbial contamination by the cooling water and increase the water spent in the processing. Huber et al. (2006) have also pointed out that an alternative is the accomplishment of cooking and cooling steps in the same vessel: after cooking with hot water or steam injection, most water is drained and the vacuum is applied, and the remaining water in the vessel evaporates, reducing the water loss of the product. Besides, water loss would not be a serious problem when the cooled mussels are canned with brine and sterilized by thermal treatment. Experiments showed that for cooling the product until 40°C, the vacuum chamber pressure needed to be reduced to about 80 mbar.

15.1.6 SPECIAL CONSIDERATIONS: RANGE OF APPLICATION

The final temperature of several vegetables when they were vacuum-cooled under similar conditions is illustrated in Figure 15.4. Process conditions were as follows: initial product temperature, 20°C to 22°C; minimum pressure, 530 Pa to 610 Pa; condenser temperature, –1.7°C to 0°C; time spent in the vacuum chamber, 0.42 h to 0.5 h.

Regarding vacuum cooler efficiency, Thompson and Chen (1988) conducted a study on the energy use at commercial scale of several cooling systems such as vacuum, hydro, water-spray vacuum, and forced-air systems, expressing the energy efficiency data as an energy coefficient as follows:

$$EC = \frac{SH}{EN} \tag{15.6}$$

where SH is the sensible heat removed from the product, assuming a specific heat of 3.8 kJ kg^{-1} °C^{-1} for fruits and 4 kJ kg^{-1} °C^{-1} for vegetables, and EN is the electrical energy consumed in operating the cooler (in kilowatts).

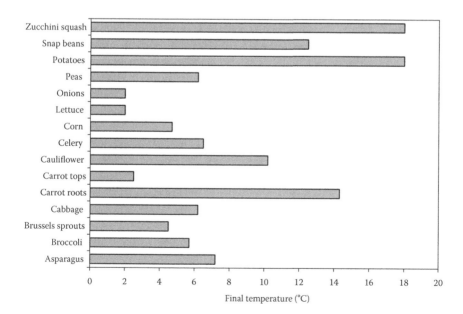

FIGURE 15.4 Comparative cooling of vegetables under similar vacuum conditions. (Adapted from ASHRAE. *ASHRAE Refrigeration Handbook*, American Society of Heating, Refrigerating and Air-Conditioning Engineers, Atlanta, GA, 2010.)

Researchers found that there were differences in energy efficiency between different cooler types; the average energy coefficient was 1.8 for VC, 1.4 for hydrocooling, 1.1 for water-spray VC, and 0.4 for forced-air cooling. Also, it was mentioned that there were large differences between coolers of the same type for vacuum coolers and hydrocoolers (up to 0.8 and 1.3, respectively). Energy use efficiency of cooling systems varied with the type of cooler used. Vacuum coolers were the most efficient, followed by hydrocoolers, water-spray vacuum coolers, and forced-air coolers. Differences were attributed to levels of nonproduct heat input and operational practices. Energy use efficiency significantly varied within coolers of the same type. The level of product throughput, the commodity type, and operational procedures have been identified as the major reasons for this fact.

15.2 MIST COOLING

15.2.1 PRINCIPLE

In processes like "hydraircooling" or "mist chilling," a two-phase flow (air and suspended water droplets) is involved. Henry and Bennett (1973) and Henry et al. (1976) described hydraircooling as a process in which a combination of chilled water and chilled air is circulated over commodities (Figure 15.5a). This technique requires less water for cooling than that for conventional hydrocooling and also reduces the requirement of keeping the cooling water clean. Cooling rates equal to or even better than those obtained in conventional unit load hydrocoolers are expected.

There are particular constraints imposed on the mist chilling process of foodstuffs: low air velocity (0.1–5 m s^{-1}), droplet diameter generally lower than 10 μm, relatively low surface temperature of the product (< 80°C), and regulated water supply to avoid excess water on the product surface (Abdul-Majeed 1981).

When spray cooling or hydraircooling is applied, a liquid film on the product surface is often observed (Abdul-Majeed 1981). Mist chilling enables water supply control that prevents excess water on the surface. This is particularly important for

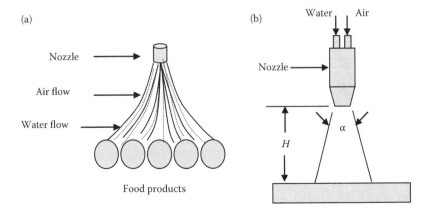

FIGURE 15.5   (a) Mist chilling of food products; (b) mist configuration.

water-sensitive products such as strawberries or bakery products. In this case, excess water can accelerate microbiological and biochemical changes responsible for food decay.

Compared to forced-air cooling, mist chilling reduces chilling time and also limits weight losses. Thus, mist chilling can limit quality decay due to dehydration.

Phenomena involved during the cooling of product stacks in a two-phase flow are complex. Heat transfer depends on aerodynamical and thermal properties of the two-phase flow, product shape, and stack arrangement (Allais and Alvarez 2001). For single products, many experimental studies were performed that allowed for a better understanding of the increase in the heat transfer and the influence of main factors on it: airflow characteristics (velocity, turbulence, and temperature) and water droplet characteristics (droplet diameter and velocity). Allais and Alvarez (2001) showed that cooling by a multiphase flow could enhance the cooling process by a maximum factor of about 2.8. The used air-mist sprays dispersed very fine droplets that were less than 10 µm in size.

15.2.2 TECHNICAL CONSIDERATIONS

Heat transfer from the spray is highly dependent on two major parameters: the impinging droplet size and the water mass flow (mass flow rate per contact surface area). The smaller the droplet size, the easier to evaporate the droplet from the surface; thus, this condition leads to heat transfer enhancement. Likewise, smaller droplets will better cool the thermal boundary layer near the surface by increasing the near-surface local vapor convection. However, a higher water mass flow or large size droplets will have a detrimental effect on cooling the surface because of surface flooding. If the spray droplets are too large, there is a risk of water running down the target surface. The detrimental effect on the quality of the processed product is due to the bleached marks that can be produced on the surface.

There are three modes of heat transfer associated with the air-mist spray chilling of surfaces slightly heated to temperatures above room temperature (Issa 2009):

1. Convection heat transfer associated with the bulk air flow
2. Evaporation heat transfer of the droplets while being airborne and also on the target surface
3. Sensible heat transfer associated with the droplet contact with the heated surface

The spray impact efficiency, η, is defined as the ratio of the actual mass flow of the droplets deposited onto the target surface to the maximum droplet mass flow leaving the nozzle exit (Issa 2009). This is expressed as

$$\eta = \frac{G}{G_{max}} \tag{15.7}$$

where G is the actual water mass flow (kg m^{-2} s^{-1}), and G_{max} is the maximum water mass flow (kg m^{-2} s^{-1}).

The spray impact on the target surface is influenced by several factors: droplet diameter, spray orientation, water flow rate, air-to-liquid loading, nozzle-to-surface distance, relative humidity, and surface temperature (Figure 15.5b).

In this sense, Issa (2009) has studied the optimal conditions for spray impact, heat transfer efficiency, and control of surface flooding. A computational fluid dynamics model using a two-phase flow composed of water droplets injected with air was developed to simulate the air-mist chilling of food products. The model took into consideration the droplet–surface interaction, water film accumulation, and surface runoff. Simulation results showed that for low-density spray, a noticeable amount of vapor was generated. This amount increased with the decrease in droplet size. The spray impact and heat transfer were shown to depend on spray droplet size and flow operating conditions. The optimal droplet size required for maximum impact was different from that required for maximum heat transfer enhancement.

15.2.3 Applications: Plant and Meat Products

Noomhorm (1991) designed and developed a precooling system for tropical fruits. The performance of the precooling systems in terms of cooling rates and half-cooling times of selected F&V were evaluated for hydrocooling, air/forced–air cooling, and hydraircooling methods. Among the products assayed, the author concluded that cucumber could be best precooled through hydraircooling. On the other hand, most adaptable precooling methods for mango and papaya were hydrocooling and forced-air cooling, respectively.

Likewise, mist chilling was used on broccoli during storage at 18°C and 4°C. During 72 h of storage at 18°C, mist-chilled broccoli significantly retained total ascorbic acid content, moisture, chlorophyll, and green color compared with non-misted samples (Barth et al. 1990, 1992). At 4°C, the increase in the microbial flora was minimized (Mohd-Som et al. 1995). Higher relative humidity in the misted samples compared with the nonmisted ones was observed; the washing effect of the misting water and possible residual chlorine effects due to the use of chlorinated tap water might explain the reduced viable counts observed in the misted samples.

Allais and Letang (2009) assessed the impact of mist chilling on high-grade strawberry postharvest quality. Strawberries were chilled at 2°C using three alternatives: air blast chilling at 0.3 m s^{-1} or 1 m s^{-1} and mist chilling at 1 m s^{-1}. Compared to air chilling, mist chilling did not reduce chilling time, but it did reduce weight loss by 20%–40%. Mist chilling had no detrimental effect on commercial loss defined as the percentage of fruit with more than 1/3 of its surface affected. According to the authors, mist chilling did not induce any major changes on strawberry quality.

Henry and Bennett (1973) reported that immersion cooling was faster than hydro-air cooling for peaches and celery, but cooling rates for sweet corn were equal to, or in some cases slightly better than, those obtained in conventional hydrocoolers.

Applying mist cooling to F&V is rather complex because, in some cases, products have to be treated in their packaging units as homogeneously as possible. Many food products being packed in bins are stacked on pallets before chilling; stack arrangement is a key parameter to take into account. In this sense, mist chilling was applied on a pallet of maize (Henry et al. 1976), on peaches (Benett and Wells 1976), and on strawberries (Allais and Alvarez 1996). Compared to air-cooled strawberries, mist-cooled strawberries at high water mass flow rates (19 g/kg strawberries) showed less commercial loss and a better appearance after 7 days of storage at 6°C. Weight loss, firmness, color, soluble sugar content, acidity, and pH did not show any significant difference between air cooling and mist cooling at low (9 g/kg strawberries) and high (19 g/kg strawberries) water flow rates.

As a complement, the adaptation of vegetables and fruits to different precooling methods is shown in Table 15.2.

Concerning meat products, in beef carcasses, the principal purpose of spray chilling is to reduce weight loss during chilling, especially during the first 24 h *postmortem* (Allen et al. 1987). Spray chilling systems are currently in use in North America and Europe for beef, lamb, poultry (Brown et al. 1993), and pork (Brown and James 1992).

Weight losses ranging from 0.75% to 2.0% of the hot carcass weight have been reported during the initial 24 h of conventional air chilling of beef, pork, and lamb (Jones and Robertson 1988; Greer and Jones 1997). Subsequent reports pointed out

TABLE 15.2
Adaptation of Vegetables and Fruits to Different Precooling Methods

Product	Vacuum Cooling	Hydrocooling	Hydro Air Cooling Saturated Air Humidity	Conventional Air
Asparagus	++	+++	++	+
Broccoli	+	+++	+	+
Carrot	–	+++	++	+
Celery	+++	+	++	+
Mushroom	+++	0	+	+
Cauliflower	–	+	+++	+
Cabbage	0	+	+++	++
Cucumber	0	0	+++	++
Endive	+	–	+++	++
Lettuce	+++	+	++	+
Sweet corn	++	++	+++	+
Melon	–	++	++	+
Leek	–	++	+++	++
Tomato	–	–	+++	++
Apricot	–	–	++	+++
Cherry	–	++	++	+
Strawberry	+	–	+++	+
Pear	–	++	++	+
Kiwi	–	++	++	+
Grape	–	–	+++	+

Source: Data from Vallespir, A. N., "El pre enfriamiento. I Parte." Available at: http://www.mapa.es/ministerio/pags/biblioteca/revistas/pdf_Hort/Hort_1990_63_56_79.pdf, 1990.

Note: +++ Very good adaptation, ++ good adaptation, + poor adaptation, – no adaptation.

that such losses could be reduced up to 75% by spraying carcasses with water (Allen et al. 1987).

However, surface drying is an important factor in limiting microbial growth. If the surface remains wet, there may be microbial problems that shorten shelf life. In this sense, some results are reported for the effect of spray chilling on the microbiological quality of meat. Hippe et al. (1991) observed an increase in bacterial counts after spray chilling for beef. On the contrary, there have also been data published showing no differences on microbial status between spray-chilled and commercially air-chilled beef (Greer and Jones 1997) or even better microbiological quality after spray chilling of pork (Greer and Dilts 1988).

Strydom and Buys (1995) suggested that these contradictory results may be due to variation in the time allowed for carcasses to dry between the moments when the spray chilling is turned off and when the chilling cycle is completed. In addition, Gill and Lander (2003) suggested that the growth of bacteria on beef carcasses during spray cooling was apparently constrained only by the temperature. Wiklund et al. (2010) evaluated the impact of spray chilling on microbiological quality of venison meat and concluded that the spray chilling did not affect the microbiological quality on deer carcasses or vacuum-packed loins.

With respect to eating quality attributes, spray-chilling system did not show differences with conventional air cooling. Meat tenderness and juiciness were unaffected by shrouding and chilling methods applied to beef roasts (Lee et al. 1990).

A variety of chilling techniques that range from using conventional air chilling systems to air-assisted chilling systems with a variety of intermittent cooling schemes have been assayed. In these studies, the amount of water used to cool the beef carcass ranged from 14 L to 28 L per carcass and the spray cooling periods range from 10 h to 17 h.

There is considerable variation in the duration of spray chilling in commercial practice, and the amount of water deposited on carcasses in a specified period of time is usually unknown. Spraying carcasses in the initial 3–12 h of carcass chilling seems to be relatively common within the commercial setting (Greer and Jones 1997).

15.2.4 HYDROAIRCOOLING WATER TREATMENT

The surface of wet commodities provides an excellent site for diseases to thrive. In addition, in some cases, hydroaircooling water is recirculated, and thus decay-producing organisms can proliferate in the hydroaircooling water and can easily spread to other commodities being hydroaircooled. Therefore, hydroaircooling water must be treated with mild disinfectants to reduce disease incidence.

Usually, hydroaircooling water is treated with chlorine to minimize the levels of decay-producing organisms (USDA 2004). Chlorine (gaseous or in the form of hypochlorous acid from sodium hypochlorite) is added to water at a level of 50 ppm to 100 ppm. However, chlorination of water only provides a surface treatment of the commodities; chlorination is not effective at neutralizing an infection below the commodity's surface.

Chlorine level must be checked at regular intervals to ensure that the proper concentration is maintained. Chlorine is volatile and disperses into the air at a rate that increases with increasing temperature.

Chlorine loses its effectiveness quickly in contact with organic matter and metals, high temperatures, and by exposure to light. Chlorine reaction with organic matter after the treatment of food products could lead to the formation of potentially mutagenic or carcinogenic compounds such as trihalomethanes (chloroform, bromodichloromethane), which constantly encourages studies for the development of new disinfectants (González et al. 2005).

REFERENCES

Abdul-Majeed, P. 1981. Analysis of heat transfer during hydrair cooling of spherical food products. *International Journal of Heat and Mass Transfer* 24, 2, 323–333.

Allais, I., Alvarez, G. 1996. Refrigeration avec brumisation des fraises fraîches. *Rev. Gen. Froid* 967, 28–33.

Allais, I., Alvarez, G. 2001. Analysis of heat transfer during mist chilling of a packed bed of spheres simulating foodstuffs. *Journal of Food Engineering* 49, 1, 37–47.

Allais, I., Letang, G. 2009. Influence of mist-chilling on post-harvest quality of fresh strawberries Cv. Mara des Bois and Gariguette. *International Journal of Refrigeration* 32, 1495–1504.

Allais, I., Alvarez, G., Flick, D. 2006. Modelling cooling kinetics of a stack of spheres during mist-chilling. *Journal of Food Engineering* 72, 197–209.

Allen, D., Hunt, M., Luchiari Filho, A., Danler, R., Goll, S. 1987. Effects of spray chilling and carcass spacing on beef carcass cooler shrink and grade factors. *Journal of Animal Science* 64, 165–170.

Ashie, I. N. A., Smith, J. P., Simpson, B. K., Haard, N. F. 1996. Spoilage and shelf-life extension of fresh fish and shellfish. *Critical Reviews in Food Science and Nutrition* 36, 87–121.

ASHRAE. 2010. Chapter 28. Methods of precooling fruits, vegetables and cut flowers. *ASHRAE Refrigeration Handbook.* Atlanta, Georgia: American Society of Heating, Refrigerating and Air-Conditioning Engineers.

Bachmann, J., Earles, R. 2000. Postharvest handling of fruits and vegetables. Horticulture Technical Note. ATTRA, Appropriate Technology Transfer for Rural Areas. Available at: http://attra.ncat.org/attra-pub/PDF/postharvest.pdf.

Barbut, S. 1998. Estimating the magnitude of the PSE problem in poultry. *Journal of Muscle Foods* 9, 35–49.

Barger, W. R. 1963. Vacuum precooling—A comparison of cooling of different vegetables. USDA, Marketing Research Report N° 600.

Barth, M., Perry, A., Schmidt, S., Klein, B. 1990. Misting effects on ascorbic acid retention in broccoli during cabinet display. *Journal of Food Science* 55, 1187–1191.

Barth, M., Perry, A., Schmidt, S., Klein, B. 1992. Misting affects market quality and enzyme activity of broccoli during retail storage. *Journal of Food Science* 57, 954–957.

Benett, A., Wells, J. 1976. Hydrair cooling: A new precooling method with special application for waxed peaches. *Journal of the American Society for Horticultural Science* 101, 428–431.

Brosnan, T., Sun, D.-W. 2001. Precooling techniques and applications for horticultural products—A review. *International Journal of Refrigeration* 24, 154–170.

Brown, T., James, S. 1992. Process design data for pork chilling. *International Journal of Refrigeration* 15, 281–289.

Brown, T., Chourouzidis, K., Gigiel, A. 1993. Spray chilling of lamb carcasses. *Meat Science* 34, 311–325.

Casp, A., Abril, J. 2003. Capítulo 10: Refrigeración. In: Procesos de conservación de alimentos. A. Madrid Vicente Ediciones, Ediciones Mundi-Prensa. Madrid. España.

Cheng, H. P. 2006. Vacuum cooling combined with hydrocooling and vacuum drying on bamboo shoots. *Applied Thermal Engineering* 26, 2168–2175.

Cheng, H. P., Hsueh, C. F. 2007. Multi-stage vacuum cooling process of cabbage. *Journal of Food Engineering* 79, 37–46.

Cheng, Q., Sun, D.-W., Scannell, A. G. M. 2005. Feasibility of water cooking for pork ham processing as compared with traditional dry and wet air cooking methods. *Journal of Food Engineering* 67, 427–433.

Davey, C. L., Gilbert, K. V. 1974. The mechanism of cold-induced shortening in beef muscle. *International Journal of Food Science and Technology* 9, 51–58.

Desmond, E., Kenny, T., Ward, P., Sun, D.-W. 2000. Effect of rapid and conventional cooling methods on the quality of cooked ham joints. *Meat Science* 56, 271–277.

Desmond, E., Kenny, T., Ward, P. 2002. The effect of injection level and cooling method on the quality of cooked ham joints. *Meat Science* 60, 271–277.

Drummond, L. S., Sun, D.-W. 2006. Feasibility of water immersion cooking of beef joints: Effect on product quality and yield. *Journal of Food Engineering* 77, 289–294.

Drummond, L., Sun, D.-W. 2008a. Immersion vacuum cooling of cooked beef—Safety and process considerations regarding beef joint size. *Meat Science* 80, 738–743.

Drummond, L., Sun, D.-W. 2008b. Temperature evolution and mass losses during immersion vacuum cooling of cooked beef joints—A finite difference model. *Meat Science* 80, 885–891.

Drummond, L., Sun, D.-W., Talens Vila, C., Scannell, A. G. M. 2009. Application of immersion vacuum cooling to water-cooked beef joints—Quality and safety assessment. *LWTM Food Science and Technology* 42, 332–337.

Everington, D. W. 1993. Vacuum technology for food processing. *Food Technology International Europe*, 71–74.

Foegeding, E., Lanier, T., Hultin, H. 1996. Characteristics of edible muscle tissues. In *Food Chemistry*, Fennema, O. R., ed.; Marcel Dekker, Inc.: New York, Chapter 15, pp. 879–942.

Gill, C., Landers, C. 2003. Effects of spray-cooling processes on the microbiological conditions of decontaminated beef carcasses. *Journal of Food Protection* 66, 1247–1252.

Gómez-López, V. M., Ragaert, P., Ryckeboer, J., Jeyachchandran, V., Johan Debevere, J., Devlieghere, F. 2007. Shelf-life of minimally processed cabbage treated with neutral electrolysed oxidising water and stored under equilibrium modified atmosphere. *International Journal of Food Microbiology* 117, 91–98.

González, R. J., Allende, A., Ruiz-Cruz, S., Luo, Y. 2005. Capítulo 12 Sanitizantes utilizados. In: *Nuevas Tecnologías de Conservación de Productos Vegetales Frescos Cortados*. González-Aguilar, G., Gardea, A. A., Cuamea-Navarro, F., Ciad, A. C., CYTED, CONACYT, COFUPRO. Logiprint Digital S. de R. L. de C. V. Guadalajara, Jalisco. México.

Greer, G. C., Dilts, B. D. 1988. Bacteriology and retail case life of spray-chilled pork. *Canadian Institute of Food Science and Technology Journal* 21, 295–299.

Greer, G., Jones, D. 1997. Quality and bacteriological consequences of beef carcass spray-chilling: Effects of spray duration and boxed beef storage temperature. *Meat Science* 45, 61–73.

He, S. Y., Feng, G. P., Yang, H. S., Wub, Y., Li, Y. F. 2004. Effects of pressure reduction rate on quality and ultrastructure of iceberg lettuce after vacuum cooling and storage. *Postharvest Biology and Technology* 33, 263–273.

Henry, F. E., Bennett, A. H. 1973. "Hydroaircooling" vegetable products in unit loads. *Transactions of the ASAE* 16, 731–733.

Henry, F., Bennett, A., Segall, R., 1976. Hydroaircooling: a new concept for precooling pallet loads of vegetables. *ASHRAE Transactions* 82(2), 541–547.

Hippe, C., Field, R., Ray, B., Russell, W. 1991. Effect of spray-chilling on quality of beef from lean and fatter carcasses. *Journal of Animal Science* 69, 178–183.

Houska, M., Sun, D.-W., Landfeld, A., Zhang, Z. 2003. Experimental study of vacuum cooling of cooked beef in soup. *Journal of Food Engineering* 59, 105–110.

Huber, E., Soares, L. P., Carciofi, B. A. M., Hense, H., Laurindo, J. B. 2006. Vacuum cooling of cooked mussels (*Perna perna*). *Food Science and Technology International* 12(1), 19–25.

Iqbal, T., Rodrigues, F. A. S., Mahajan, P. V., Kerry, J. P. 2009. Mathematical modeling of the influence of temperature and gas composition on the respiration rate of shredded carrots. *Journal of Food Engineering* 91, 325–332.

Issa, R. J. (2009). Numerical investigation of the chilling of food products by air-mist spray. *International Journal of Mechanical, Industrial and Aerospace Engineering* 3, 130–139.

James, S. J. 2001. Rapid chilling of food—A wish or a fact? Bulletin 2001.6 IIR.

Jin, T. X., Zhu, H. M., Xu, L. 2006. Moisture movement characteristics and their effect on the ultrastructure of cooked meat during vacuum cooling. *Biosystems Engineering* 95, 111–118.

Jones, S. D., Robertson, W. 1988. The effects of spray-chilling carcasses on the shrinkage and quality of beef. *Meat Science* 24, 177–188.

Kader, A. A. 2002. Chapter 4. Postharvest biology and technology: An overview. In *Postharvest Technology of Horticultural Crops*, 3rd edition. Adel A. Kader, Technical Editor. University of California, Agriculture and Natural Resources. Publication 3311. California, USA.

Kinsella, K., Sheridan, J., Rowe, T., Butler, F., Delgado, A., Ramirez, Q., Blair, I., McDowell, D. 2006. Impact of a novel spray-chilling system on surface microflora, water activity and weight loss during beef carcass chilling. *Food Microbiology* 23, 483–490.

Koukounaras, A., Siomos, A. S., Sfakiotakis, E. 2007. Postharvest CO_2 and ethylene production and quality of rocket (*Eruca sativa* Mill.) leaves as affected by leaf age and storage temperature. *Postharvest Biology and Technology* 46, 167–173.

Kuitche, A., Daudin, J. 1996. Modeling of temperature and weight loss kinetics during meat chilling for time-variable conditions using an analytical-based method—I. The model and its sensitivity to certain parameters. *Journal of Food Engineering* 28, 55–84.

Landfeld, A., Houska, M., Kyhos, K., Qibin, J. 2002. Mass transfer experiments on vacuum cooling of selected pre-cooked solid foods. *Journal of Food Engineering* 52, 207–210.

Lee, L. M., Hawrysh, Z. J., Jeremiah, L. E., Hardin, R. T. 1990. Shrouding, spray-chilling and vacuum-packaged aging effects on processing and eating quality attributes of beef. *Journal of Food Science* 55, 1270–1273.

Mallikarjunan, P., Mittal, G. 1994. Heat and mass transfer during beef carcass chilling—Modeling and simulation. *Journal of Food Engineering* 23, 277–292, 1994.

Marsh, B. B. 1977. The Basis of quality in muscle foods. *Journal of Food Science* 42, 295–297.

Martínez, J. A., Artés, F. 1999. Effect of packaging treatments and vacuum-cooling on quality of winter harvested iceberg lettuce. *Food Research International* 32, 621–627.

McDonald, K., Sun, D.-W. 2000. Vacuum cooling technology for the food processing industry: A review. *Journal of Food Engineering* 45, 55–65.

McDonald, K., Sun, D.-W. 2001. The formation of pores and their effects in a cooked beef product on the efficiency of vacuum cooling. *Journal of Food Engineering* 47, 175–183.

McDonald, K., Sun, D.-W., Kenny, T. 2000. Comparison of the quality of cooked beef products cooled by vacuum cooling and by conventional cooling. *Lebensmittel Wissenschaft und Technologie* 33, 21–29.

Mohd-Som, F., Art-Spomer, L., Martin, S., Schmidt, S. 1995. Microflora changes in misted and nonmisted broccoli at refrigerated storage temperatures. *Journal of Food Quality* 18, 279–293.

Noomhorm, A. 1991. Designed and developed a pre cooling system for tropical fruits. Available at: http://www.lib.ku.ac.th/KUCONF/KC2611018.pdf.

Ozturk, H. M., Ozturk, H. K. 2009. Effect of pressure on the vacuum cooling of iceberg lettuce. *International Journal of Refrigeration* 32, 402–410.

Schmidt, F. C., Aragão, G. M. F., Laurindo, J. B. 2010. Integrated cooking and vacuum cooling of chicken breast cuts in a single vessel. *Journal of Food Engineering* 100, 219–224.

Solomon, M. B., Van Luck, R. L. J. M., Eastridge, J. S. 1998. Biophysical basis of pale, soft, exudative (PSE) pork and poultry muscle: A review. *Journal of Muscle Foods* 9, 1–11.

Strydom, P., Buys, E. M. 1995. The effects of spray-chilling on carcass mass loss and surface associated bacteriology. *Meat Science* 39, 265–276.

Sun, D.-W., Hu, Z. 2002. CFD predicting the effects of various parameters on core temperature and weight loss profiles of cooked meat during vacuum cooling. *Computers and Electronics in Agriculture* 34, 111–127.

Sun, D.-W., Hu, Z. 2003. CFD simulation of coupled heat and mass transfer through porous foods during vacuum cooling process. *International Journal of Refrigeration* 26, 19–27.

Sun, D.-W., Wang, L. 2000. Heat transfer characteristics of cooked meats using different cooling methods. *International Journal of Refrigeration* 23, 508–516.

Sun, D.-W., Wang, L.-J. 2001. Vacuum cooling. In: *Advances in Food Refrigeration*. Da-Wen Sun (ed.). Leatherhead Publishing, LFRA Limited, Surrey, England.

Suslow, T. V., Cantwell, M. 2009. Mushroom: Recommendations for Maintaining Postharvest Quality. Available at: http://postharvest.ucdavis.edu/Produce/ProduceFacts/Veg/mushroom.shtml.

Tao, F., Zhang, M., Hangqing, Y., Jincai, S. 2006. Effects of different storage conditions on chemical and physical properties of white mushrooms after vacuum cooling. *Journal of Food Engineering* 77, 545–549.

Tao, F., Zhang, M., Hang-qing, Y. 2007. Effect of vacuum cooling on physiological changes in the antioxidant system of mushroom under different storage conditions. *Journal of Food Engineering* 79, 1302–1309.

Thompson, F., Chen, Y. L. 1988. Comparative energy use of vacuum, hydro, and forced air coolers for fruits and vegetables. *ASHRAE Transactions* 92, 1427–1433.

Thompson, J. F., Chen, Y. L., Rumsey, T. R. 1987. Energy use in vacuum coolers for fresh market vegetables. *Applied Engineering in Agriculture*, 3, 196–199.

Thompson, J. M., Perry, D., Daly, B., Gardner, G. E., Johnston, D. J., Pethick, D. W. 2006. Genetic and environmental effects on the muscle structure response post-mortem. *Meat Science* 74, 59–65.

USDA. 1999. *Performance Standards for the Production of Certain Meat and Poultry Products*. Washington, DC: Office of Federal Register, National Archives and Records Administration.

USDA. 2004. The Commercial Storage of Fruits, Vegetables, and Florist and Nursery Stocks. Washington, DC: Agricultural Research Service, U.S. Department of Agriculture.

Vallespir, A. N. 1990. El pre enfriamiento. I Parte. Available at: http://www.mapa.es/ministerio/pags/biblioteca/revistas/pdf_Hort/Hort_1990_63_56_79.pdf.

Van Laack, R. L. J. M., Liu, C.-H., Smith, M. O., Loveday, H. D. 2000. Characteristics of pale, soft, exudative broiler breast meat. *Poultry Science* 79, 1057–1061.

Viña, S. Z., Chaves, A. R. 2006. Antioxidant responses in minimally processed celery during refrigerated storage. *Food Chemistry* 94, 68–74.

Viña, S. Z., Cerimele, E. L. 2009. Quality changes in fresh chives (*Allium schoenoprasum* L.) during refrigerated storage. *Journal of Food Quality* 32, 747–759.

Wang, L., Sun, D.-W. 2002a. Modeling vacuum cooling processes of cooked meat—Part 1: Analysis of vacuum cooling system. *International Journal of Refrigeration* 25, 854–861.

Wang, L., Sun, D.-W. 2002b. Modeling vacuum cooling processes of cooked meat—Part 2: Mass and heat transfer of cooked meat under vacuum pressure. *International Journal of Refrigeration* 25, 862–871.

Wiklund, E., Kemp, R. M., le Roux, G., Li, Y., Wu, G. 2010. Spray chilling of deer carcasses. Effects on carcass weight, meat moisture content, purge and microbiological quality. *Meat Science* 86, 926–930.

Woelfel, R. L., Owens, C. M., Hirschler, E. M., Martínez-Dawson, R., Sams, A. R. 2002. The characterization and incidence of pale, soft, and exudative broiler meat in a commercial processing plant. *Poultry Science*, 81, 579–584.

Zheng, L., Sun, D.-W. 2004. Vacuum cooling for the food industry—A review of recent research advances. *Trends in Food Science and Technology* 15, 555–568.

16 Special (Emerging) Freezing Techniques (Dehydrofreezing, Pressure-Shift Freezing, Ultrasonic-Assisted Freezing, and Hydrofluidization Freezing)

Laura A. Ramallo, Miriam N. Martino,
and Rodolfo H. Mascheroni

CONTENTS

16.1   INTRODUCTION

As any other evolving technology, freezing tries to develop new applications of known procedures, as in combination with partial dehydration in dehydrofreezing or using fluidization principles in hydrofluidization (HF) freezing, or testing promising developing (emerging) technologies such as very high pressures in pressure-shift freezing or ultrasound in ultrasonic-assisted freezing.

The first two applications are much more advanced in their concepts and applications, and the last two have promising possibilities that must be studied much more thoroughly.

16.2   DEHYDROFREEZING

Although freezing is a classic procedure to preserve foods, freezing preservation should not be applied directly for most fruits and vegetables because it tends to cause deteriorative changes in texture and other properties. Fruits and vegetables have a high water content (ranging from 75% to 95%); consequently, they can be susceptible to forming large ice crystals during freezing. Water expands when frozen, and large ice crystals that formed cause cell injury. Even if an increasing freezing rate can reduce the possibilities of the formation of large ice crystals, the tissue damage is still inevitable due to the large water content (Li and Sun 2002b). Dehydrofreezing is a variant of freezing in which a food is dehydrated to a desirable moisture content and then frozen (Robbers et al. 1997; Spiazzi et al. 1998; Maestrelli et al. 2001; Li and Sun 2002a; Sun 2006). So partial dehydration constitutes the first stage of the dehydrofreezing process, affecting the freezing process mainly through the freezing time and the product quality. In fact, each operation (dehydration and freezing) may affect, partially or totally, the properties of food, and those changes can be linked to each other.

Diminution of moisture content can reduce the amount of water to be frozen and therefore lowers the refrigeration load necessary for the freezing process (Spiazzi et al. 1998). Consequently, a partial dehydration of fruits and vegetables can reduce the costs of the later stage of freezing, as well as the costs of packaging, distribution, and storage.

Nevertheless, the treatment designer not only must minimize process costs but also must provide all the conditions to assure appropriate product quality. In the continual development of technological processes necessary to preserve sensory characteristics and physical structure of fruits and vegetables, novel studies that analyze the application of different partial dehydration procedures prior to freezing have been published. These dehydration techniques are as follows:

- *Osmotic dehydration.* This pretreatment, as well as the advantage of the water removal and consequent reduction in the ice crystal formation, can increase the content of cryoprotective solutes (Tregunno and Goff 1996; Dermesonlouoglou et al. 2007). Osmo-dehydrofreezing has been studied in many vegetables: apricots (Forni et al. 1997), apples (Tregunno and Goff 1996; Marani et al. 2007; Blanda et al. 2008), kiwi fruits (Forni et al. 1990;

Robbers et al. 1997; Talens et al. 2002; Chiralt et al. 2001; Marani et al. 2007), melons (Maestrelli et al. 2001), beans (Biswal et al. 1991), strawberries (Chiralt et al. 2001; Marani et al. 2007; Blanda et al. 2009), mangoes (Nunes et al. 1996; Chiralt et al. 2001), and pears (Marani et al. 2007), among others. The osmotic dehydration treatment and the effects related to the improvement of nutritional, sensorial, and functional properties of the products are analyzed in Chapter 10.

- *Air drying.* Conventional convection drying constitutes a technique that is more used in the dehydration of plant-based foods, in which the solid to be dried is exposed to a continuously flowing stream of hot air in order to induce moisture loss. Diverse changes in physical and chemical characteristics of fruits and vegetables can occur during drying. Several researches have shown that a partial dehydration prior to freezing may improve the quality of the final product since, during thawing, the drip loss is smaller than in the fresh product (Maestrelli et al. 2001; Moraga et al. 2006; Ramallo and Mascheroni 2010). Another advantage of drying prior to freezing of some fruits is that mechanical properties (texture) of a dried–frozen–thawed product resemble more the fresh food texture than that of a frozen–thawed product (Bolin and Huxoll 1993). Nevertheless, for other fruits, the mechanical properties of fresh fruit were very different than those of dried–frozen–thawed fruits (Ramallo and Mascheroni 2010).
- *Combination pretreatments.* The combination of osmotic dehydration and air drying, applied before freezing, has reduced exudate losses and improved the texture and the sensory acceptability after thawing of muskmelons (Maestrelli et al. 2001), pears (Bolin and Huxoll 1993), strawberries (Maestrelli et al. 1997; Torreggiani et al. 2000), and kiwifruits (Robbers et al. 1997), among others.

In recent years, other techniques with a promising future for its application in dehydrofreezing have been developed.

- *Microwave drying.* It is a rapid dehydration method that can be applied to fruits and vegetables, in which the heat is generated inside the food. Microwave energy may be combined with hot air drying (Zhang et al. 2006) and with osmotic dehydration (Torringa et al. 2001) in order to reduce the drying time and improve the final product quality. Advantages for the application of microwaves in vegetable dehydration were investigated (Nijhuis et al. 1998), but there are no data of its application prior to the freezing process. In recent times, microwave heating has been successfully used in freeze drying (Wang and Chen 2007; Duan et al. 2010).
- *Pulsed electric field* (PEF). It was successfully used in accelerating convective mass transfer. PEF induced inactivation of enzymes and permeabilization of plant membranes (Knorr and Angersbach 1998), which make PEF attractive for use in further processing such as dehydration (Ade-Omowaye et al. 2001). PEF treatment prior to air-blast freezing of potato produced a

decrease in the freezing time. The PEF pretreatment gave significant structural damage in freeze-thawed potato samples. This is an undesirable phenomenon and can deteriorate the quality of thawed potatoes.

Among pretreatments prior to fruit freezing (osmotic and air dehydration), osmotic dehydration has been studied more than air drying because of two fundamental reasons: less energy requirement and maintaining the fruit quality. Nevertheless, the results are contradictory in some cases, mainly for the drip loss during thawing. Data reported by Maestrelli et al. (2001) indicated that air dehydration applied before freezing of melon spheres significantly reduced exudate losses after thawing and improved the texture at thawing, demonstrating that the reduction of water content protects the tissues from freezing damage. Nevertheless, osmotic dehydration treatment alone did not reduce the exudate loss.

As is well known, physicochemical characteristics of dehydrated food are affected by drying conditions. Food properties also change during the freezing–thawing process, and these changes are linked to the water content of the partially dehydrated product.

16.2.1 FREEZING PROCESS

The materials dehydrofrozen are normally whole little fruits or vegetables or bigger ones diced or cut into slices, with a thickness lower than 1 cm. With the aim of obtaining high final quality through a rapid freezing process, these materials are individually quick-frozen in standard high-freezing-rate industrial freezing equipment (fluidized bed freezers or cryogenic freezers), which are also described in this book in Chapters 11 and 12.

16.2.2 DEHYDROFREEZING EFFECTS ON FRUIT AND VEGETABLE QUALITY

16.2.2.1 Color Changes

In fruits and vegetables, color is an important quality parameter and it is closely linked with the acceptability of the product. Usually, during dehydrofreezing operation, the most important color changes take place during the air-drying process, even though the freezing–thawing process also induces additional changes in the color of plant-based foods. These changes are linked to the type of fruits or vegetables and with process conditions.

The individual effect of air drying on apple color can be observed in several works. Thus, redness and yellowness increased significantly during air drying at 55°C (Mandala et al. 2005) and during drying at 70°C (Krokida et al. 2001), which is indicative of the browning reaction. Also, changes of color were significant after freezing and thawing of fresh apples (Chassagne-Berces et al. 2010) as a consequence of no pretreatment. In this study, the experimental data indicated that lightness decreased, and redness and yellowness increased after apple freezing–thawing. Nevertheless, for apple samples previously immersed in $NaHSO_3$ solution, the instrumental color shows no appreciable differences among samples partially dried and fresh fruits before and after being frozen–thawed (Namor et al.

1974). It was observed in some fruits such as apple that color change depends on preliminary chemical treatment rather than on the drying–freezing–thawing process conditions.

Color coordinates a^*, b^*, and L^* of pineapples did not significantly change during the pretreatment by air drying: chroma did not change, while hue decreased (increase in yellowness), so the color of pineapple slices appeared more "concentrated" after drying (Ramallo and Mascheroni 2006). Results reported by Maestrelli et al. (2001) show melon air drying, as a pretreatment to freezing, caused an increase in chroma values. But the subsequent freezing process induced a decrease in chroma values of all samples (fresh and dried fruits). Thus, after being frozen–thawed, the color of melon returned close to the fresh fruit color (Maestrelli et al. 2001). Differences in pears' color were undetectable visually in the combined process of partial drying and freezing–thawing (Bolin and Huxoll 1993), although the instrumental measures of the color registered some differences between the color of fresh and dried fruits.

The color results obtained during dehydrofreezing of kiwifruits show some differences in relation to the behavior of other fruits: significant changes in color parameters were observed during air drying where the kiwifruits turned darker (Robbers et al. 1997). Additionally, in sensory tests of dehydrofrozen fruits, panelists judged the appearance of apples, pears, and peaches after dehydrofreezing as very acceptable; only the appearance of kiwifruit after dehydrofreezing was considered unacceptable (Torreggiani et al. 1987).

16.2.2.2 Mechanical Properties Changes

Generally, in compression tests, the plant tissue mechanical response shows an initially linear stress–strain relationship, measured as elasticity modulus, and afterward, the stress falls as a consequence of sample fracture or failure. Air drying causes drastic changes in the mechanical properties of fruits and vegetables, and the texture parameter variation is strongly linked to the characteristics of the fruit: it can bring about an increase in the values of failure stress and a decrease in elasticity modulus, as in the case of pineapples (Ramallo and Mascheroni 2010). Kotwaliwale et al. (2007) also reported that hardness increased at the initial stage of mushroom drying. In contrast, failure stress (F_{max}) decreased during potato drying; consequently, the drying of potato slices causes the softening of the tissue (Troncoso and Pedreschi 2007).

Nevertheless, freezing always reduces the resistance to the deformation of vegetable tissue (Sousa et al. 2007; Marani et al. 2007).

These changes in mechanical properties vary according to the product and to the operation conditions, mainly with the dehydration process. So during air drying, a softening of kiwi compared with fresh fruits was observed (Robbers et al. 1997). Softening was independent of the moisture eliminated (between 40% and 60% wet basis). A decrease in the maximum force values after freezing of fresh fruits was also registered.

The freezing–thawing process can hasten firmness and turgor loss of vegetable tissue, mainly in nondehydrated fruits such as pineapples (Ramallo and Mascheroni 2010), strawberries (Torreggiani et al. 2000), pears (Bolin and Huxoll 1993), and kiwifruits (Robbers et al. 1997). This effect may be due to the cellular damage produced by ice crystal formation. Pineapple samples hot-air-dried followed by freezing–thawing treatment showed greater failure resistance than fresh fruits.

16.2.2.3 Rehydration/Reconstitution Behavior

Limited literature concerning rehydration capacity of dehydrofrozen fruits and vegetables is available. Additionally, there is no consistency in measuring procedures and calculations of rehydration indices for dried foods (Lewicki 1998), which makes the comparison of results difficult.

It is generally accepted that the level of rehydration is dependent on the degree of cellular damage (McMinn and Magee 1997; Krokida and Marinos-Kouris 2003); this injury to the material can be caused by drying treatment, due to which the variables of the drying process have significant influence upon rehydration ability.

There are a large number of studies for fruit and vegetable frozen storage, but in these reports, the product is not rehydrated because it was not previously dried. Because of this, little knowledge of the freezing effect on the rehydration ability of dehydrofrozen plant-based foods is available. The rehydration capacity of the dehydrofrozen apple varies progressively with the extent of the preceding dehydration (Namor et al. 1974).

16.2.2.4 Nutritional Changes: Loss of Vitamins

Several researches on chemical degradation of vitamins, mainly ascorbic acid, during fruit and vegetable frozen storage are available (Lisiewska and Kmiecik 1996; Sahari et al. 2004; Volden et al. 2009). Also, numerous researches have studied the vitamin retention during the drying process (Zanoni et al. 1999; Asami et al. 2003; Nicoleti et al. 2004; Erenturk et al. 2005; Orikasa et al. 2008; Santos and Silva 2009).

Due to the direct influence that temperature has on nutritional component degradation, it is reasonable to predict that the most important changes in the vitamin content will take place during fruit and vegetable drying.

In many cases, fruit frozen storage did result in a noticeable L-ascorbic acid decrease (Volden et al. 2009; Lisiewska and Kmiecik 1996). Cieślik et al. (2007) investigated the effects of freezing for 48 h at −22°C on several different vegetables; they found inconsistent effects.

Furthermore, the freezing–thawing step might affect the nutritional quality indirectly because vitamins and other water-soluble compounds may dissolve in liquid exuding from fruits. At this stage, it is crucial to diminish the losses by exudation to preserve the nutritional quality of the dehydrofrozen product. Ramallo and Mascheroni (2010) evaluated the ascorbic acid concentration in the exudates of pineapple slices. They found that pineapple samples with high moisture evidenced a marked decline in ascorbic acid content due to the freezing–thawing effect: the difference in the ascorbic acid content after being frozen–thawed was 25% for fresh fruits, whereas when fruits were dried–frozen–thawed, they lost approximately 20% of their ascorbic acid content.

Moraga et al. (2006) studied the effect of partial dehydration prior to freezing strawberries on the final quality of the product. These authors found that air drying markedly reduces the exuded liquid amount. Quite similar results have been published by Maestrelli et al. (2001) in a study concerning partial water removal prior to melon freezing.

16.3 PRESSURE-SHIFT FREEZING

It is a well-known anomalous thermodynamic characteristic of water that it is less dense in its solid state (ice) than in liquid state. Directly related to this property is the fact that at increasing pressures (above the atmospheric one), there is a steady descent in the freezing point, as can be seen in Figure 16.1 that presents a scheme of the pressure–temperature equilibrium for the water–ice system.

This feature has been considered as an important option to get instantaneous and uniform ice crystal nucleation over the whole volume of food and not only on the refrigerated surface, which would induce much less tissue damage, because ice crystals would be smaller, and the simultaneous nucleation in all the material would avoid cell dehydration and damage typical of normal freezing processes.

To attain this objective, one of the possible thermodynamic paths (see scheme in Figure 16.1) is to compress the food to very high pressures, higher than 100 MPa (path a–b in the scheme), then to refrigerate it in the unfrozen state to a sufficiently low temperature (path b–c), and finally to release pressure (path c–d), which provokes nucleation over all the volume. The type of ice crystals formed is ICE I, which is normal at ambient pressures. During this rapid ice crystal growth, there is an instantaneous temperature increase within the frozen temperature range (not shown), due to the heat of solidification released during crystallization, to near the initial freezing point. This heat must be removed by the refrigeration system.

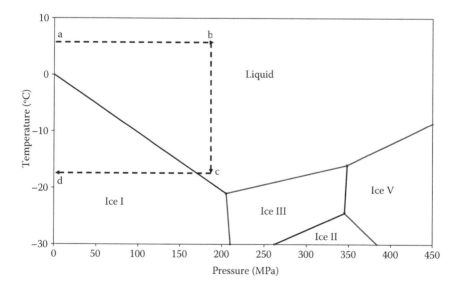

FIGURE 16.1 Partial scheme of the pressure–temperature equilibrium diagram for the water–ice system, including a possible path for high-pressure shift freezing.

The release of pressure can be either slow over several minutes (Fuchigami et al. 1997; Levy et al. 1999) or quick in a matter of seconds (Kanda et al. 1992; Otero et al. 1998; Levy et al. 1999), inducing uniform supercooling throughout the sample. The faster the pressure release, the lower the nucleation pressure and the greater the degree of supercooling caused. The higher the pressure and the lower the temperature before expansion, the more ice formed and hence the shorter the plateau time for a given cooling temperature (Otero and Sanz 2000).

At present, this technique is in the developmental stage, and its possible advantages related to higher final quality cannot overcome the disadvantages of its very high fixed costs.

16.4 ULTRASONIC-ASSISTED FREEZING

Ultrasound can be considered as an emerging and promising technology for the food industry for both analysis and modification of foods, although early reports on its application in fish freezing systems have been found since 1842 (Kissam et al. 1981). And even more, it is becoming feasible to consider commercial opportunities based on industrial-scale ultrasonic systems with worthwhile economic gains (Zheng and Sun 2006; Soria and Villamiel 2010).

Two avenues are opened for ultrasound applications. Mc Clements (1995) summarized those showing examples of both low- and high-intensity ultrasound techniques. Low-intensity ultrasound is mainly used as a nondestructive technique. The transmission characteristics of the ultrasonic waves, as affected by material properties, can provide information on composition, structure, and quality properties of foods (Saggin and Coupland 2001). Low-intensity ultrasound is used to measure material thickness, particle size, and flow rate, for material cleaning, phase transitions, creaming, and sedimentation, among others. On the other hand, high-intensity ultrasound can produce permanent physical or chemical changes in food products. It can generate emulsions, disrupt cells, promote chemical reactions, inhibit enzymes, tenderize meat, and modify crystallization processes.

Also, Mason et al. (2005) divided these applications in nondestructive testing as those used in diagnostic medicine, employing high frequency (5–10 MHz) or low power (<1 W/cm^2), and those that cause physical disruption of the material, known as low frequency (20 kHz to 1 MHz) or power ultrasound (>5 W/cm^2), generally used at around 40 kHz. Human hearing is within the 16- to 18-kHz range.

Like any sound wave, ultrasound is propagated via a series of compression and rarefaction waves induced in the molecules of the medium through which it propagates. The major difference between low and power ultrasound is produced when the power is sufficiently high to generate cavitation bubbles in the medium, which, in turn, will cause mechanical and even chemical effects.

16.4.1 LOW INTENSITY ULTRASOUND

The most common application of low-intensity ultrasound in food technology is as a noninvasive analytical technique with particular reference to quality control. Ultrasonic velocity, attenuation coefficient, and acoustic impedance are the parameters most frequently used for providing information about the physicochemical properties of foods, such as composition, structure, and physical state. According to McClements (1995), the velocity at which an ultrasonic wave travels through a material depends on its elastic modulus and density. The elastic modulus of the material varies whether a compression or shear wave is used. The ultrasonic velocity of a material can be determined either by the wavelength of the ultrasound at a known frequency or by the time taken for a wave to travel a known distance. The attenuation coefficient is a measure of the decrease in amplitude of an ultrasonic wave as it travels through a material. The major causes of attenuation are adsorption and scattering. Adsorption is caused by physical mechanisms converting energy stored as ultrasound into heat, for example, fluid viscosity, thermal conduction, and molecular relaxation. Scattering occurs in heterogeneous materials, such as emulsions, suspensions, and foams, when an ultrasonic wave is incident on a discontinuity (e.g., a particle). Unlike in the case of adsorption, the energy is still stored as ultrasound, but it is not detected because its propagation direction and phase have been altered. Measurements of the adsorption and scattering of ultrasound provide valuable information about the physicochemical properties of food materials, including concentration, viscosity, molecular relaxation, and microstructure. The acoustic impedance is a measure of the resistance a medium imposes to a sound wave. It depends on sound velocity, density, and material elastic properties. Ultrasonic imaging techniques can detect internal boundaries between different materials, since high percentage of ultrasound is reflected when two materials have very different acoustic impedances. Besides, a change in the material state from liquid to solid and therewith changing elastic properties will affect the acoustic impedance.

One of the major reasons why low-intensity ultrasound has not been used more widely in the food industry has been the lack of commercial ultrasonic instrumentation specifically designed to characterize foods. Thus, researchers have had to design and set up their own experiments, which required a fairly good understanding of the physical principles of ultrasound. This situation is changing, and a number of instrument manufacturers have recently developed ultrasonic devices that are suitable for characterizing foods (McClements 1995; Kress-Rogers and Brimelow 2001). Most ultrasonic instruments utilize either pulsed or continuous wave ultrasound. Pulse techniques are by far the most widely used because they are easy to operate, measurements are rapid and noninvasive, and the technique can easily be automated. Continuous wave techniques are used when highly accurate measurements are needed and tend to be found in specialized research laboratories (McClements 1995). Recently, more convenient and rapid methods have been developed that use Fourier transform analysis of broadband ultrasonic pulses to measure the frequency dependence of the ultrasonic velocity, attenuation coefficient, and acoustic impedance.

16.4.1.1 Applications of Low-Intensity Ultrasound in Food Freezing

Ultrasonic sensors have the advantages of propagating through opaque materials such as equipment walls and also being relatively cheap and suitable for automation (Gülseren and Coupland 2007). Although ultrasonic transmittance devices are easy to implement in a laboratory, it is often problematic for online applications, where existing equipment may have to be replaced or modified to allow the fitting of transducers. This potential disadvantage of ultrasonic transmittance measurements, which require the propagation of sound across a known and fixed distance, can be overcome by using ultrasonic reflectance. Saggin and Coupland (2001) stressed that ultrasonic reflectance is a precise and reliable method in measuring the concentration of simple food solutions because it is a function of both ultrasonic velocity and density. Also, with appropriate calibration, one can estimate the solid content of solutions within approximately 3% standard error of the true value. These researchers designed a transducer that can simply be attached to the outer wall of the equipment either by physical coupling or merely pressing the sensor in place by hand. Successful applications to determine sugar content of fruit juices, solid content of fat, and phase transitions in lipids were reported (Gülseren and Coupland 2007). However, few research groups are facing the challenge to develop a sensitive ultrasound device to measure formation of ice *in situ* to optimize freezing processes (Sigfusson et al. 2004; Gülseren and Coupland 2007, 2008; Aparicio et al. 2008; Sanz et al. 2008). Once set, the operating parameters of a continuous freezer are rarely altered during a process run. This is inherently inefficient as the process is optimized for the worst-case sample (i.e., high initial temperature or large size) and so most of the food will be overfrozen. If it were possible to perform measurements online, it would be feasible to continuously alter the freezer operation to respond to the changing product properties (Sigfusson et al. 2004). Thus, researchers have been motivated to develop ultrasonic sensors for characterizing partially frozen foods and to adapt them for industrial applications, based on the large difference between the ultrasonic velocity in ice (approximately 3900 m/s) and in water (approximately 1400 m/s). Variations in ice content were highly manifested by changes in speed of sound during freezing of orange juice with different concentrations (Figure 16.2). However, high ultrasonic attenuation may appear during freezing due to different causes, decreasing the transmitted signal. One possible mechanism for ultrasonic losses in frozen solutions is scattering by the crystals themselves or by the grain boundaries between the crystals. Another factor is the presence of air bubbles formed on freezing. Since the solubility of gases is significantly higher in water than in ice, bubbles are known to form upon freezing. Bubbles pulsate when the ultrasonic frequency corresponds to a critical bubble radius causing large losses of acoustic energy. Some of these factors were analyzed by Gülseren and Coupland (2008). They measured ultrasonic velocity and attenuation as a function of temperature in partially frozen sucrose solutions, which had either been degassed under vacuum or not prior to freezing. The temperature of the sudden changes in ultrasonic velocity and attenuation on melting corresponded to the thermodynamic melting point on the phase diagram (Figure 16.3). These changes could be used to easily establish the phase boundary. Ultrasonic velocity increased approximately linearly with ice content in all cases and was somewhat affected by the initial sucrose concentration and degassing. Their experiments showed that the high ultrasonic attenuation in frozen systems was determined mainly by the presence of air bubbles and only weakly by ice content.

Several approaches have been used to estimate the amount of ice (or the amount of unfrozen solution) as a function of temperature. Carcione et al. (2007)

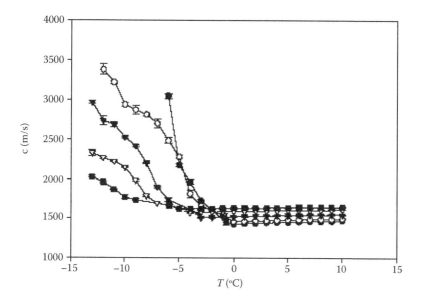

FIGURE 16.2   Ultrasonic velocity in orange juice solutions (● 5%, ○ 10%, ▼ 20%, △ 30%, and ■ 40%) as a function of temperature. Curve discontinuity indicates the formation of ice. (From Gülseren, I. and Coupland, J.N., *Journal of Food Engineering*, 79, 1071–1078, 2007. With permission.)

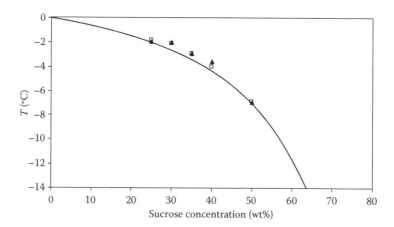

FIGURE 16.3   Sucrose–water phase diagram (calculated from literature data of Young and Jones, 1949). Points show the temperature of the discontinuity in the ultrasonic velocity (□) or attenuation (▲) temperature curves at the freezing point obtained by Gülseren and Coupland (2008).

approached the subject based on a poroelastic model to predict the degree of freezing of orange juice by means of the wave velocity and attenuation factor. The model showed an excellent agreement with the experimental data, particularly the ultrasonic velocities.

Aparicio et al. (2008) developed an easy and feasible method to determine a representative overall temperature and the ice content of NaCl model solutions and real food such as fish. They even patented this quick method, suitable for online monitoring of frozen, freezing, and thawing systems, which can also be adapted to a large variety of containers, geometrical situations, and water contents (Sanz et al. 2008). The invention relates to a noninvasive device for determining the temperature and ice content of frozen foods using ultrasound, comprising a 1- to 10-MHz frequency receiver/reflector and transmitter. The method uses a program to perform calculations as a function of the thickness of the sample and the flight time of the wave, which is determined using conventional methods, with the type of material (fruit–vegetable or meat–fish) and the known initial freezing point.

16.4.2 POWER ULTRASOUND

Mason et al. (2005) clearly explained the rationale behind the profound modifications that ultrasound can produce in liquid systems by propagating compression and rarefaction waves. At sufficiently high power, rarefaction cycle may exceed the attractive forces of the molecules of a liquid, and cavitation bubbles will form. Such bubbles grow by a process known as rectified diffusion, that is, small amounts of vapor (or gas) from the medium enter the bubble during its expansion phase and are not fully expelled during compression. Since the acoustic field is not uniform, it will cause the cavitation bubble to become unstable and to collapse (Figure 16.4). It is this collapse that generates the energy for chemical and mechanical effects. For example, in aqueous systems at an ultrasonic frequency of 20 kHz, each cavitation bubble collapse acts as a localized "hotspot" generating temperatures of about 4000 K and pressures above 1000 atmospheres. This bubble collapse, distributed through the medium, has a variety of effects within the system depending upon the type of material involved. In the case of solid–liquid systems, unlike cavitation bubbles collapsing in the bulk liquid, collapse of a cavitation bubble on or near a surface is asymmetrical. The result is an inrush of liquid predominantly from the side of the bubble remote from the surface, resulting in a powerful liquid jet targeted at the surface. The effect is equivalent to high-pressure jetting and is the reason that ultrasound is used for cleaning. It can also increase mass and heat transfer to the surface. Surface imperfections or trapped gas can act as the nuclei for cavitation bubbles. The collapse of a cavitation bubble in the liquid phase near a particle can force it into rapid motion. Under these circumstances, the general dispersive effect is accompanied by interparticle collisions that can lead to erosion, surface cleaning, and wetting of the particles and particle size reduction. Besides, the mechanical effects of cavitation close to a liquid–liquid interface are used to produce effective emulsification or homogenization. Applications of power ultrasound on drying, emulsification, freezing and thawing, cutting, meat tenderization, and sono-crystallization, among others, are summarized by Bhaskaracharya et al. (2009).

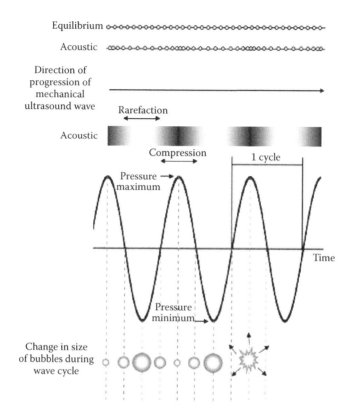

Equilibrium

Acoustic

Direction of progression of mechanical ultrasound wave

Rarefaction

Acoustic

Compression

1 cycle

Pressure → maximum

Time

Pressure minimum

Change in size of bubbles during wave cycle

FIGURE 16.4 Ultrasonic cavitation. (Reprinted from *Trends in Food Science and Technology*, 21, Soria, A.C., Villamiel, M., 323–331, Copyright 2010, with permission from Elsevier.)

16.4.2.1 Applications of Power Ultrasound in Food Freezing

High power ultrasound has also been applied on freezing to produce primary nucleation or accelerate the freezing rate. Primary nucleation rate highly depends on supersaturation, the number of nuclei, and its distribution with time. Thus, batch crystallization from a solution by cooling or evaporation generally needs a marked decrease in temperature or an increase in concentration to trigger primary nucleation to start and proceed over an often not well-defined period of time. It is also difficult to obtain reproducible crystal size distributions in batch crystallization. These problems can largely be avoided by using pressure as an additional driving force for nucleation, as can be achieved by fluid dynamic or ultrasonic cavitation, during a predetermined time interval (Virone et al. 2006). The already existing supersaturation that is low compared to the pressure-induced one preserves the developed nuclei and allows their further outgrowth. The total number of nuclei depends on the cavitator design, the power input, and the physical properties of the solute. Since cavitation can create a very high supersaturation, it can induce solid nuclei in systems that will otherwise only give an oily substance upon cooling. Due to the large number of nuclei so rapidly created, cavitation can also be applied to obtain a large number of ultrafine particles or to create nuclei of another polymorphism (Virone et al. 2006). Inada et al. (2001) demonstrated that ultrasonic vibration can actively control supercooling of water, a critical issue in cold-energy storage and transport systems using ice slurries. Experimental results and a simulated model based on cavitation phenomena indicated that ultrasonic vibrations strongly promoted the phase transition from supercooled water to ice for both pure and tap water. Virone et al. (2006) designed a novel cavitator where a standing wave was created to obtain a well-defined pressure profile in a reactor volume. From these experiments, acoustic cavitation appears to be an adequate tool to induce a step function for reproducible nucleation with only a negligible sensitivity for the initial supersaturation. A strong reduction in induction times with high reproducibility was found in batches of ammonium sulfate solutions. They reported that the median induction time from insonated experiments was 39.07 ± 8.2 min, while it was 150.07 ± 29.9 min from the blanks. Chow et al. (2003) used a novel optical microscope stage that permits temperature control and simultaneous excitation of the sample with alternating pressures at ultrasonic frequencies. They have stressed the fact that both primary and secondary nucleations of ice may occur consecutively, within a sucrose solution that was continuously sonicated. The primary nucleation of ice in sucrose solutions was achieved at higher nucleation temperatures and with a lower standard deviation than those obtained without ultrasound. Their results showed that increasing both the ultrasonic output level and the duty cycle can increase nucleation temperature. After fine dendritic structures and crystal breakup, secondary nucleation was observed to form new smaller crystal nucleating sites. Cavitation bubbles appear to be important in the fragmentation of the dendrites, although ultrasonic streaming was also reported to be significant (Chow et al. 2003).

Besides, ultrasound-assisted processes can have promising applications in freeze drying of high-value food (vitamins, nutrients, natural additives, etc.) and pharmaceutical products (drugs, vaccines, etc.). Freeze drying is a convenient drying process for thermally sensitive samples. The whole process begins with freezing, followed by the primary drying step or sublimation, then the secondary drying or desorption step, and finally storage. Power ultrasound has been used by several authors to optimize the freezing step (Nakagawa et al. 2006; Hottot et al. 2007). Freezing is a key step because it fixes the morphology of the frozen material and, in turn, the final morphology of the freeze-dried material. It can also be a damaging step since it can lead to biological activity losses. Hottot et al. (2007) stressed the compromise to optimize the freeze-drying process. The ice crystal sizes must be large enough to obtain the shortest primary drying times. However, in the secondary period, these sizes must be small enough to offer a large surface area of the dried product to allow an easy desorption of unfrozen water from the amorphous matrix. For a given formulation, three independent factors control the freezing process: the cooling rate, the supercooling degree, and the ice nucleation temperature. Nakagawa et al. (2006) set up an ultrasound system to optimize an industrial freeze-drying process by controlling the freezing step. They can efficiently trigger ice crystal nucleation of model solutions of mannitol, sucrose, and bovine serum albumin. No significant differences in the ice morphology of frozen sample prepared with or without ultrasound were found, if they were nucleated at the same temperature or operating conditions (cooling rate, sample geometry, etc.). The overall primary drying rate during freeze drying increased with nucleation temperature for both spontaneously and controlled nucleated samples; however, ultrasound allowed nucleation to occur at much higher temperatures (Figure 16.5). A great enhancement of the sublimation rate was achieved by increasing nucleation temperature from average nucleation values to around −2°C with ultrasound. A step forward was obtained by analyzing the operating conditions of temperature and acoustic power to freeze 10% w/w mannitol aqueous solutions contained in vials (Saclier et al. 2010). These authors found that increasing supercooling and acoustic power resulted in decreasing ice crystals' mean size and increasing their mean circularity, therefore allowing management of the size and morphology of ice crystals by this technique. These data as a whole confirmed the adequacy of ultrasound to control the nucleation processes. Passot et al. (2009) obtained higher activity recovery of a model protein (catalase) in a prototype freeze dryer when investigating the impact on the primary drying stage of an ultrasound-controlled ice nucleation technology and compared it with usual freezing protocols. The ultrasound technology made it possible to decrease the sublimation time by 14%, compared with the freezing method involving a constant shelf cooling rate of 1°C/min, and to improve intrabatch homogeneity. Summarizing, ultrasound accelerates notably the primary sublimation rates during the freeze-drying process, improves the quality factors related to structural properties (rehydration facility, water vapor permeability, etc.), and reduces the operating costs related to drying times.

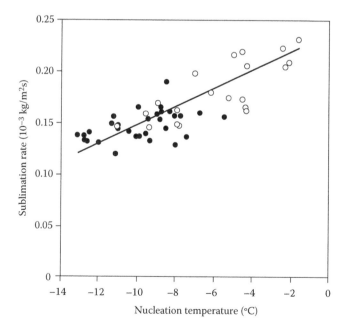

FIGURE 16.5 Primary drying rates as function nucleation temperatures (10% mannitol). (○) controlled nucleation by ultrasound; (●) spontaneous nucleation. (From Nakagawa, K., et al., *Chemical Engineering and Processing*, 45, 783–791, 2006. With permission.)

Frost deposition is a well-known and undesirable phenomenon in numerous fields of industry such as aeronautics, air conditioning, and cryogenics. In the near future, ultrasound could also improve heat exchanger performance by decreasing frost deposits associated with the increase in air pressure drop and decrease in thermal efficiency. Li et al. (2010) carried out a series of experiments concerning the initial frost nucleation and growth in atmospheric flow with a 20-kHz ultrasound laboratory device. The visual results showed that the size of deposited freezing droplets on the cold surface during frost nucleation with the effect of ultrasound was much smaller and sparser than without ultrasound. Furthermore, the frost grew along the horizontal direction compared with the normal vertical direction and had feather-like appearance without the effect of ultrasound. A reduction in the frost thickness of about 75% was obtained. According to Li et al. (2010), ultrasound has a mechanical effect that can cause the periodic intensive vibration of water vapor around a flat surface. The acceleration of the water vapor due to intensive vibration is 104 times the gravity acceleration.

Such a large acceleration can cause violent disturbances of vapor along propagation direction, which will modify directional migration of the vapor to the cold surface due to the concentration gradient.

Although most of the applications were carried out at laboratory scale, there are some examples at the pilot plant level. The design challenge for scaled-up units is to arrange multiple transducers to give a reasonably uniform intensity distribution throughout a realistic working volume. According to Luque de Castro and Priego-Capote (2007), to achieve high-intensity fields in large volumes, it is preferable to operate at lower face intensity (approx. 100 W m^{-2}) over an extended area and place multiple transducers around the medium that is being sonicated. Accordingly, it is likely that in the near future, industries involved with batch crystallization, freeze drying, or frost deposition will be using ultrasound techniques to solve their problems.

16.5   HF FREEZING

Food freezing by HF is an advanced method of immersion freezing, developed with the aim of increasing heat transfer rate, reducing power costs, lowering ambient pollution, and attaining high-quality final products. It was developed and described by Fikiin (1985, 1992) and Fikiin and Fikiin (1998). Figure 16.6 is a scheme of this HF system.

In brief, HF can be viewed as a combination of immersion freezing with forced liquid fluidization (Peralta 2009). In food freezing by HF, a concentrated aqueous solution at low temperature is used as a liquid refrigerant (as in immersion freezing). This liquid is pumped upward through orifices or nozzles into a vessel where the food products to be frozen are charged, thereby creating a turbulent agitation due to the ascending liquid jets. Thus, a fluidized bed of food products and a highly turbulent liquid are produced, obtaining high heat transfer coefficients (Fikiin 1992; Fikiin and Fikiin 1998; Scheerlinck et al. 2002; Fikiin et al. 2003; Verboven et al. 2003; Peralta et al. 2008). In particular, Verboven et al. (2003) determined convection heat transfer to spherical particles inside a pilot HF freezing unit that contained a food tank with a perforated bottom plate to create agitating jets. An aqueous solution of

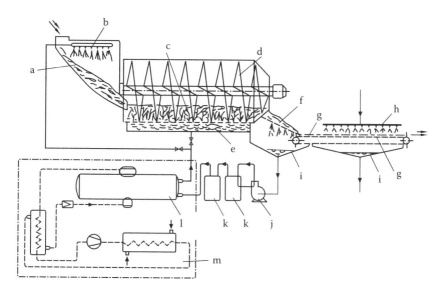

FIGURE 16.6 HFM-based freezing system: (a) charging funnel; (b) sprinkling tubular system; (c) refrigerating cylinder; (d) perforated screw; (e) double bottom; (f) perforated grate for draining; (g) netlike conveyor belt; (h) sprinkling device for glazing; (i) collector vats; (j) pump; (k) rough and fine filters; (l) cooler of refrigerating system; (m) refrigeration plant. (Adapted from Fikiin, K.A., and Fikiin, A.G., *IIR Proceedings Series Refrigeration Science and Technology* 6, 319–326, 1998.)

30% ethanol + 20% glucose was used as the refrigeration medium in a temperature range of –20°C to 0°C and flow rates from 5 to 15 L min^{-1}. They measured the cooling profiles of aluminum spheres of 5–50 mm to obtain surface heat transfer coefficients. Coefficients were within a range of 154–1548 W m^{-2} C^{-1} and depended on diameter, flow rate, refrigeration temperature, and fluid agitation level. Peralta et al. (2008, 2009) developed a pilot-scale prototype HF and studied heat transfer to stationary copper spheres of different diameters. Measurements were carried out using a solution of NaCl–water as a refrigerant. Re, Pr, the orifice–sphere distance, and the curvature of the sphere surface were considered in a correlation for Nusselt number for the HF system. Finally, Peralta et al. (2008) and Peralta (2009) developed a mathematical model to predict heat and mass transfer during HF taking into account liquid fluid dynamics and thermodynamic properties.

With respect to final food quality, if the HF process is adequately controlled, it appears as having important potential advantages (Fikiin and Fikiin 1998; Flair-Flow 2003; Mascheroni 2007) because high transfer rates are attained with low temperature gradients. The critical crystallization zone of water in the food is quickly surpassed, assuring that a fine ice crystal structure is produced. This prevents part of the damage to cell tissues inherent to ice crystallization; food surface is immediately frozen to a solid crust, which lowers osmotic dehydration and gives the product excellent appearance. HF uses secondary refrigerants that are environmentally compatible; fluidization is attained with jets of low flow rates and pressures due to the high buoyancy forces, with the consequent savings in energy and lower mechanical stresses on food pieces; besides, equipment operation is continuous, with low maintenance costs, convenient for automation and reduction in labor expenses (Peralta 2009).

Notwithstanding, as in immersion freezing, one of the limitations is to find freezing solutions that simultaneously reach low temperatures and do not contaminate foods with undesired or abnormal tastes. Most of the tested solutions include salts and/or sugars. Fikiin et al. (2003) proposed the use of ice slurries based on sugar–ethanol aqueous solutions to freeze fruit pieces.

Another critical point is the type and size of foods that can be frozen. The system is especially useful for little fishes or shrimps, vegetables, and fruits (whole or in pieces), but not for other foods that could degrade during the fluidization process (such as meat balls).

In spite of the alleged advantages of HF, up to now, it must be considered as an emerging technology that needs further studies on heat and mass transfer, final product quality, and continuous equipment development.

REFERENCES

Ade-Omowaye, B.I.O., Angersbach, A., Taiwo, K.A., Knorr, D. (2001). Use of pulsed electric field pretreatment to improve dehydration characteristics of plant based foods. *Trends in Food Science and Technology* 12:285–295.

Aparicio, C., Otero, L., Guignon, B., Molina-García, A.D., Sanz, P.D. (2008). Ice content and temperature determination from ultrasonic measurements in partially frozen foods. *Journal of Food Engineering* 88(2):272–279.

Asami, D.K., Hong, Y.-J., Barrett, D.M., Mitchell, A.E. (2003). Comparison of the total phenolic and ascorbic acid content of freeze-dried and air-dried marionberry, strawberry, and corn grown using conventional, organic, and sustainable agricultural practices. *Journal of Agricultural and Food Chemistry* 51:1237–1241.

Bhaskaracharya, R.K., Kentish, S., Ashokkumar, M. (2009). Selected applications of ultrasonics in food processing. *Food Engineering Reviews* (1):31–49.

Biswal, R., Bozorgmehr, K., Tompkins, F., Liu, X. (1991). Osmotic concentration of green beans prior to freezing. *Journal of Food Science* 56:1008–1012.

Blanda, G., Cerretani, L., Bendini, A., Cardinali, A., Scarpellini, A., Lercker, G. (2008). Effect of vacuum impregnation on the phenolic content of Granny Smith and Stark Delicious frozen apple cvv. *European Food Research and Technology* 226:1229–1237.

Blanda, G., Cerretani, L., Cardinali, A., Barbieri, S., Bendini, A., Lercker, G. (2009). Osmotic dehydrofreezing of strawberries: polyphenolic content, volatile profile and consumer acceptance. *Lebensmittel Wissenschaft und Technologie* 42:30–36.

Bolin, H.R., Huxoll, C.C. (1993). Partial drying of cut pears to improve freeze/thaw texture. *Journal of Food Science* 58:357–360.

Carcione, J.M., Campanella, O.H., Santos, J.E. (2007). A poroelastic model for wave propagation in partially frozen orange juice. *Journal of Food Engineering* 80:11–17.

Chassagne-Berces, S., Fonseca, F., Citeau, M., Marin, M. (2010). Freezing protocol effect on quality properties of fruit tissue according to the fruit, the variety and the stage of maturity. *LWT—Food Science and Technology* 43:1441–1449.

Chiralt, A., Martínez-Navarrete, N., Martínez-Monzó, J., Talens, P., Moraga, G., Ayala, A., Fito, P. (2001). Changes in mechanical properties throughout osmotic processes: cryoprotectant effect. *Journal of Food Engineering* 49:129–135.

Chow, R., Blindt, R., Chivers, R., Povey, M. (2003). The sonocrystallisation of ice in sucrose solutions: primary and secondary nucleation. *Ultrasonics* 41:595–604.

Cieślik, E., Leszczyńska, T., Filipiak-Florkiewicz, A., Sikora, E., Pisulewski, P.M. (2007). Effects of some technological processes on glucosinolate contents in cruciferous vegetables. *Food Chemistry* 105:976–981.

Dermesonlouoglou, E., Giannakourou, M., Taoukis, P. (2007). Stability of dehydrofrozen tomatoes pretreated with alternative osmotic solutes. *Journal of Food Engineering* 78:272–280.

Duan, X., Zhang, M., Mujumdar, A.S., Wang, S. (2010). Microwave freeze drying of sea cucumber (*Stichopus japonicus*). *Journal of Food Engineering* 96:491–497.

Erenturk, S., Gulaboglu, S., Gultekin, S. (2005). The effects of cutting and drying medium on the vitamin C content of rosehip during drying. *Journal of Food Engineering* 68:513–518.

Fikiin, A.G. (1985). Method and system for immersion cooling and freezing of foodstuffs by hydrofluidisation. Invention Certificate No. 40164, Bulgarian Patent Agency.

Fikiin, A.G. (1992). New method and fluidized water system for intensive chilling and freezing of fish. *Food Control* 3:153–160.

Fikiin, K.A., Fikiin, A.G. (1998). Individual quick freezing of foods by hydrofluidisation and pumpable ice slurries. *IIR Proceedings Series Refrigeration Science and Technology* 6:319–326.

Fikiin, K., Tsvetkov, O., Laptev, Y., Fikkin, A., Kolodyaznaya, V. (2003). Thermophysical and engineering issues of the immersion freezing of fruits in ice slurries based on sugar-ethanol aqueous solution. *Ecolibrium* (August):10–15.

Flair-Flow (2003). Novelties of Food Freezing Research in Europe and Beyond. Flair-Flow Europe Synthetic Brochure for SMEs No.10, INRA, France.

Forni, E., Sormani, A., Scalise, S., Torreggiani, D. (1997). The influence of sugar composition on the colour stability of osmodehydrofrozen intermediate moisture apricots. *Food Research International* 30:87–94.

Forni, E., Torreggiani, D., Crivelli, G., Maestrelli, A., Bertolo, G., Santelli, F. (1990). Influence of osmosis time on the quality of dehydrofrozen kiwi fruit. In: *International Symposium of Kiwifruit*, Padova, Italy; published in *Acta Horticulturae* 282:425–433.

Fuchigami, M., Miyazaki, K., Kato, N., Teramoto, A. (1997). Histological changes in high pressure-frozen carrots. *Journal of Food Science* 62(4):809–812.

Gülseren, I., Coupland, J.N. (2007). Ultrasonic velocity measurements in frozen model food solutions. *Journal of Food Engineering* 79:1071–1078.

Gülseren, I., Coupland, J.N. (2008). Ultrasonic properties of partially frozen sucrose solutions. *Journal of Food Engineering* 89:330–335.

Hottot, A., Vessot, S., Andrieu, J. (2007). Freeze drying of pharmaceuticals in vials: Influence of freezing protocol and sample configuration on ice morphology and freeze-dried cake texture. *Chemical Engineering and Processing* 46:666–674.

Inada, T., Zhang, X., Yabe, A., Kozawa, Y. (2001). Active control of phase change from supercooled water to ice by ultrasonic vibration 1. Control of freezing temperature. *International Journal of Heat and Mass Transfer* 44:4523–4531.

Kanda, Y., Aoki, M., Kosugi, T. (1992). Freezing of tofu (soybean curd) by pressure-shift freezing and its structure. *Nippon Shokuhin Kogyo Gakkaishi* 39:608–614.

Kissam, A.D., Nelson, R.W., Ngao, J., Hunter, P. (1981). Water-thawing of fish low with frequency acoustics. *Journal of Food Science* 47:71–75.

Knorr, D., Angersbach, A. (1998). Impact of high intensity electric field pulses on plant membrane permeabilisation. *Trends in Food Science and Technology* 9:185–191.

Kotwaliwale, N., Bakane, P., Verma, A. (2007). Changes in textural and optical properties of oyster mushroom during hot air drying. *Journal of Food Engineering* 78:1207–1211.

Kress-Rogers, E., Brimelow, C.J.B. (2001). Ultrasonic instrumentation in the food industry. In: *Instrumentation and Sensors for the Food Industry*, 2nd ed., Woodhead Publishing Limited and CRC Press, Boca Raton, FL. Chapter 12, pp. 326–402.

Krokida, M.K., Marinos-Kouris, D. (2003). Rehydration kinetics of dehydrated products. *Journal of Food Engineering* 57:1–7.

Krokida, M.K., Maroulis, Z.B., Saravacos, G.D. (2001). The effect of the method of drying on the colour of dehydrated products. *International Journal of Food Science and Technology* 36:53–59.

Levy, J., Dumay, E., Kolodziejczyk, E., Cheftel, J.C. (1999). Freezing kinetics of a model oil-in-water emulsion under high pressure or by pressure release. Impact on ice crystals and oil droplets. *Lebensmittel- Wissenschaft und- Technologie* 32:396–405.

Lewicki, P.P. (1998). Some remarks on rehydration of dried foods. *Journal of Food Engineering* 36:81–87.

Li, B., Sun, D.-W. (2002a). Novel methods for rapid freezing and thawing of foods—a review. *Journal of Food Engineering* 54:175–182.

Li, B., Sun, D.-W. (2002b). Effect of power ultrasound on freezing rate during immersion freezing of potatoes. *Journal of Food Engineering* 55:277–282.

Li, D., Chen, Z., Shi, M. (2010). Effect of ultrasound on frost formation on a cold flat surface in atmospheric air flow. *Experimental Thermal and Fluid Science* 34:1247–1252.

Lisiewska, Z., Kmiecik, W. (1996). Effects of level of nitrogen fertilizer, processing conditions and period of storage of frozen broccoli and cauliflower on vitamin C retention. *Food Chemistry* 57(2):267–270.

Luque de Castro, M.D., Priego-Capote, F. (2007). Ultrasound-assisted crystallization (sono-crystallization). *Ultrasonics Sonochemistry* 14:717–724.

Maestrelli, A., Giallonardo, G., Forni, E., Torreggiani, D. (1997). Dehydrofreezing of sliced strawberries: A combined technique for improving texture. In *Engineering and Food*, Jowitt, R. (ed.), pp. F37–40. Sheffield, UK: Sheffield Academic Press.

Maestrelli, A., Lo Scalzo, R., Lupi, D., Bertolo, G., Torreggiani, D. (2001). Partial removal of water before freezing: cultivars and pre-treatments as quality factors of frozen muskmelon (*Cucumis melo*, cv *reticulates Naud.*). *Journal of Food Engineering* 49:255–260.

Mandala, I.G., Anagnostaras, E., Oikonomou, C.K. (2005). Influence of osmotic dehydration conditions on apple air-drying kinetics and their quality characteristics. *Journal of Food Engineering* 69:307–316.

Marani, C.M., Agnelli, M.E., Mascheroni, R.H. (2007). Osmo-frozen fruits: mass transfer and quality evaluation. *Journal of Food Engineering* 79:1122–1130.

Mascheroni, R.H. (2007). Overview on new freezing technologies. ALCUEFOOD Seminar, Sao Paulo, March 2007. Available at: http://www.docstoc.com/docs/40456538/ALCUEFOOD-Seminar-in-São-Paulo-Brasil, accessed March, 22, 2011.

Mason, T.J., Riera, E., Vercet, A., Buesa-Lopez, P. (2005). Application of ultrasound. In *Emerging Technologies for Food Processing*, Sun, D.-W. (ed.), London: Elsevier Ltd. Chapter 13, pp. 325–350.

McClements, J. (1995). Advances in the application of ultrasound in food analysis and processing. *Trends in Food Science and Technology* 61:293–299.

McMinn, W.A., Magee, T.R.A. (1997). Physical characteristics of dehydrated potatoes—part I. *Journal of Food Engineering* 33:37–48.

Moraga, G., Martínez-Navarrete, N., Chiralt, A. (2006). Compositional changes of strawberry due to dehydration, cold storage and freezing–thawing processes. *Journal of Food Processing and Preservation* 30:458–474.

Nakagawa, K., Hottot, A., Vessot, S., Andrieu, J. (2006). Influence of controlled nucleation by ultrasounds on ice morphology of frozen formulations for pharmaceutical proteins freeze-drying. *Chemical Engineering and Processing* 45:783–791.

Namor, M.S.S., Cowell, N.D., Rolfe, E.J. (1974). Dehydrofreezing of apple slices: a critical study of the process. In: *Proceedings of the IV International Congress Food Science and Technology*, Madrid, Spain: Consejo Superior de Investigaciones Cientificas. Vol. IV, pp. 59–67.

Nicoleti, J., Silveira-Junior, V., Telis-Romero, J., Telis, V. (2004). Ascorbic acid degradation during convective drying of persimmons with fixed temperature inside the fruit. In: *Proceedings of the 14th International Drying Symposium (IDS 2004)*, São Paulo, Brazil: State University of Campinas. vol. C, pp. 1836–1843.

Nijhuis, H.H., Torringa, H.M., Muresan, S., Yuksel, D., Leguijt, C., Klock, W. (1998). Approaches to improving the quality of dried fruit and vegetables. *Trends in Food Science Technology* 9:13–20.

Nunes, M.H., Miguel, M.H., Kieckbusch, T.G. (1996). Influencia dos solutos na conservacao de fatias de manga no processo combinado desidratacao osmótica e congelamento. Anais do VI Congresso Brasileiro de Engenharia e Ciências Térmicas (VI ENCIT) e VI Congreso Latinoamericano de Transferencia de Calor y Materia (LATCYM 96), Santa Catarina, Brazil: Universidade Federal de Santa Catarina. vol. I, pp. 647–650.

Orikasa, T., Wu, L., Shiina, T., Tagawa, A. (2008). Drying characteristics of kiwifruit during hot air drying. *Journal of Food Engineering* 85:303–308.

Otero, L., Sanz, P.D. (2000). High-pressure shift freezing. Part 1. Amount of ice instantaneously formed in the process. *Biotechnology Progress* 16:1030–1036.

Otero, L., Sanz, P.D. (2003). Modelling heat transfer in high pressure food processing: a review. *Innovative Food Science and Emerging Technologies* 4:121–134.

Otero, L., Solas, M.T., Sanz, P.D., de Elvira, C., Carrasco, J.A. (1998). Contrasting effects of high pressure-assisted freezing with conventional air freezing on eggplants tissue microstructure. *Lebensmittel-Untersuchung und Forschung (A)* 206(5):338–342.

Passot, S., Tréléa, I.C., Marin, M., Galan, M., Morris, G.J., Fonseca, F. (2009). Effect of controlled ice nucleation on primary drying stage and protein recovery in vials cooled in a modified freeze-dryer. *Journal of Biomechanical Engineering* 131(7): 074511.

Peralta, J.M. (2009). Congelación de alimentos por hidrofluidización. Tesis Doctoral, Universidad Nacional del Litoral, Santa Fe, Argentina.

Peralta, J.M., Rubiolo, A.C., Zorrilla, S.E. (2008). CFD modeling of a liquid round jet impinging on a sphere in a hydrofluidization system. In: *10th International Congress on Engineering and Food*.

Peralta, J.M., Rubiolo, A.C., Zorrilla, S.E. (2009). Design and construction of a hydrofluidization system. Study of the heat transfer on a stationary sphere. *Journal of Food Engineering* 90:358–364.

Ramallo, L.A., Mascheroni, R.H. (2006). Cambios de color, volumen y vitamina C durante el secado de ananá. TRABAJOS DEL X CONGRESO CYTAL Tomo IV: 1278–1287.

Ramallo, L.A., Mascheroni, R.H. (2010). Dehydrofreezing of pineapple. *Journal of Food Engineering* 99:269–275.

Robbers, M., Singh, R.P., Cunha, L.M. (1997). Osmotic–convective dehydrofreezing process for drying kiwifruit. *Journal of Food Engineering* 62:1039–1047.

Saclier, M., Roman Peczalski, R., Andrieu, J. (2010). Effect of ultrasonically induced nucleation on ice crystals' size and shape during freezing in vials. *Chemical Engineering Science* 65:3064–3071.

Saggin, R., Coupland, J.N. (2001). Oil viscosity measurement by ultrasonic reflectance. *Journal of the American Oil Chemists' Society* 78(5):509–511.

Saggin, R., Coupland, J.N. (2002). Measurement of solid fat content by ultrasonic reflectance in model systems and chocolate. *Food Research International* 35(10):999–1005.

Sahari, M.A., Boostani, F.M., Hamidi, E.Z. (2004). Effect of low temperature on the ascorbic acid content and quality characteristics of frozen strawberry. *Food Chemistry* 86:357–363.

Santos, P.H.S., Silva, M.A. (2009). Kinetics of L-ascorbic acid degradation in pineapple drying under ethanolic atmosphere. *Drying Technology* 27:947–954.

Sanz, P., Aparicio, C., Molina, A., Otero, L., Guignon, B. (2008). Device for determining temperature and ice content of frozen foods, e.g., fruit, determines thickness of sample, and flight time of ultrasound wave by type of material and initial freezing point of material. Patent. WO2008090242-A1

Scheerlinck, N., Jancsók, P., Verboven, P., Nicolaï, B. (2002). Influence of shape on the fast freezing of small fruits by means of hydrofluidisation. AgEng, N° 02-PH-034, Budapest, Hungary.

Sigfusson, H., Ziegler, G.R., Coupland J.N. (2004). Ultrasonic monitoring of food freezing. *Journal of Food Engineering* 62(3):263–269.

Soria, A.C., Villamiel, M. (2010). Effect of ultrasound on the technological properties and bioactivity of food: a review. *Trends in Food Science and Technology* 21:323–331.

Sousa, M.B., Canet, W., Alvarez, M.D., Fernandez, C. (2007). Effect of processing on the texture and sensory attributes of raspberry (cv. *Heritage*) and blackberry (cv. *Thornfree*). *Journal of Food Engineering* 78:9–21.

Spiazzi, E.A., Raggio, Z.I., Bignone, K.A., Mascheroni, R.H. (1998). Experiments in dehydrofreezing of fruits and vegetables: mass transfer and quality factors. In: *Advances in the Refrigeration Systems, International Institute of Refrigeration Proceeding Series* 6:401–408.

Sun, D.-W. (2006). *Handbook of Frozen Food Processing and Packaging*, 1st ed., Chapter 8. CRC Press, Florida, USA.

Talens, P., Martínez-Navarrete, N., Fito, P., Chiralt, A. (2002). Changes in optical and mechanical properties during osmodehydrofreezing of kiwi fruit. *Innovative Food Science and Emerging Technologies* 3:191–199.

Torreggiani, D., Forni, E., Crivelli, G., Bertolo, G., Maestrelli, A. (1987). Researches of dehydrofreezing of fruits. Part 1: Influence of dehydration levels on the products quality. In: *Proceedings of the XVII International Congress of Refrigeration*; Vienna, Austria, Vol. C, pp. 461–467.

Torreggiani, D., Rondo Brovetto, B., Maestrelli, A., Bertolo, G. (2000). High quality strawberry ingredients by partial dehydration before freezing. In: *Proceedings of the 20th International Congress of Refrigeration*, Sydney, Australia, 19–24 September 1999, Vol. IV, pp. 2100–2105.

Torringa, E., Esveld, E., Scheewe, I., van der Berg, R., Bartels, P. (2001). Osmotic dehydration as a pre-treatment before combined microwave-hot-air drying of mushrooms. *Journal of Food Engineering* 49:185–191.

Tregunno, N., Goff, H. (1996). Osmodehydrofreezing of apples: structural and textural effects. *Food Research International* 29:471–479.

Troncoso, E., Pedreschi, F. (2007). Modeling of textural changes during drying of potato slices. *Journal of Food Engineering* 82:577–584.

Verboven, P., Scheerlinck, N., Nicolai, B.M. (2003). Surface heat transfer coefficients to stationary spherical particles in an experimental unit for hydrofluidisation freezing of individual foods. *International Journal of Refrigeration* 26:328–336.

Virone, C., Kramera, H.J.M., van Rosmalena, G.M., Stoopb, A.H., Bakker, T.W. (2006). Primary nucleation induced by ultrasonic cavitation. *Journal of Crystal Growth* 294:9–15.

Volden, J., Bengtsson, G.B., Wicklund, T. (2009). Glucosinolates, l-ascorbic acid, total phenols, anthocyanins, antioxidant capacities and colour in cauliflower (*Brassica oleracea* L. ssp. *botrytis*); effects of long-term freezer storage. *Food Chemistry* 112:967–976.

Wang, W., Chen, G. (2007). Freeze drying with dielectric-material-assisted microwave heating. *AIChE Journal* 53:3077–3088.

Young, F.E., Jones, F.T. (1949). Sucrose hydrates. The sucrose–water phase diagram. *Journal of Physical Chemistry* 53(9):1334–1350.

Zanoni, B., Peri, C., Nani, R., Lavelli, V. (1999). Oxidative heat damage of tomato halves as affected by drying. *Food Research International* 31:395–401.

Zhang, M., Tang, J., Mujumdar, A.S., Wang, S. (2006). Trends in microwave-related drying of fruits and vegetables. *Trends in Food Science and Technology* 17:524–534.

Zheng, L., Sun, D.-W. (2006). Innovative applications of power ultrasound during food freezing processes: a review. *Trends in Food Science and Technology* 17:16–23.

17 Thawing

Quang Tuan Pham

CONTENTS

17.1   INTRODUCTION

Except for ice cream, frozen desserts, and the like, frozen food is usually thawed before eating, cooking, or further processing. Sometimes tempering (raising the temperature of frozen food to just below the freezing point) is carried out, as tempered food is firmer than completely thawed food and can be more easily cut or flaked. This chapter covers both the thawing and tempering processes.

Thawing is a much more difficult process to carry out safely than chilling or freezing. It usually takes much longer than freezing, and there is a high danger of subjecting the surface of the food to high temperature and humidity, which favor microbial growth. The surface is often the most vulnerable part since it is exposed to contamination from the environment, and it is the first part to heat up during thawing. It is desirable that the product is uniformly treated, and exposure to high temperature is stopped as soon as the desirable end point has been reached. In the home and at small establishments, thawing is often carried out using procedures that are rated unsafe (Lacroix et al. 2003; Mitakakis et al. 2004).

This chapter will first examine what happens to food during thawing or tempering, examine the factors that influence the choice and design of thawing processes, review the available thawing/tempering methods, and describe the various calculation and design methods.

17.2   PHYSICAL BEHAVIOR DURING THAWING

17.2.1   THAWING CURVE

Due to the latent heat peak, the inner parts of a large piece of frozen food will rise to and then remain at just below the freezing point for a period of time, known as the freezing plateau. On the other hand, those parts of the food that thaw earliest (such as the corners) will rapidly heat up toward ambient temperature and stay at those temperatures for a long time, causing nonuniformity of treatment and increasing the risk of local microbial growth (Figure 17.1). The nonuniformity is greatest for large products and when the heat transfer rate is fast. A compromise must be chosen between speed of thawing and uniformity of treatment.

17.2.2   BIOT NUMBER

The Biot number Bi is an important concept in understanding the physics of freezing and thawing processes and in solving practical problems. Consider, for example, a slab of frozen food being thawed in air or water: heat must first cross from the environment into the outer layer of the food through an "external resistance" $1/h$ formed by the boundary layer of film around the food and possibly some wrapping. Then it must travel to the center of the food through a layer with thickness R, which forms an "internal resistance" R/k_u. The ratio of internal to external resistances is proportional to Bi:

$$Bi \equiv \frac{R/k_u}{1/h} = \frac{hR}{k_u}. \qquad (17.1)$$

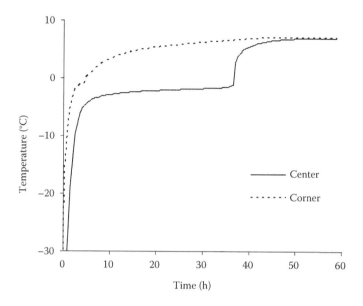

FIGURE 17.1 Evolution of temperature at the center and corner of a block of solid food, showing the fast rise toward ambient temperature at the corner and the thawing plateau just under the freezing point.

If the Biot number is much greater than 1 (internal resistance dominating), the engineer must concentrate on reducing internal resistance (e.g., by reducing the thickness or by using heat pipes), and it is of little use trying to increase air/water velocity or reducing the wrapping's thickness. The reverse is true when the Biot number is small (much less than 1). A large Biot number will also mean that the temperature will be more uneven (Figure 17.2), and microbial risks may be greater at the surface, so greater care is necessary, such as using a cooler thawing medium and controlling the process time better.

17.3 QUALITY CHANGES DURING THAWING

17.3.1 Microbial Growth

With meat and fish, microbial growth is usually the most important concern during thawing. In contrast to freezing, where the most vulnerable part is the center of the product, during thawing, it is the surface and especially the corners that are at risk. For uncut meat, since it is the surface that is most contaminated, the potential for microbial growth is much greater during thawing than during freezing.

Microbial growth is affected by temperature, water activity, pH, and the concentrations of nutrients. During meat and fish thawing, conditions for microbial growth are highly favorable (high water activity due to melted ice and/or water condensing on the surface, near-neutral pH, high availability of nutrients), and it is important to control growth by restricting temperature rise.

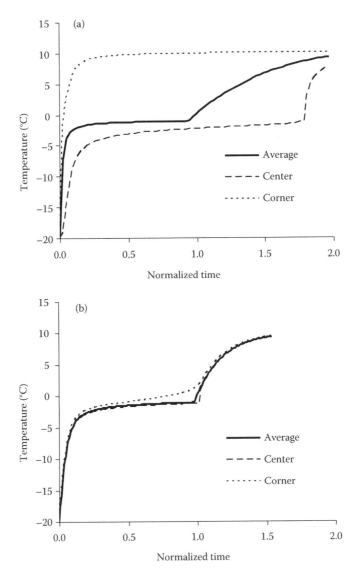

FIGURE 17.2 Product temperatures during thawing of a block of food for (a) $Bi = 11.5$ (internal resistance dominating) and (b) $Bi = 0.1$ (external resistance dominating). Time has been normalized so that the average temperature is 0°C at unit time.

Below 5°C, most pathogenic bacteria will not grow, and therefore, it is often recommended or required that food temperature does not rise above 5°C at any time, especially for ready-to-eat food (Food and Drug Administration [FDA] 2004). However, microorganisms must first pass through a lag phase during which various repair and adjustment mechanisms take place before growth can start (Figure 17.3). Therefore, when operating conditions are rigorously controlled, and especially when the food will be cooked, the 5°C limit does not always need to be adhered to. For

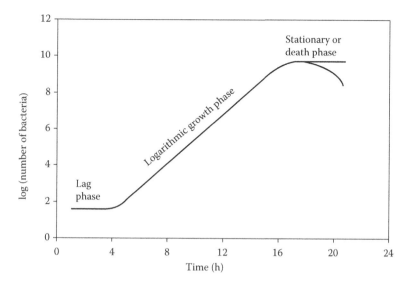

FIGURE 17.3 Typical bacterial growth curve at constant temperature.

example, thawing plastic-wrapped lamb in air at 10°C and 0.75 m/s and unwrapped lambs at 7.5°C and 0.75 m/s cause insignificant changes in bacterial number (Creed et al. 1979). The FDA (2004) permits animal food that is due to undergo cooking after thawing to exceed this temperature for up to 4 hours. It also states in relation to eating establishments that "any procedure" is permitted if a portion of frozen ready-to-eat food is "thawed and prepared for immediate service in response to an individual consumer's order," since in these situations, the lag time is unlikely to be exceeded.

Flexible industrial thawing schedules can be designed for specific situations by calculating or monitoring the temperature profiles in the food throughout the process. For example, Pham et al. (1994) carried out a finite difference method (FDM) simulation of the air thawing of meat blocks and compared predicted growth of *Escherichia coli* at the corners with measured values. They found that the predicted values are generally conservative, that is, equal to or higher than measured values, due to the simulation being better at locating the critical spots. In poorly controlled situations such as in the home or when thawing is done commercially by untrained operators, it is best that (air) thawing be carried out at 5°C or below, that is, in a refrigerator or cool room. If the frozen food is wrapped, the wrapping should be removed before thawing as it will hinder heat transfer, slowing down the process, as well as trap melted ice and raise the water activity at the surface of the product, creating an ideal environment for bacterial growth.

17.3.2 WATER LOSS

Water loss in the form of drip is a major concern during thawing (Pham and Mawson 1997). Freezing is accompanied by water migration on both the cellular scale and the macroscale. On the cellular scale, ice usually forms in the extracellular space. On the macroscale, when food freezes, the surface is colder than the inside, and this temperature gradient also causes moisture to migrate toward the surface, especially in porous foods such as bread.

On thawing, some of this water will be reabsorbed (Spiess 1979) while the rest will be lost as drip. In meat, this is particularly severe if the freezing conditions are such that large intracellular crystals have disrupted the cell membrane, which happens when freezing rates are such that the meat freezes in about 10 minutes (Añón and Calvelo 1980). For meat, drip loss may also be greater at shorter thawing time, as this gives less time for resorption (Gonzalez-Sanguinetti et al. 1985). Surface drip can be minimized by ensuring that $t_d > 50$ minutes at the surface, where t_d is the time for product temperature to rise from $-5°C$ to $-1°C$. However, experimental data on water loss during thawing are conflicting and inconclusive because drip loss may be influenced by many factors prior to thawing: postmortem treatment and pH (for meat), cut-surface-to-volume ratio, orientation of cut surface, rate of freezing, duration and temperature of frozen storage, and temperature fluctuations during frozen storage.

17.3.3   ICE CRYSTAL GROWTH

In cryogenically frozen foods, and to a lesser extent, immersion frozen foods, crystal size is very small, which is often claimed to lead to better texture. However, some of this advantage may be lost during thawing due to ice crystal growth, especially if the food spends a long time at temperatures not too far below the freezing point. At temperatures above the glass transition temperature, the rate factor for crystal growth follows the Arrhenius relationship. In the range $-16.5°C$ to $-5°C$, Martino and Zaritzky (1989) found the activation energy to be about 11.6×10^4 J mol^{-1} for 0.28 N NaCl and 42.4×10^3 J mol^{-1} for beef muscle. This activation energy implies a doubling of the recrystallization rate approximately every $9°C$ rise in temperature. Some quick-frozen foods are stored below the glass transition temperature, and the increase in the recrystallization rate as this temperature is exceeded may be more dramatic.

17.3.4   APPEARANCE, NUTRITIONAL, AND EATING QUALITY

The appearance and eating quality (aroma, tenderness, juiciness) of thawed meat do not seem to depend greatly on the thawing method, as long as the process does not take unusually long (one day or more). The exception is that meat thawed by direct contact with water may gain a bleached appearance. With microwave thawing, if local runaway heating occurs, the food may have burn marks and smell burnt. In fish and other seafoods, the loss of eating quality due to changes in odor and flavor often occurs much faster than bacterial growth at low temperatures, so thawing should be as fast as possible. For bread and some other foods, ice melting and condensation may cause the food to become soggy. Concerning fruit and vegetables, Ibanez et al. (1996) tested air (5°C and 24°C), water (0°C and 40°C), and microwave thawing on blanched broccoli, blanched spinach, lemon juice, and strawberries. Vegetables showed no significant effects of the thawing method on ascorbic acid retention, while fruit products exhibited a large loss after thawing for 24 hours at room temperature.

17.4 THAWING METHODS

17.4.1 EXTERNAL HEATING METHODS

17.4.1.1 Air Thawing

Air thawing has a low rate of heat transfer and hence a long process time, especially when there is no forced air circulation. Air has low thermal conductivity and low volumetric heat capacity, resulting in a low surface heat transfer coefficient (of the order of 10 W m^{-2} K^{-1}, although this may be increased slightly by radiative heat transfer). On the other hand, low surface heat transfer can be an advantage as it ensures that the product is heated relatively uniformly (large external resistance, low Biot number), and the surface stays cooler than in water. For that reason, air is better than water for tempering, where a uniform final product temperature is desirable.

Industrial thawing may be carried out at air temperatures higher than 5°C without microbial growth (Creed et al. 1979) due to the lag phase in microbial growth and the surface staying cooler than the air, but the process must be carefully designed and controlled. To accelerate the process, thawing may be carried out in two stages: initially a high temperature (typically 20°C) is used, perhaps in combination with high air velocities, to increase the heat input. This is then followed by a gentler thawing period where temperature is reduced to a sage level, usually 5°C or colder. The extent to which the surface temperature rises depends on the geometry and composition of the product and the temperature and velocity of the air. Therefore, careful experimentation or numerical modeling must be carried out to verify that temperature rises are acceptable.

In the home and small commercial premises, where there is practically no monitoring and control, it is less clear what approach is safer: to thaw quickly at room temperature or slowly in the refrigerator. Although the FDA stipulates the latter, the United States Department of Agriculture (USDA) will permit the former (Jimenez et al. 2000).

17.4.1.2 Water Thawing

Water thawing is much faster than air thawing due to the high thermal conductivity and volumetric heat capacity of water. In still water, the heat transfer coefficient is about ten times that in air. Higher heat transfer coefficients are still obtained by circulating the water with a pump, spraying it, or using jet impingement. Water circulation is also important to avoid microbial growth that might occur in stagnant water.

Because water is a good conductor of heat, the surface of the food is almost at the same temperature as the water. Therefore, it is most important that water should be cold enough to avoid microbial growth, except when thawing is very fast (such as for peas and other small solids). In uncut meat, microbial growth occurs only at the surface, so hot water may be used for thawing since most microorganisms will not grow above 40°C or so.

When unwrapped food is thawed in water, nutrients will be leached out of the product while water gets absorbed, giving some food such as meat a bleached, unattractive appearance. This is worse with hot water than with cold water. Furthermore, in commercial premises, nutrients and pollutants such as salts, blood, fats, and oils quickly accumulate in the liquid and may cause cross contamination, and the cost of treating and/or discharging the water may be very large.

17.4.1.3   Steam Vacuum Thawing

Vacuum thawing, or steam vacuum thawing, takes place in an evacuated chamber. Steam is introduced and condenses rapidly on the surface, giving a very high heat transfer rate. The aim of using vacuum is twofold: (1) to lower the condensation temperature of the steam and (2) to enhance the heat transfer rate by removing air molecules, which tend to insulate the product from the steam. Vacuum thawing is faster than water thawing, and there is less pollution and absorption of water. Carver (1975) found that frozen blocks of headed and gutted whiting and shrimp can be thawed in 1/2 to 1 hour. However, the equipment for vacuum thawing is expensive.

17.4.1.4   Direct Contact Thawing

In direct contact thawing, the frozen food is sandwiched between hollow metal plates that are heated by circulating hot fluid. The heat transfer coefficient is high, but the food must be in a suitable form, that is, rectangular blocks or slabs. Since the equipment is expensive and the advantage of high heat transfer coefficient is small in the thawing of thick slabs (due to the high Biot number), this method is little used in practice.

17.4.2   INTERNAL HEATING METHODS

17.4.2.1   Ohmic Thawing

Ohmic or resistance heating requires electrodes to be in contact with the product. Ohmic heating at 60 Hz and up to 480 V was tested on shrimp blocks by Balaban et al. (1994). Yun et al. (1998) thawed meat in combination with conventional water immersion at 60 to 210 V, 60 Hz to 60 kHz. They found that frequency changes do not significantly affect thawing time, while reduced drip loss and improved water-holding capacity are obtained at lower voltages.

17.4.2.2   Dielectric Thawing

Dielectric thawing comprises the use of radio frequency (RF) and microwave. Heat is generated inside the product by the rapid reorientation of dipolar molecules of water or other molecules under the influence of an alternating electromagnetic field. RF operates at 3 kHz to 300 MHz, while microwave operates at 300 MHz to 300 GHz (wavelengths of 1 mm to 1 m). The frequencies allowed for industrial, scientific, and medical use are 13.56, 27.12, and 40.68 MHz in the RF range and 915 MHz, 2450 MHz, 5.8 GHz, and 24.124 GHz in the microwave range (Piyasena et al. 2003). Domestic microwave ovens work at 2450 MHz, while industrial microwave ovens also use 915 MHz, which is more penetrating.

The longer wavelength of RF means that it is more penetrating than microwave and therefore may be more suitable for large foods, while microwave gives better energy absorption and hence higher heating rates (Ohlsson et al. 1974; Bengtsson

1963). RF heating of foods has been successfully applied to various foods such as eggs, fruit, vegetables, fish, and meat (Piyasena et al. 2003) and has also been used for the tempering of meat. Compared to air thawing, for both fish blocks (Jason and Sanders 1962a,b) and meat (Bengtsson 1963; Sanders 1966), RF gave much faster thawing with less drip. Fish also retains flavor better with RH thawing (Jason and Sanders 1962a,b). Myung (1998) found that frozen cooked rice thawed by conventional heating or microwave heating has similar quality characteristics, better than that thawed at room temperature or in a pressure cooker.

A common problem with all internal thawing methods is uneven heat distribution, often leading to runaway heating. With ohmic heating, the conductivity of water or thawed food is much higher than that of ice or frozen food; therefore, the electrical current tends to channel along regions that are already thawed. With RF and microwave heating, again the thawed food absorbs more energy than the frozen parts and may cause runaway heating, resulting in a mixture of cooked and frozen regions. For example, the penetration depth δ_p for 2450 MHz microwave in tuna flesh decreases drastically from 150 mm at −30°C to only 6 mm at −1°C (Figure 17.4; Liu et al. 2005).

With microwave, the surface and corners of a rectangular block of food tend to absorb heat first, especially since these parts are already warmed by air. The use of cold air to refrigerate the surface has been suggested (Bialod et al. 1978). Curved surfaces may serve to focus the microwave as in a lens, so that the center of a cylindrical or spherical product may heat up much faster than the rest (Oliveira and Franca 2002). Nonuniform product composition will also cause problems. Fat is particularly effective at absorbing microwaves and may start to fry before the water melts. Merabet (2000) patented a process whereby an edible water-in-oil microemulsion is incorporated into food before freezing to ensure uniform energy absorption during microwave thawing. Various other methods are used to improve heat distribution: turntables, microwave stirrers, and multiple antennas. Intermittent

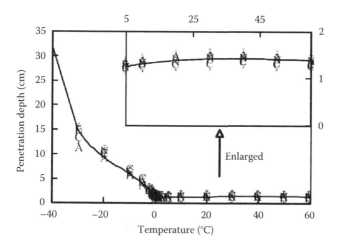

FIGURE 17.4 Variation of penetration depth with temperature for tuna flesh. (From Liu, C.M., et al., *Int. J. Food Sci. Technol.* 40, 14, 2005. With permission.)

interruptions or continuous power at a reduced level will allow time for the heat to be conducted away from the hot spots (Chamchong and Datta 1999). Gong et al. (1995) patented a method for feedback control of power input by sensing the amount of evolved water vapor.

Due to the difficulties in obtaining uniform heating, dielectric heating methods have not been widely used for thawing in industry, except for microwave tempering (not thawing), which is quite successful. In microwave tempering, no part of the product is allowed to rise above melting point. It is important to ensure that the product is not prewarmed by air before tempering but should be taken straight from cold storage; else the corners and edges will warm up first, which would lead to runaway heating. Carton-sized meat blocks can be tempered by microwave in around 10 minutes, as opposed to several days by air. Tempering can be done by itself, to prepare food for slicing and cutting (which is easier and less messy when the food is not too soft), or as a first step prior to air thawing.

17.4.3 Miscellaneous Methods

17.4.3.1 Power Ultrasound Thawing

Miles et al. (1999) tested the use of power ultrasound thawing on meat and fish. The technique was successful only within a narrow frequency band around 500 kHz and intensities around 0.5 W cm^{-2}. Beef, pork, and cod samples were thawed to a depth of 7.6 cm within about 2.5 hours. Surface heating was a problem both at higher frequencies due to an increase in attenuation and at lower frequencies ($\leq$430 kHz) due to cavitation.

17.4.3.2 Low-Frequency Acoustics Thawing

Low-frequency acoustics has been tested for thawing by Kissam et al. (1981). Blocks of Pacific cod 91 mm thick weighing 12.7 kg were thawed in 18°C water and simultaneously exposed to 1500-Hz acoustic energy not exceeding 60 W. The acoustic input causes a 71% time reduction.

17.4.3.3 Pressure-Assisted Thawing

Water has the unusual property that it expands upon freezing and thus a pressure increase will cause the freezing point to fall. The minimum freezing point that is obtainable by pressurization is about −22°C at 210 MPa (Figure 17.5). If frozen food is pressurized and heated, it starts to thaw at a lower temperature, where the risk of microbial growth is low. Furthermore, the temperature driving force, that is, the difference between environmental and product temperatures, is increased, leading to faster process time. Makita (1992) found that pressure-assisted thawing (PAT) takes about a third of the time required for conventional thawing. Mussa and LeBail (2000) found that high-pressure thawing reduced the number of *Listeria innocua* by 1.5 and 3 log cycles at 100 and 200 MPa, respectively.

PAT has been applied to pollock whiting fillets (Chourot 1997), beef (Zhao et al. 1998), and whiting fillets. Meat thawed at high pressure had sensory qualities comparable to those of conventionally thawed meat (Makita 1992). Drip could be reduced by maintaining pressure for longer than necessary (Chevalier et al. 1999).

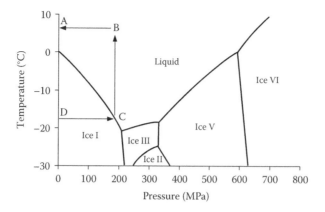

FIGURE 17.5 PAT on the phase diagram.

There is some protein denaturation and protein loss in the drip, especially at pressures above 150 MPa. With salmon, pressures above 150 MPa cause noticeable color changes, while at 200 MPa, the product texture was significantly modified (Zhu et al. 2004).

17.5 MATHEMATICAL DESIGN OF THAWING PROCESSES

17.5.1 THERMAL PROPERTY CHANGES DURING THAWING

Foods do not freeze or thaw at a sharp freezing point but do so gradually over a temperature range. The temperature at which the melting of the ice in the foods is completed in thawing is the *initial freezing point*, T_f. It is also the temperature at which ice starts to appear during freezing if no supercooling occurs.

Due to the presence of solutes, the initial freezing point of foods is lower than that of pure water. As ice freezes out of the solution and temperature falls, the remaining solution becomes more and more concentrated but is always in equilibrium with the ice. For dilute aqueous solutions, it can be shown that the ice fraction at temperature T below the freezing point is given by Raoult's law of freezing point depression (Miles et al. 1983; Fikiin 1998):

$$F = 1 - \frac{T_0 - T_f}{T_0 - T} \tag{17.2}$$

where $T_0 = 273.15$ K. Thus, for food that begins to freeze at $-1°C$, about half the freezable (unbound) water in the food remains unfrozen at $-2°C$ (Figure 17.6). In other words, there is as much ice to melt between $-20°C$ and $-2°C$ as between $-2°C$ and $-1°C$. If the food is being thawed in air, the second part of the process (from $-2°C$ to $-1°C$) may take longer than the first part. Hence, tempering a food takes much less time than complete thawing, even though the final tempering temperature may be only a few degrees below the freezing point.

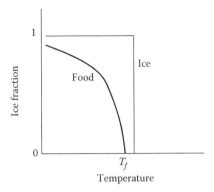

FIGURE 17.6   Frozen fraction versus temperature.

For frozen food, the apparent specific heat (the amount of heat required to raise the food temperature by a degree) increases gradually with temperature and is greatest just under the initial freezing point T_f because that is where most of the ice thaws (Figure 17.7). The peak is known as the "latent heat peak."

The thermal conductivity of ice is about four times that of liquid water, and that of frozen food may be two or three times that of unfrozen food. The transition occurs within a few degrees and is particularly sharp at the initial freezing point (Figure 17.8).

17.5.2   ANALYTICAL THAWING TIME PREDICTION METHODS

Analytical methods are exact formulas, obtained by making simplifying assumptions about the physical situation. All analytical methods are solutions of the Fourier heat conduction equation, which govern temperatures inside a solid food:

$$\rho c \frac{\partial T}{\partial t} = \nabla \cdot \left(k \nabla T \right) + q. \tag{17.3}$$

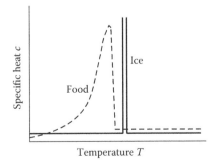

FIGURE 17.7   Apparent specific heat of food around the freezing point.

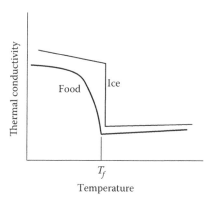

FIGURE 17.8 Thermal conductivity behavior of ice and water-rich foods.

Several analytical solutions have been found for idealized cases of freezing and thawing based on different sets of simplifying assumptions (Lind 1991; Pham 2006). The most important is Plank's equation, which often serves as the starting point for more practically useful empirical formulas. By assuming zero specific heat (i.e., only the latent heat of freezing has to be considered) and a sharp freezing point, we obtain Plank's equation (Plank 1913), which can be written in the form (Pham 1986a)

$$t_{Plank} = \frac{1}{E} \frac{\rho \Delta H_f R}{h(T_a - T_f)} \left(1 + \frac{Bi}{2}\right) \tag{17.4}$$

where E is a shape factor (1 for slabs, 2 for infinite cylinders, 3 for spheres). When $Bi \gg 1$ (internal resistance dominating), Plank's equation reduces to

$$t_{Plank} = \frac{1}{E} \frac{\rho \Delta H_f R^2}{2k(T_a - T_f)} \tag{17.5}$$

that is, the thawing time is proportional to the square of thickness or radius. When $Bi \ll 1$ (external resistance dominating), Plank's equation becomes

$$t_{Plank} = \frac{1}{E} \frac{\rho \Delta H_f R}{h(T_a - T_f)}. \tag{17.6}$$

Plank's equation underpredicts freezing and thawing times and is not recommended for practical use. However, it illustrates well the effect of various physical factors, and it is the starting point for several more accurate empirical equations, which we will come to next.

17.5.3 EMPIRICAL AND MODIFIED ANALYTICAL METHODS

Empirical calculation methods are those obtained by regression (curve fitting) of experimental data. Most empirical methods start from an analytical expression that assumes certain idealized conditions and then add empirical correction terms; hence, they can be termed modified analytical methods. The prerequisite for obtaining a reliable empirical method is a set of carefully planned experimental data that cover a wide range of operating conditions. The best known such dataset for thawing is that of Cleland et al. (1986, 1987), who carried out hundreds of tests on Tylose slabs, cylinders, spheres, bricks, and other shapes, plus 18 on lean minced beef. The range of material is limited to beef and Tylose gel, which are quite similar. However, the experimental data are usually supplemented by numerically predicted results.

Cleland (1990) and Lind (1991) reviewed several empirical methods. The equations of Cleland et al. (1986) were found to be most accurate. One of these is reproduced here in a form that highlights the relationship with Plank's equation:

$$t_T = \frac{1}{E} \frac{\rho \Delta H_{10} R}{h\left(T_a - T_f\right)} \left(p + r\frac{Bi}{2} \right) \quad (17.7)$$

where
$p = 0.7754 + 2.2828 Ste\ Pk$
$r = 0.4271 + 2.1220 Ste - 1.4847 Ste^2$
$Ste = c_u(T_a - T_f)/\Delta H$
$Pk = c_f(T_f - T_i)/\Delta H$
ΔH_{10} = enthalpy change from $-10°C$ to $0°C$ (kJ kg^{-1})

The similarity with Plank's equation (Equation 17.4) may be observed. Approximate expressions for the shape factor E have been proposed for ellipses and ellipsoids (Pham 1991; Ilicali et al. 1999) and for the more general case (Hossain et al. 1992a,b,c). For multidimensional regular shapes (rectangular rods, finite cylinders, rectangular bricks), analytical expressions for E in the form of infinite series are also available (McNabb et al. 1990a,b).

17.5.4 NUMERICAL SOLUTION OF THE FOURIER EQUATION

17.5.4.1 Overview

The numerical solution of the Fourier heat conduction equation (Equation 17.3) involves discretizing the space domain to obtain a set of ordinary differential equations (ODEs) relating the temperatures at a finite number of nodes (points in space). The ODEs can be written in matrix form as

$$\mathbf{C}\frac{d\mathbf{T}}{dt} + \mathbf{KT} = \mathbf{f} \quad (17.8)$$

where $\mathbf{T}$ is a vector of nodal temperatures, $\mathbf{C}$ is termed the global capacitance matrix and contains the specific heat c, $\mathbf{K}$ is termed the global conductance matrix and contains the thermal conductivity k, and $\mathbf{f}$ is termed the global forcing matrix and contains known terms arising from heat generation and boundary conditions. The exact form of the matrices $\mathbf{C}$, $\mathbf{K}$, and $\mathbf{f}$ in Equation 17.8 depends on which method is used for discretizing space. The most common such methods are FDM, finite element method (FEM), and finite volume method (FVM). FDM is most convenient and efficient for problems involving simple geometries, while FEM and FVM are better suited to complex geometries and composite objects. While an FDM program can be written by most researchers with scientific or engineering mathematics training, FEM or FVM tends to be available as commercial software. Further details and comparison of the three methods may be found in the work of Pham (2006).

17.5.4.2   Dealing with Latent Heat Peak

In the numerical modeling of both freezing and thawing, the evolvement of a large amount of (latent) heat poses special problems. Pham (2006) reviewed methods for dealing with these difficulties. The main methods that have been proposed are the *apparent specific heat* method (Bonacina et al. 1973), the *enthalpy* method (Eyres et al. 1946; Voller and Swaminathan 1991), and the *quasi-enthalpy* (or *temperature correction*) method (Pham 1985, 1986b). The apparent specific heat method solves Fourier equation in its original form (Equation 17.3). The enthalpy method solves a modified heat conduction equation, which does not involve the specific heat c:

$$\rho \frac{\partial H}{\partial t} = \nabla \cdot \left(k \nabla T \right) + q. \tag{17.9}$$

The quasi-enthalpy method solves the original Fourier equation (as in the apparent specific heat method) but applies a correction at each step to ensure that heat balance is retained. Occasional programmers should use an explicit version of the enthalpy method, which is very easy to program. The quasi-enthalpy combines speed and ease of programming (Pham 1995; Voller 1996), while commercial FEM packages use an iterative version of the apparent specific heat method.

17.5.4.3   Dealing with Frost and Condensation

In air thawing, frosting and condensation release latent heat, which is often a large component of the total heat gain, especially at the beginning of thawing. The condensation flux is given by $k_y(Y_a - Y_s)$. For an unwrapped product being thawed in air, the total heat flux Q at the surface is the sum of latent, sensible, and radiative heats:

$$Q = h_c \left(T_a - T_s \right) + \Delta H_v k_y \left(Y_a - Y_s \right) + \varepsilon \sigma \left(T_R^4 - T_s^4 \right). \tag{17.10}$$

Y_s is the saturation humidity at the surface temperature:

$$Y_s = Y^{sat}(T_s). \tag{17.11}$$

The temperature gradient underneath the surface is proportional to the rate of heat gain:

$$k_u \frac{\partial T}{\partial z}\bigg|_s = Q. \tag{17.12}$$

Combining these equations, we get

$$k_u \frac{\partial T}{\partial z}\bigg|_s = h_c \left(T_a - T_s\right) + H_v k_y \left[Y_a - Y^{sat}\left(T_s\right)\right] + \varepsilon\sigma\left(T_R^4 - T_s^4\right) \tag{17.13}$$

which may serve as the boundary condition of the heat conduction equation (Equation 17.3). The convective heat and mass transfer coefficients may be related by the psychrometric relationship:

$$\frac{h_c}{k_y} = c_a \, (Sc/\mathrm{Pr})^{2/3} \approx 0.90 c_a. \tag{17.14}$$

The heat transfer coefficient depends on the temperature, airflow conditions, and geometry of the product. It has to be measured, estimated from empirical relationships (Kondjoyan 2006), or calculated by computational fluid dynamics (CFD). A layer of frost or condensation will form an extra resistance to heat transfer while at the same time disturbing the boundary layer and causing an increase in the heat transfer coefficient. These effects are complex and have been studied by Mannapperuma and Singh (1988, 1989) and Lind (1991).

17.5.5 COMPUTATIONAL FLUID DYNAMICS

CFD is a class of numerical methods that computes the temperature and velocity fields in a fluid. It can be used to calculate the surface heat transfer coefficient from more easily measurable environmental conditions such as fluid velocity and turbulence. Otherwise, the heat transfer coefficient has to be estimated experimentally or calculated from available empirical formulas. However, the advantages of CFD are not always fully realized. While velocity can easily be measured or calculated, turbulence data are rarely available and need special intruments. Yet turbulence may have a very strong influence on heat transfer (Kondjoyan 2006).

Turbulent flow involves very fast random fluctuations in the velocity field. The effect of these random fluctuations cannot be solved from first principles (except by direct numerical solution of the flow equations, which requires too much computer power for the vast majority of problems), so they must be solved approximately by using one of a number of so-called turbulence models, such as the well-known k–ε model. Since these models are semiempirical, the results are only approximate at best, particularly in highly swirling flow.

A full CFD computation of a transient process involving turbulent flow can be extremely time consuming. Therefore, CFD is usually used to calculate an approximate value of the heat transfer coefficient, which is then applied to a conduction-only numerical program.

17.5.6 CALCULATION OF MICROWAVE HEAT ABSORPTION

Microwave heating contributes the internal heat generation q in the heat conduction equation (Equation 17.3). To calculate this term, two approaches may be taken: Lambert's law or Maxwell's equations (Ayappa and Davis 1991). Lambert's law states that equal fractions of the microwave energy are absorbed by equal thicknesses of a given material:

$$P = P_0 e^{-2\alpha z} = P_0 e^{-2z/\delta_E} \tag{17.15}$$

where α is the absorbance of the material and $\delta_E \equiv 1/\alpha$ is the penetration depth of the beam, or the depth at which the *intensity* of the radiation falls to $1/e$ or about 37% of the original value at the surface. [Regier and Schubert (2005) distinguish the *power dissipation penetration depth* $\delta_P \equiv 1/(2\alpha)$ and the *electric field penetration depth* $\delta_E \equiv 1/\alpha = 2\delta_P$.]

A more rigorous solution is obtained by applying Maxwell's equations. The power absorption profile predicted by Lambert's law decreases exponentially with distance from the surface, while that calculated from Maxwell's equations often show wave-like fluctuations. If those fluctuations are too large, then heat will not be conducted away fast enough from the high absorption locations, and temperature nonuniformities and perhaps runaway heating will result. Ayappa and Davis (1991) compare the two approaches and found that Lambert's law is a good approximation when the thickness of the absorbing material exceeds a certain critical value:

$$2R > 5.4\delta_P - 0.08 \text{ cm} \tag{17.16}$$

while for cylindrical samples, Oliveira and Franca (2002) found that the corresponding criterion is

$$R > 14.06\delta_P \tag{17.17}$$

The greater critical depth in cylindrical samples is attributed to focusing effect. Liu et al. (2005) found that for the thawing of tuna slabs with thicknesses from 2 to 32 cm, temperatures calculated from Lambert's are similar to those predicted by Maxwell's equations, and both are compatible with experimental results.

17.6 CONCLUSIONS

Thawing is an important and sensitive process but often performed in a doubtful manner, especially in the home and in small establishments. Fortunately, the existence of a lag phase before the resumption of microbial growth usually ensures that the end product remains safe, especially when subjected to further cooking.

It is difficult to devise general rules for thawing, each product and each packaging requiring a different approach. The usual recommendation for the consumer is to thaw in air in the refrigerator at 5°C or below; however, this is not necessarily optimal for microbial growth or eating quality. Numerical methods will allow the design of faster and safer multistage thawing processes with varying air temperature. More validation tests are required to justify this approach, however.

Conventional air and water thawing are likely to remain the predominant thawing methods, with RF and microwave used in some establishment where time is seen as critical, and for tempering before cutting and communition. PAT and, to a lesser extent, steam vacuum thawing are unlikely to become widely applied commercially in the near future due to the high equipment cost.

NOMENCLATURE

Bi	Biot number, hR/k_u
c	Specific heat of food, J K^{-1} kg^{-1}
$\mathbf{C}$	Global capacitance matrix
c_a	Specific heat of air, J K^{-1} kg dry air^{-1}
c_f	Specific heat of frozen food, J K^{-1} kg^{-1}
c_u	Specific heat of unfrozen food, J K^{-1} kg^{-1}
E	Shape factor
$\mathbf{f}$	Global forcing matrix
F	Fraction of the freezable water that has solidified
h	Overall heat transfer coefficient, W m^{-2} K^{-1}
H	Specific enthalpy, J kg^{-1}
h_c	Convection heat transfer coefficient, W m^{-2} K^{-1}
ΔH_f	Latent heat of freezing, kJ kg^{-1}
ΔH_v	Latent heat of vaporization, kJ kg^{-1}
ΔH_{10}	Enthalpy change of food from -10°C to 0°C, kJ kg^{-1}
k	Thermal conductivity, W m^{-1} K^{-1}
$\mathbf{K}$	Global conductance matrix
k_u	Thermal conductivity of unfrozen food, W m^{-1} K^{-1}
k_y	Mass transfer coefficient based on dry basis moisture content, kg m^{-2} s^{-1}
P	Electromagnetic energy flux, W m^{-2}
P_0	Electromagnetic energy flux at surface, W m^{-2}
Pr	Prandtl number of air
q	Heat generated, W m^{-3}
Q	Total heat flux at surface, W m^{-2}
R	Radius or half-dimension of food, m
Sc	Schmidt number of air
t	Time, s
T	Temperature, K
$\mathbf{T}$	Vector of nodal temperatures, K
T_0	Initial freezing point of water, K

T_a	Environment temperature, K
T_f	Initial freezing point, K
T_R	Radiating temperature, K
T_s	Surface temperature, K
Y_a	Dry basis moisture content of air, (kg moisture) (kg air)$^{-1}$
Y_s	Dry basis moisture content of air near surface, (kg moisture) (kg air)$^{-1}$
Y^{sat}	Saturation moisture content, (kg moisture) (kg air)$^{-1}$
z	Distance from food surface, m
α	Absorptivity
δ_E	Electric field penetration depth, $1/\alpha$
δ_P	Power dissipation penetration depth, $1/(2\alpha)$
ε	Surface emissivity
σ	Stefan–Boltzmann constant, W m^{-2} K^{-2}
ρ	Density of food, kg m^{-3}

REFERENCES

Añón, M.C. and Calvelo, A. 1980. Freezing rate effects on the drip losses of frozen beef. *Meat Science* 4:1–14.

Ayappa, K.G. and Davis, E. 1991. Microwave heating: an evaluation of power formulations. *Chemical Engineering Science* 46(4):1005–1016.

Balaban, M., Henderson, T., Teixeira, A. and Otwell, W.S. 1994. Ohmic thawing of shrimp blocks. In *Developments in Food Engineering*, eds. T. Yano, R. Matsuno and K. Nakamura, vol. 1, pp. 307–309. London: Blackie Academic and Professional.

Bengtsson, N. 1963. Electronic Defrosting of Meat and Fish at 35 MHz and 2,450 MHz: a Laboratory Comparison. *5th International Congress on Electro-Heat 5th International Congress on Electro-Heat, Wiesbaden*, Presentation No. 414.

Bialod, D., Jolion, M. and Legoff, R. 1978. Microwave thawing of food-products using associated surface cooling. *Journal of Microwave Power and Electromagnetic Energy* 13(3):269–274.

Bonacina, C., Comini, G., Fasano, A. and Primicerio, M. 1973. Numerical solution of phase-change problems, *International Journal of Heat and Mass Transfer* 16:1825–1832.

Carver, J.H. 1975. Vacuum cooling and thawing fishery products. *Marine Fisheries Review* 37(7):15–21.

Chamchong, M. and Datta, A.K. 1999. Thawing of foods in a microwave oven, I. Effect of power levels and power cycling. *Journal of Microwave Power and Electromagnetic Energy* 34(1):9–21.

Chevalier, D., Le Bail, A., Chourot, J.M. and Chantreau, P. 1999. High pressure thawing of fish (whiting): influence of the process parameters on drip losses. *Lebensmittel-Wissenschaft Und-Technologie* 32(1):25–31.

Cleland, A.C. 1990. *Food Refrigeration Processes: Analysis, Design and Simulation*. Chapter 7. London: Elsevier Applied Science.

Cleland, D.J., Cleland, A.C., Earle, R.L. and Byrne, S.J. 1986. Prediction of thawing times for foods of simple shape. *International Journal of Refrigeration* 9:220–228.

Cleland, D.J., Cleland, A.C., Earle, R.L. and Byrne, S.J. 1987. Experimental data for freezing and thawing of multi-dimensional objects. *International Journal of Refrigeration* 10:22–31.

Chourot, J.M. 1997. Contribution a l'etude de la decongelation par haute pression. Doctorate Thesis, Universite de Nantes, France.

Creed, P.G., Bailey, C., James, S.J. and Harding, C.D. 1979. Air thawing of lamb carcasses. *International Journal of Food Science and Technology* 14(2):181–191.

Eyres, N.R., Hartree, D.R., Ingham, J., Jackson, R., Sarjant, R.J. and Wagstaff, J.B. 1946. The calculation of variable heat flow in solids. *Transactions of the Royal Society* A240:1.

Fikiin, K.A. 1998. Ice content prediction methods during food freezing: a survey of the Eastern European literature. *Journal of Food Engineering* 38:331–339.

Food and Drug Administration (FDA). 2004. Food Code (2001, Updated April 2004) Chapter 3: Food. Available at: http://www.cfsan.fda.gov/~dms/fc01-3.html.

Gong, C.S., Kim, S.T. and Chai, E.S. 1995. Automatic thawing device of microwave oven and control method thereof. *U.S. Patent* US5436433.

Gonzalez-Sanguinetti, S., Añon, M.C. and Calvelo, A. 1985. Effect of thawing rate on the exudate production of frozen beef. *Journal of Food Science* 50:607–700, 706.

Hossain, Md.M., Cleland, D.J. and Cleland, A.C. 1992a. Prediction of freezing and thawing times for foods of regular multidimensional shape by using an analytically derived geometric factor. *International Journal of Refrigeration* 15(4):227–234.

Hossain, Md.M., Cleland, D.J. and Cleland, A.C. 1992b. Prediction of freezing and thawing times for foods of two-dimensional irregular shape by using a semi-analytical geometric factor. *International Journal of Refrigeration* 15(4):235–240.

Hossain, Md.M., Cleland, D.J. and Cleland, A.C. 1992c. Prediction of freezing and thawing times for foods of three-dimensional irregular shape by using a semi-analytical geometric factor. *International Journal of Refrigeration* 15(4):241–246.

Ibanez, E., Foin A., Cornillon P. and Reid D.S. 1996. Study of different thawing methods for frozen fruits and vegetables. *1996 IFT Annual Meeting: Book of Abstracts*. Chicago: Institute of Food Technologists.

Ilicali, C., Teik, T.H. and Shian, L.P. 1999. Improved formulations of shape factors for the freezing and thawing time prediction of foods. *Lebensmittel-Wissenschaft and Technologie* 32(5):312–315.

Jason, A.C. and Sanders, H.R. 1962a. Dielectric thawing of fish. I. Experiments with frozen herrings. *Food Technology* 16(6):101–106.

Jason, A.C. and Sanders, H.R. 1962b. Dielectric thawing of fish. II. Experiments with frozen white fish. *Food Technology* 16(6):107–112.

Jimenez, S.M., Pirovani, M.E., Salsi, M.S., Tiburzi, M.C. and Snyder, O.P. 2000. The effect of different thawing methods on the growth of bacteria in chicken. *Dairy, Food, and Environmental Sanitation* 20(9):678–683.

Kissam, A.D., Nelson, R.W., Ngao, R. and Hunter, P. 1981. Water thawing of fish using low frequency acoustics. *Journal of Food Science* 47(1):71–75.

Kondjoyan, A. 2006. A review on surface heat and mass transfer coefficients during air chilling and storage of food products. *International Journal of Refrigeration* 29:863–875.

Lacroix, B.J., Li, K.W.M. and Powell, D.A. 2003. Consumer food handling recommendations: is thawing of turkey a food safety issue? *Canadian Journal of Dietetic Practice and Research* 64:59–61.

Lind, I. 1991. Mathematical modelling of the thawing process. *Journal of Food Engineering* 14(1):1–23.

Liu, C.M., Wang, Q.Z. and Sakai, N. 2005. Power and temperature distribution during microwave thawing, simulated by using Maxwell's equations and Lambert's law. *International Journal of Food Science and Technology* 40(1):9–21.

Makita, T. 1992. Application of high pressure and thermophysical properties of water to biotechnology. *Fluid Phase Equilibrium* 76:87–95.

Mannapperuma, J.D. and Singh, R.P. 1988. Thawing of frozen foods in humid air. *International Journal of Refrigeration* 11:173–186.

Mannapperuma, J.D. and Singh, R.P. 1989. A computer-aided method for the prediction of properties and freezing/thawing times of foods. *Journal of Food Engineering* 9(4):275–304.

Martino, M.N. and Zaritzky, N.E. 1989. Ice recrystallization in a model system and in frozen muscle tissue. *Cryobiology* 26:138–148.

McNabb, A., Wake, G.C., Hossain, Md.M. and Lambourne, R.D. 1990a. Transition times between steady states for heat conduction, Part I: General theory and some exact results, *Occasional Pubs in Mathematics and Statistics* No.20, Massey University, New Zealand.

McNabb, A., Wake, G.C., Hossain, Md.M. and Lambourne, R.D. 1990b. Transition times between steady states for heat conduction, Part II: Approximate solutions and examples, *Occasional Pubs in Mathematics and Statistics* No.21, Massey University, New Zealand.

Merabet, M. 2000. Microwave thawing using micro-emulsions. *U.S. Patent* US6149954.

Miles, C., van Beek, G., and Veerkamp, C. 1983. Calculation of the thermophysical properties of foods. In *Physical Properties of Foods*, eds. R. Jowitt et al., pp. 269–312. London: Applied Science Publishers.

Miles, C.A., Morley, M.J. and Rendell, M. 1999. High power ultrasonic thawing of frozen foods. *Journal of Food Engineering* 39(2):151–159.

Mitakakis, T.Z., Sinclair, M.I., Fairley, C.K., Lightbody, P.K., Leder, K. and Hellard, M.E. 2004. Food safety in family homes in Melbourne, Australia. *Journal of Food Protection* 67:818–822.

Mussa, D.M. and LeBail, A. 2000. High pressure thawing of fish: evaluation of the process impact on *Listeria innocua*. 2000 IFT Annual Meeting, Institute of Food Technologists.

Myung, S.O. 1998. Effect of thawing methods and storage periods on the quality of frozen cooked rice. *Journal of Food Science and Nutrition* 3(3):234–240.

Ohlsson, T., Bengtsson, N.E. and Risman, P.O. 1974. The frequency and temperature dependence of dielectric food data as determined by a cavity perturbation technique. *Journal of Microwave Power and Electromagnetic Energy* 9(2):129–145.

Oliveira, M.E.C. and Franca, A.S. 2002. Microwave heating of foodstuffs. *Journal of Food Engineering* 53:347–359.

Pham, Q.T. 1985. A fast, unconditionally stable finite-difference method for heat conduction with phase change. *International Journal of Heat Mass Transfer* 28:2079–2084.

Pham, Q.T. 1986a. Simplified equation for predicting the freezing time of foodstuffs. *Journal of Food Technology* 21:209–221.

Pham, Q.T. 1986b. The use of lumped capacitances in the finite-element solution of heat conduction with phase change. *International Journal of Heat Mass Transfer* 29:285–292.

Pham, Q.T. 1991. Shape factor for the freezing time of ellipses and ellipsoids. *Journal of Food Engineering* 13:159–170.

Pham, Q.T. 1995. Comparison of general purpose finite element methods for the Stefan problem. *Numerical Heat Transfer Part B - Fundamentals* 27:417–435.

Pham, Q.T. 2006. Modelling heat and mass transfer in frozen foods: a review. *International Journal of Refrigeration*, 29(6):876–888.

Pham, Q.T. and Mawson, R. 1997. Moisture Migration and Ice Recrystallization in Frozen Foods. In *Quality in Frozen Foods*, eds. M.C. Erickson and Y.C. Hung, pp. 67–91. New York: Chapman and Hall.

Pham, Q.T., Lowry, P.D., Fleming, A.K., Willix, J. and Fitzgerald, C. 1994. Temperature and microbial growth in meat blocks undergoing air thawing. *International Journal of Refrigeration* 17:280–287.

Piyasena, P., Dussault, C., Koutchma, T., Ramaswamy, H.S. and Awuah, G.B. 2003. Radio frequency heating of foods: principles, applications and related properties—a review. *Critical Reviews in Food Science and Nutrition* 43(6):587.

Plank, R. 1913. Die Gefrierdauer von Eisblocken. *Zeitschrift fur die gesamte Kalte Industrie* 20(6):109–114.

Regier, M. and Schubert, H. 2005. Introducing microwave processing of food: principles and technologies. In *The Microwave Processing of Foods*, eds. H. Schubert and M. Regier, pp. 3–21. Cambridge: Woodhead Publishing Ltd.

Sanders, H.R. 1966. Dielectric thawing of meat and meat products. *International Journal of Food Science and Technology* 1(3):183–192.

Spiess, W.E.L. 1979. Impact of freezing rates on product quality of deep frozen foods. In *Food Process Engineering 1979—Proceedings of the Second International Congress on Engineering and Food, Helsinki*, eds. P. Linko and J. Larinkari, vol. 1, pp. 689–694. London, U.K.: Applied Science Publishers.

Voller, V.R. 1996. An overview of numerical methods for solving phase change problems. In *Advances in Numerical Heat Transfer*, eds. W.J. Minkowycz and E.M. Sparrow, vol. 1, pp. 341–375. London: Taylor & Francis.

Voller, V.R. and Swaminathan, C.R. 1991. Generalized source-based method for solidification phase change. *Numerical Heat Transfer Part B* 19:175–189.

Yun, C.G., Lee, D.H. and Park, J.Y. 1998. Ohmic thawing of a frozen meat chunk. *Journal of Food Science and Technology (Korean)* 30(4):842–847.

Zhao, Y., Flores, R.A. and Olson, D.G. 1998. High hydrostatic pressure effects on rapid thawing of frozen beef. *Journal of Food Science* 63(2):272–275.

Zhu, S., Ramaswamy, H.S. and Simpson, B.K. 2004. Effect of high-pressure versus conventional thawing on color, drip loss and texture of Atlantic salmon frozen by different methods. *Lebensmittel-Wissenschaft Und-Technologie* 37(3):291–299.

18 Freeze-Drying Equipment

Antonello A. Barresi and Davide Fissore

CONTENTS

18.1 INTRODUCTION

Freeze drying (FD) is a process where water is removed from a frozen product by sublimation, working at low temperature and pressure. As it has been discussed in Chapter 3, FD has a lot of advantages in comparison with traditional methods used for drying foodstuffs:

- Original flavor and aroma are perfectly preserved.
- Original product shape, color, and texture are maintained.
- Very little shrinkage occurs.
- Long shelf life is obtained.
- The freeze-dried product has excellent rehydratation characteristics.

An FD process is carried out in a special equipment, whose cost can be very high: savings realized by stabilizing an otherwise unstable product at ambient temperature, thus eliminating the need for refrigeration, can compensate for the investment in FD equipment and processing. As a consequence, FD is mainly used when

- We deal with a relatively expensive raw material.
- Maintenance of color, texture, and taste is extremely important.
- Proper and instant rehydratation cannot be achieved through other drying methods.
- Storage and distribution costs become too high if other preservation methods are used.
- Weight reduction is an important characteristic of the dried product.

An FD process consists of three steps, namely, freezing and primary and second-ary drying (see Chapter 3 for more details); thus, the plant where the FD process is carried out consists of:

- A refrigerating equipment for freezing the product.
- A transport system for product loading and unloading.
- The FD cabinet.

In some cases, freezing and drying are carried out in the same apparatus. The aim of this chapter is to discuss how an FD process of foodstuffs can be realized, thus showing and discussing the main elements of a typical FD plant.

18.2   FREEZING

The first step of an FD cycle is freezing. When the product temperature is lowered, ice crystals start appearing at about $-2°C$; as the cooling continues, water gradually crystallizes. However, even at very low temperatures, some of the water does not crystallize because it is too closely bound to other molecules in the product. The size and shape of the ice crystals are affected by the freezing rate: a slow freezing rate gives large ice crystals. Solid products like vegetables, fruits, and meat have a natu-ral cell structure, whereas liquids like coffee extract and fruit juices do not have such cell structure; thus, the freezing method must be adapted to the different product types: for solid products, freezing can be performed by vacuum freezing within the freeze dryer (see Chapter 3), while for liquid products, freezing can be obtained by placing the product over a cooled shelf inside the freeze dryer. For most productions, freezing is performed in a separate freezing room.

If cooled shelves are used, they are maintained at $-40°C$ or $-50°C$, and in special plants, $-60°C$ or slightly lower temperatures can be reached. The containers can be loaded onto precooled or room temperature shelves: if the shelves are precooled, the loading must be done quickly in order to minimize the condensation of water vapor from the air on the shelves. For the freezing of food, stainless-steel belt conveyors can be used, and a cold gas cools and freezes the product. The design of such convey-ors is difficult due to problems of sealing of the moving belt and of product abrasion (Oetjen and Haseley 2004). Another possibility for freezing foods is to use a flow of cold air in a fluidized bed freezer if the product is granulated or in small pieces, thus obtaining a very rapid freezing. Quick freezing of liquid products can be obtained also using rotating cylinders.

The possible freezing rate can be estimated using the methods described in Chapters 2 and 3, while a more detailed description of the pretreatments and of the freezing equipment can be found in Chapters 4 through 16.

18.3   PRIMARY AND SECONDARY DRYING

Primary drying is the second step of the FD cycle: most of the ice is sublimated under low pressure, and heat is supplied to the product as this phase change is endo-thermic. The residual moisture content immediately after the primary drying may be in the range of 10%–15% due to the water that did not previously crystallize and that is strongly bound by adsorption phenomena to the partially dried cake. A further drying phase is thus required (secondary drying): using a low vapor pressure and a moderate temperature (20°C and higher), it is usually possible to bring down the residual moisture content to below 1%–2% without exceeding the maximum temperature allowed by the product. Both primary and secondary drying are carried out in the FD cabinet, which has thus to comprise:

- A heat supply system to perform the sublimation.
- A refrigerating system to condense the sublimated water vapor.
- A vacuum system to remove the noncondensable gases.
- A control system to ensure efficient operation as well as high and uniform product quality.

Heat can be supplied by means of heated shelves: the trays containing the frozen food are placed between fixed, hollow shelves, which are heated internally by steam, heated water, or other thermal fluids, and thus heat is supplied by radiation. If the tray is placed over the shelf, heat is also supplied by conduction (Brennan 2006). Modern FD systems for foodstuffs are based on heat transfer by radiation. Figure 18.1 shows the amount of heat transferred by radiation and by conduction between parallel surfaces (each at a constant temperature) at different distances in a vacuum chamber: the radiation heat transfer between the parallel surfaces is constant, independent of the distance between the surfaces; the conduction heat transfer depends to a large extent on the contact between the surfaces; the total heat transfer is the sum of the two fluxes. Even slightly warped trays do not contact heating surfaces uniformly, thus causing large temperature differences within the trays. To avoid heat damage to the product in the areas of high tray temperature, the entire heating surface must be maintained at a lower temperature, thus reducing the overall drying rate. The predictable and precise transfer of heat by radiation over large surfaces is the very reason why radiation is preferred.

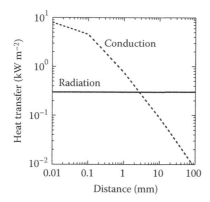

FIGURE 18.1 Heat transfer by radiation and conduction between parallel surfaces each at a constant temperature at different distances in a vacuum chamber.

The trays can be either open flat trays, with shallow walls, or finned trays, as those shown in Figure 18.2: the distance between the fins can be modified to meet the dimension of the product. Rolfgaard (1987) compared these types of trays, showing that the finned trays have an uneven temperature distribution because the distances between the shelf and the tray can vary between 0.1 and 1 mm; the presence of the ribs could compensate for this only partially. There is a major difference between the two forms of trays: with a flat tray and mostly radiation energy, the density of the heat flow is limited, while it can be significantly larger with finned trays standing on the heated shelf. The ice sublimation rate of finned trays can thus be about three to four times larger than that of flat trays. Finned trays can be useful for granular products, while all product forms and sizes can be filled into an open, flat tray, but careful tray filling is required to avoid any difference in layer thickness (Oetjen and Haseley 2004).

The volumes of water vapor produced during the drying are too large to be pumped by mechanical vacuum pumps in the operating pressure range typical of FD: 1 kg of ice at 0.4 mbar represents a volume of about 2800 m^3, which becomes about 25,000 m^3 at 0.04 mbar. In early freeze dryers, the water vapor was pumped out of the vacuum chamber by large steam ejectors, but these need large quantities of cooling water and steam. Due to the high energy costs, this method has been replaced by vapor traps (see Figure 18.3): water vapor condenses to ice on their refrigerated surfaces, and only the noncondensable gases are pumped out by mechanical vacuum pumps.

Condensers must fulfill some requirements (Oetjen and Haseley 2004):

– The surface area has to be large enough to condense the ice and to maintain the ice thickness below a maximum value, so that the heat transfer from the tube surface to the condensing surface of ice is not reduced. To condense 1 kg of water in 1 h on 1 m^2 surface on the top of an existing ice layer of 1 cm, the temperature difference required between the cooling surface and the ice surface should be about 4.5°C. To reduce this temperature difference to 2°C, the condenser would have to be defrosted every 30 min. Therefore, condensing surface is usually designed large enough to take up the total amount of ice of one charge in a layer of about 1 cm.

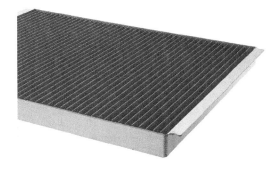

FIGURE 18.2   Finned tray used in freeze drying of foodstuffs.

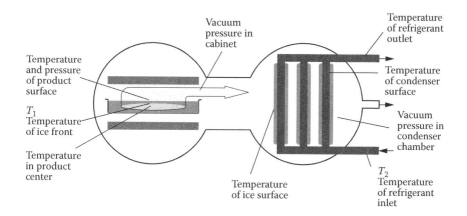

FIGURE 18.3   Scheme of equipment used for freeze drying of foodstuffs.

- If the condenser is separated from the drying chamber, the geometry of the connections between the two devices must be designed so that large volumes of water vapor can be transported, without the occurrence of chocked flow.
- The temperature difference of the refrigerant between the inlet and the outlet of the condenser should be small in order to ensure a uniform condensation (for large surfaces, it can be necessary to use several coils or plates in parallel and to control separately the temperature of each of them).
- The permanent gases have to be pumped off at the lowest possible position in the condenser as their density is higher than that of water vapor, and thus, they concentrate at the bottom of the condenser.

An efficient vapor removal system is characterized by a small total difference between the ice front temperature (T_1) in the product and the refrigerant temperature in the vapor trap (T_2). The refrigerant temperature (T_2) must be so low that it ensures the right sublimation temperature (T_1), besides avoiding any risk of melting the product or damaging it. A good design of a freeze dryer is thus determined by a low resistance to the vapor flow, that is, a big inlet area to the vapor trap, and by a low temperature drop from the condenser surface to the refrigerant, that is, a thin ice layer. For a given production, the refrigeration requirement may vary from a temperature of –35°C to –40°C, dependent on the design. The electricity consumption will be about 25% higher for the –40°C operation.

18.4   FREEZE-DRYER CABINET

It has been stated that one of the main advantages of FD is the product quality, while the main disadvantage is the process cost. Let us now look into some of the major considerations regarding equipment selection with respect to safeguarding and optimizing the product quality, besides minimizing the process cost.

The drying process is fundamentally the same for any type of product, but the way in which different products are prepared for drying varies considerably. Products for FD fall into three basic categories: liquids, individually quick frozen (IQF) products, and combined products. Liquids include coffee, tea, juices, and other extracts. IQF products include products such as segments of or whole fruit, berries, seafood, meat, and vegetables. Combined products include soup blocks, rice dishes, baby foods, camping foods, etc. Preparation for the FD of liquids is the most complex part of the process. First, the liquids must be treated to achieve the required density and color characteristics. The concentrate is then frozen, and the product is granulated and sieved to produce granules of optimum size depending on requirement. The granules are loaded onto the trays to pass through the freeze dryer. Everything is weighed to ensure the correct volume on each tray for perfect FD. The trays are automatically loaded using special tray feeders to secure a uniform and accurate filling. IQF products are loaded by weight or volume directly onto the trays. Combined products are normally loaded into special plastic molds placed onto the trays and defining the block size of the product.

Within its vacuum chamber, a freeze dryer contains the product trays, the heating plates, and, possibly, also the vapor trap (Dalgleish 1990; Snowman 1997; Fellows 2000). Figure 18.4 shows an example of how flat or finned trays in a tray carrier are suspended from an overhead rail. The trays are carried freely between the heating plates for radiation heating from both sides. The water vapor can flow in the interspace down into the vapor trap at the base of the cabinet. The vertical vapor velocity

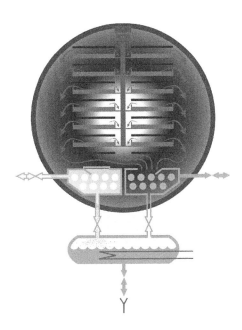

FIGURE 18.4 Cross section of a freeze-drying cabinet. (Courtesy of GEA-Niro A/S, Søborg, Denmark.)

is moderate (below 1 m s^{-1} for flat trays and below 1.5 m s^{-1} for finned trays), thus eliminating almost completely the carryover of product particles.

The refrigeration plant for the vapor trap accounts for the major part of the energy consumption in the FD process. So the efficiency of the vapor trap dictates the energy demand. Figure 18.4 shows a modern vapor trap that ensures an efficient operation—the freeze dryer contains a pair of condensers: while one is in operation, the other, shut off by a vacuum tight sliding gate, performs the deicing (Brennan 2006). Deicing takes place without breaking the vacuum in the drying chamber: a deicing tank containing water at about 20°C is connected to the condenser through automated valves. During deicing, one side of the condenser is connected to the deicing tank by opening the appropriate valve: due to the lowering of the vapor pressure caused by the cold condenser, the deicing tank water starts boiling and the warm vapor enters the condenser, thus melting the ice. The resulting water is drained back to the deicing tank. The energy consumption used for the deicing is very low, allowing frequent deicing. This feature is known as the continuous deicing (CDI).

The vapor flow is directed into the condenser in such a way that the vapor condenses as ice and not as snow. Thus, a thin layer thickness (2–3 mm) and a good conductivity ensure the lowest possible ΔT_{ice}. Vapor pressure drop is maintained at a minimum due to the short distance of the vapor flow and the large opening area to the condenser chamber, which yields a relatively low vapor velocity.

Production plants for food can be either discontinuous or continuous. In the next two paragraphs, the main features of both configurations are discussed.

18.4.1 BATCH FREEZE DRYER

The vacuum chamber is cylindrical, mounted horizontally, and fitted with doors at front and back. Once the frozen food is sealed into the chamber, the pressure must be reduced rapidly in order to avoid the melting of the ice; the low pressure has to be maintained for the time required to complete the drying. The refrigerated condenser, which may be a plate or a coil, is located inside the drying chamber or in a smaller chamber connected to the main chamber by a duct. The water vapor freezes onto the surface of the plate or coil. The temperature of the refrigerant must be below the saturation temperature corresponding to the pressure in the chamber. As drying proceeds, ice builds up on the surface of the condenser, thus reducing its effectiveness. If it is not to be defrosted during the drying cycle, then a condenser with a large surface area has to be used; otherwise, some deicing has to be foreseen.

Figure 18.5 shows the sketch of the cross section of a batch freeze dryer with cabinet, heating plates, and vapor condenser built as individual units. A water-flushing deicing system is used for the smaller cabinets: at the end of every FD cycle, the condenser is flushed with preheated water. Ice in the condenser melts within 10 min, and the water is drained. This method is ideal for small systems, ensuring minimum investment cost and simple operation.

Large cabinets incorporate the CDI system previously described (see Figure 18.4). During deicing, vapor at 25°C from the deicing vessel condenses at the cold condenser surface, thus melting the ice. In order to restore the condenser to operating condition, the condenser chamber is closed off from the deicing vessel. The

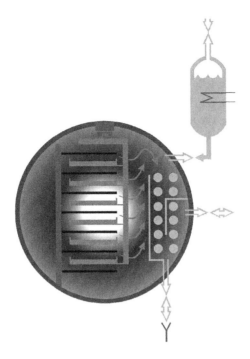

FIGURE 18.5 Cross section of a small-scale batch freeze-drying cabinet (ATLAS-RAY™). (Courtesy of GEA-Niro A/S, Søborg, Denmark.)

condenser is cooled to operating temperature, resulting in the condensation of any remaining vapor. As the vapor condenses, the pressure in the condenser decreases until operating vacuum is achieved, thus eliminating any loss of operating vacuum at switch-over between vapor condenser chambers. The CDI system is fully automatic; it ensures:

- A maximum ice buildup to 5 mm, resulting in a negligible temperature drop over the ice and low energy consumption in the refrigeration plant.
- A constant condenser capacity.
- A high FD capacity per square meter of tray surface.
- A short time from one charge to another.

Usually two pumps are used in series for the noncondensable gases. The first may be a root blower or an oil-sealed rotary pump, while the second is a gas-ballasted, oil-sealed rotary pump.

18.4.2 CONTINUOUS FREEZE DRYER

Figure 18.6 shows the sketch of the cross section of a continuous freeze dryer. The internal vapor condenser, with built-in deicing system, allows saving space;

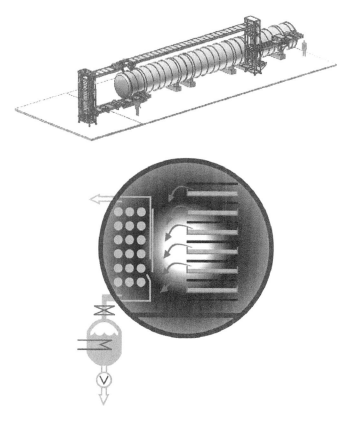

FIGURE 18.6 Scheme of continuous equipment for freeze drying and cross section of the cabinet (ATLAS-CONRAD™). (Courtesy of GEA-Niro A/S, Søborg, Denmark.)

moreover, it is more reliable, as it does not rely on large external vacuum valves with pressure drops that are difficult to secure: using the internal system, deicing is performed under vacuum, thus avoiding the need to seal the chamber against large pressure differentials. No product abrasion and low vapor velocities within the dryer guarantee as little as 0.1% product loss during the process. Deicing under vacuum, rather than at atmospheric pressure, eliminates the need to reestablish vacuum, thus strongly reducing the energy consumption. Trays with frozen product are loaded into the dryer through an efficient airlock system to an inlet elevator inside the drying chamber. When the elevator has a full stack of product trays, the entire stack is pushed into the first drying zone. More stacks follow and are pushed in turn through the various subsequent drying zones of the dryer, each adjusted to provide the drying characteristics required. When trays arrive at the exit, they are unloaded, again through an airlock, and the trays are emptied.

There are a number of other designs of freeze dryers suitable for processing granular materials. In one design, a stack of circular heaters is located inside a vertical, cylindrical vacuum chamber. The frozen granules enter the top of the chamber through an entrance lock and fall onto the top plate. A rotating central vertical shaft carries arms, which sweep the top surface of the plates: the arm rotating on the top plate pushes the granules outward and over the edge of the plate, onto the plate below, which has a larger diameter than the top plate. The arm on the second plate pushes the granules inward, toward a hole in the center of the plate, through which they fall onto the third plate, which has the same diameter of the top plate. In this way, the granules travel down to the bottom plate of the stack, from which they fall through an exit lock and are discharged from the chamber (Brennan 2006).

In another design, the frozen granules enter at one end of a horizontal, cylindrical vacuum chamber via an entrance lock, onto a vibrating deck that carries them to the other end of the chamber. Then they fall onto a second deck, which transports them to the front end of the chamber, where they fall onto another vibrating deck. Thus, the granules move back and forth inside the chamber until the drying is complete. Then they are discharged from the chamber through a vacuum lock. Heat is supplied by radiation from heated plates located above the vibrating decks. Conveyor belts can be used to move the granules instead of vibrating decks (Lorentzen 1975).

18.5   VACUUM SPRAY FREEZE DRYER

A prototype of this freeze dryer, designed for instant coffee and tea, is described by Mellor (1978). The preconcentrated extract is sprayed into a tall cylindrical vacuum chamber surrounded by a refrigerated coil. Before FD, preconcentration of the initial product up to 30% solids can be useful. The droplets freeze by evaporative cooling, losing about 15% of their moisture. A central hopper collects the partially dry powder as it falls freely to the bottom of the tower, which is connected to a tunnel, where the drying process is completed on a stainless-steel belt using radiant heaters. Generally, sprayed freeze-dried coffee has less flavor than normal freeze-dried coffee; concentration before spraying can improve flavor retention in the dried product.

18.6   MONITORING AND CONTROL

In order to safeguard a quality and uniform production from an FD operation, it is essential that a monitoring and control system makes the working conditions within the freeze dryer reliable and consistent. The FD process responds in fact very quickly to any interruption or discontinuity, with reduced end product quality as a consequence. The simultaneous and automatic monitoring and control of the equipment can be obtained using modern computer or programmable logic control (PLC) based control systems. As a minimum, the following actions should be required:

- An appropriate time/temperature program securing that a particular product is always exposed to the same thermal history
- Recording of maximum product temperature through the drying to ensure that no heat damage will occur
- Control of the exact vacuum level to ensure that the sublimation conditions remain constant at a preset level for each particular production

Currently, even the most advanced industrial freeze dryers have control systems that are no more than data acquisition systems for certain key variables. Monitored data and information obtained in previous runs, carried out with the same product, are used to manage the process. In some cases, the working temperature is selected by the operator; in some other cases, the working temperature is obtained from the mathematical simulation of the process, with the goal of optimizing the process, using either semiempirical models, which do not properly account for the heat and mass transfer mechanisms occurring during the process, or detailed models (see Chapter 3).

Poor process control is a consequence of the limitations of the current technology, mainly due to the impossibility, in a production process, of measuring the parameters of interest, namely, the product temperature and the residual moisture content. Moreover, at least in the pharmaceutical field, regulatory guidance, up to now, imposes to operate the manufacturing process in open loop, so that only an activity of monitoring is allowed during production. Nevertheless, at least during the phase of process development carried out at laboratory or pilot scale, it would be very useful to have an inline control system to minimize the drying time, taking into account the final quality of the product: process development can be expensive and highly time consuming, and thus the use of an efficient control system can give significant advantages.

The chamber pressure and the temperature of the heating shelf are the two variables that can be manipulated for control purposes in order to reduce the time required to achieve the desired residual moisture content, besides maintaining the product temperature below the maximum allowable value. Few papers appeared in the past about the inline control of an FD process, and almost all of them dealt with the control of liquid formulations containing drugs and proteins. Liapis and Litchfield (1979) proposed to manipulate the radiator energy output and the total pressure in the drying chamber in order to minimize the time required to complete the primary drying; the application of this control strategy to secondary drying was investigated by Sadikoglu et al. (1998). In both papers, a simplified model of the process is used, and the temperature profile in the product is assumed to be fully known, thus allowing to discriminate between a process whose dynamics is controlled by the heat transfer (for which the manipulation of the shelf temperature is effective) and a process whose dynamics is controlled by the mass transfer (for which the manipulation of the chamber pressure is effective). As the full temperature profile cannot be measured, the proposed control policies should be interpreted with caution, as it is highlighted in the conclusions of the papers of Liapis and Litchfield (1979) and of Litchfield and Liapis (1982). Tang et al. (2005) and Pikal et al. (2005) proposed to manipulate the shelf temperature and the chamber pressure using a simplified model and the results obtained by means of the noninvasive measurement of the product temperature proposed by Milton et al. (1997); a similar approach, named thermodynamic lyophilization control, was proposed by Oetjen and Haseley (2004). Galan et al. (2007) and Pisano et al. (2010) exploited another type of noninvasive measurement of the product temperature (dynamic parameters estimation) to calculate the control action that optimizes the performance of the system. Fissore et al. (2008) proposed to use the estimation of the whole temperature profile in the product provided by a Kalman filter in the framework of a model-based predictive control to calculate the control action that maintains the product temperature close to the maximum allowed value.

Most of these control policies have been designed for managing the FD of pharmaceuticals in vials because of the high value of the product and the rigid constraint on the maximum temperature allowed by the product. The FD of foodstuff is less demanding in terms of control: shelf temperature and chamber pressure can be optimized off-line (as it has been shown in Chapter 3); during the operation, anyway, care must be paid to maintain the product temperature below the maximum allowed value. Various devices can be used to monitor the process (Barresi et al. 2009): Figure 18.7 shows the results that can be obtained in a typical FD cycle; spinach samples are considered as test product. The blanched product, prepared in samples of almost cubic shape, was frozen before FD and dried in a small-scale plant at constant shelf temperature and chamber pressure. Various variables can be measured:

- *Product temperature* is monitored using some thermocouples placed inside the samples (Figure 18.7a). The main drawback is that the thermocouple allows to monitor only the temperature in a fixed point, generally the bottom (or the center) of the sample, and not the whole temperature profile. In some cases, in particular, when radiation flux from the upper tray is relevant, the maximum product temperature can be at the sublimating interface, and this cannot be measured by any thermocouples as the position of

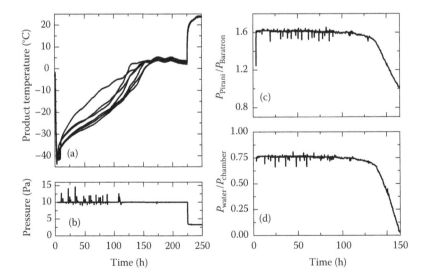

FIGURE 18.7 Freeze drying of spinach (in a Lyobeta freeze dryer by Telstar, Terrassa, Spain): cycle run using 32 samples having an almost cubic shape, with a side of 44 mm, and a total weight of 1.935 kg (chamber volume = 0.2 m³). The time evolution of the following variables is shown: (a) product temperature measured by some thermocouples placed in the center of different samples; (b) total pressure in the chamber measured by the capacitive sensor (the spikes are caused by the PRT); (c) ratio between the Pirani and Baratron signals; (d) ratio between partial pressure of water and total pressure in the chamber.

the sublimation front is moving during the drying. Moreover, in case of liquid products, the thermocouples produce a modification of the elementary phenomena of nucleation and of ice crystal growth as the tip of the thermocouple acts as a heterogeneous nucleation site: this can result in an increase in the average crystal size of the ice that leads to a lower mass transfer resistance. Furthermore, a preferential path for the vapor molecules is created by the presence of the body of the probe, and this also causes a reduction in the mass transfer resistance in the dried layer. All these effects can lead to a significant faster drying kinetics for the monitored samples, which, as a consequence, cannot be considered representative of the whole batch. Temperature measurement can be used to detect the end of the primary drying as, at that point, the temperature sharply increases as the heat supplied by the shelf is no longer used to sublimate water, but directly heats the product.

- *Chamber pressure* (Figure 18.7b) can be detected using either a thermal conductivity gauge (Pirani), whose working principle is a Wheatstone bridge, or a capacitance manometer (Baratron). The former is much cheaper, but the signal depends on the gas type and, in case of mixture, like water and inert, on the composition; thus, the use of Pirani gauge should be discouraged for monitoring FD because the chamber gas composition continuously changes during each run and is generally different in different cycles, depending on setup, loading, and product features. On the other hand, taking into account the known dependence of the Pirani response on the water vapor fraction, it is possible to evaluate the partial pressure of water into the drying chamber elaborating the different signals obtained from Baratron and Pirani sensors. Moreover, it is possible to detect the end of the primary drying as at that point, the concentration of water into the drying chamber becomes very low, so Pirani and Baratron sensors measure almost the same value of pressure and their ratio approaches 1 (Figure 18.7c). Other methods are available to monitor the time evolution of the *water concentration in the chamber* (Figure 18.7d), for example, moisture sensors (Roy and Pikal 1989; Trelea et al. 2007) and mass spectrometers (Jennings 1980; Connelly and Welch 1993): while the former is largely used for food applications, the latter has a very high cost that prevents its use except for research purposes or when high value products are processed.

- *Product mass* can be monitored by weighing some samples inside the drying chamber. Various balances can be used to this purpose: the capacitive balance proposed by Rovero et al. (2001) works efficiently in case of radiative heating, while, when heat is transferred through the shelf by conduction, it has the drawback that the heat transfer to the product is limited by the volumetric gap that acts as an additional resistance and thus the results obtained are not representative of the whole batch. Other balances have been proposed for the FD of pharmaceuticals in vials, but they could also be used with liquid products in glass containers as well as with products in a tray over a shelf. Among others, Christ (1994) proposed a balance that requires glass vials with a specific geometry that not always corresponds to that of the vials of the batch, and thus the measurement cannot be representative of the whole system, while Vallan et al. (2005) and Vallan (2007) proposed a novel balance that is able to monitor some vials that have the same geometry of those of the batch and that remain almost always in contact with the cooling shelf: they are lifted just before the measurement, and thus, the thermal exchange between the glass vials and the cooling surface is not significantly modified and the measurement can be considered representative of the whole system. A series of tests has shown that the weighing frequency can be chosen in a wide range without affecting the process, but since the insertion of the balance itself in the drying chamber can interfere with the process, both the monitored vials and the balance case must be properly shielded to avoid systematic errors due to radiation effects. A mathematical model takes into account temperature affection and adjusts the final outcome.

Noninvasive monitoring techniques have been proposed as valuable alternatives to the use of thermocouples; these techniques are able to give information about the state of the whole system. Among others, Milton et al. (1997) proposed a nonintrusive method [the manometric temperature measurement (MTM)], which is useful for monitoring product temperature throughout almost all the primary drying. MTM uses the inline measure of the pressure rise due to the shut-off of the valve placed between the drying chamber and the condenser for a short time interval, about 30 s: the value of the chamber pressure is related to the temperature of the sublimating interface by means of a mathematical model. By this way, performing some pressure rise tests (PRTs) throughout all the drying, it is possible to evaluate the evolution of the product temperature, and since this test also gives information about the entity of the sublimation flux of the solvent vapor (obtained from the initial slope of the pressure rise curve), it can be used to detect the end point of the primary drying phase. Figure 18.8 shows some examples of curves of pressure rise obtained during the experimental run with spinach samples previously described, as well as the evolution of the water flow rate obtained from the sublimation and calculated from the pressure rise curves; it is possible to point out that when the primary drying is almost finished, according to the product temperature, to the ratio between the Pirani and Baratron gauges, and to the measurement of the water concentration in the chamber, the flow rate of water is also very low; moreover, when secondary drying is started, and thus the shelf temperature is increased and the chamber pressure is set to the lowest possible value (see Figure 18.7), the water flow rate increases at the beginning and then approaches zero. Again from Figure 18.8, it is also possible to see that the pressure rise during the primary drying is much higher than that obtained during secondary drying, as it is expected because of the larger sublimation fluxes; moreover, the value of the sublimation flux calculated from the PRT curve during the secondary drying can be affected by error due to the difficulties in evaluating the slope of this curve when the pressure increase is very small.

Several approaches, based on simplified mathematical models of the process, have been proposed in literature to interpret the PRT, and they have been referred with different names, for example, the barometric temperature measurement (Oetjen

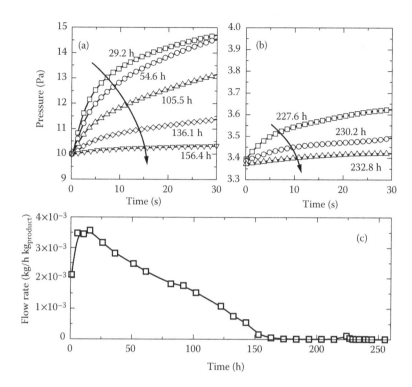

FIGURE 18.8 Example of results obtained using the PRT during the same experimental run of Figure 18.7. In the upper graphs, the curves of pressure rise obtained at various time instant during the primary drying (a) and during the secondary drying (b) are shown, while in the bottom graph (c), the flow rate of sublimation calculated from the PRT during the whole cycle is given.

and Haseley 2004) and the pressure rise analysis (Chouvenc et al. 2004). Dynamic parameters estimation (DPE) is based on a similar approach, but uses a more detailed model, taking into account the heat accumulation in the product during the PRT. DPE allows estimating the product temperature, the moving front position, and the heat and mass transfer coefficients. This algorithm, looking for the best fitting between the measured values of the pressure rise during the PRT and the values calculated by means of the model, also provides the estimate of the temperature profile in the whole product (Velardi et al. 2008; Pisano et al. 2011).

Figure 18.9 shows an example of the results that can be obtained with DPE in the experimental run with spinach samples: a good agreement is obtained between the measured values of chamber pressure and the estimated ones; the DPE outcomes are an interface temperature of 232 K and an ice thickness of 36.9 mm. Both data are useful as they allow monitoring the progress of the drying as well as verifying that the product temperature remains below the maximum value characteristic of the product.

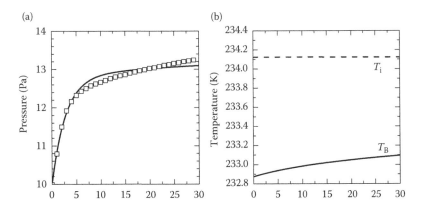

FIGURE 18.9 Example of results obtained using the DPE to interpret the results of a PRT in the same experimental run of Figure 18.7: comparison between measured (symbols) and estimated (solid line) pressure rise curve (a) and estimated product temperature (b) at the interface (dashed line) and at the bottom of the samples (solid line). The PRT was done 12 h after the beginning of the primary drying.

ACKNOWLEDGMENT

Valuable contributions of Dr. Salvatore Velardi and of Dr. Roberto Pisano (Politecnico of Torino) are gratefully acknowledged.

REFERENCES

Barresi, A. A., Pisano, R., Fissore, D., Rasetto, V., Velardi, S. A., Vallan, A., Parvis, M., and Galan, M. 2009. Monitoring of the primary drying of a lyophilization process in vials. *Chemical Engineering and Processing* 48:408–423.

Brennan, J. G. 2006. Evaporation and dehydration. In *Food Processing Handbook* (J. G. Brennan, ed.). Weinheim: Wiley-WCH.

Chouvenc, P., Vessot, S., Andrieu, J., and Vacus, P. 2004. Optimization of the freeze-drying cycle: a new model for pressure rise analysis. *Drying Technology* 22:1577–1601.

Christ, M. 1994. Freeze-drying plant. International Patent Classification n. WO1994EP02426 19940722.

Connelly, J. P., and Welch, J. V. 1993. Monitor lyophilization with mass spectrometer gas analysis. *Journal of Parenteral Science and Technology* 47:70–75.

Dalgleish, J. Mc N. 1990. *Freeze-Drying for the Food Industries*. London: Elsevier Applied Science.

Fellows, P. 2000. Freeze drying and freeze concentration. In *Food Processing Technology: Principles and Practice*. Cambridge: Woodhead Publishing.

Fissore, D., Velardi, S. A., and Barresi, A. A. 2008. In-line control of a freeze-drying process in vial. *Drying Technology* 26:685–694.

Galan, M., Velardi, S., Pisano, R., Rasetto, V., and Barresi, A. A. 2007. A gentle PAT approach to in-line control of the lyophilisation process. *Proceedings of the Conference of the International Institute of Refrigeration—New Ventures in Freeze-Drying*, Strasbourg, France, November 7–9. Vol. 2007-3, p. 17, ISBN: 978-2-913149.

Jennings, T. A. 1980. Residual gas analysis and vacuum freeze drying. *Journal of Parenteral Drug Association* 34:62–69.

Liapis, A. I., and Litchfield, R. J. 1979. Optimal control of a freeze dryer—I. Theoretical development and quasi steady-state analysis. *Chemical Engineering Science* 34:975–981.

Litchfield, R. J., and Liapis, A. I. 1982. Optimal control of a freeze dryer—II. Dynamic analysis. *Chemical Engineering Science* 37:45–55.

Lorentzen, J. 1975. Industrial freeze drying plants for food. In *Freeze Drying and Advanced Food Technology* (S. A. Goldblith, L. Rey, H. H. Rothmayr, eds.). London: Academic Press.

Mellor, J. D. 1978. *Fundamentals of Freeze-Drying.* London: Academic Press.

Milton, N., Pikal, M. J., Roy, M. L., and Nail, S. L. 1997. Evaluation of manometric temperature measurement as a method of monitoring product temperature during lyophilization. *PDA Journal of Pharmaceutical Science Technology* 51:7–16.

Oetjen, G. W., and Haseley, P. 2004. *Freeze-Drying.* Weinheim: Wiley-VHC.

Pikal, M. J., Tang, X., and Nail, S. L. 2005. Automated process control using manometric temperature measurement. United States Patent, n° US6,971,187 B1.

Pisano, R., Fissore D., and Berresi, A. A. 2010. In-line optimization and control of an industrial freeze-drying process for pharmaceuticals. *Journal of Pharmaceutical Sciences* 99:4691–4709.

Pisano, R., Fissore, D., and Barresi A. A. 2011. Innovation in monitoring food freeze-drying. *Drying Technology* 29:1920–1931.

Rolfgaard, J. 1987. Industrial freeze-drying for the food and coffee industry. Ballerup (Denmark): Atlas Industries.

Rovero, G., Ghio, S., and Barresi, A. A. 2001. Development of a prototype capacitive balance for freeze-drying studies. *Chemical Engineering Science* 56:3575–3584.

Roy, M. L., and Pikal, M. J. 1989. Process control in freeze drying: determination of the end point of sublimation drying by an electronic moisture sensor. *Journal of Parenteral Science and Technology* 43:60–66.

Sadikoglu, H., Liapis A. I., and Crosser, O. K. 1998. Optimal control of the primary and secondary drying stages of bulk solution freeze drying in trays. *Drying Technology* 163:399–431.

Snowman, J. W. 1997. Freeze dryers. In *Industrial Drying of Foods* (C. G. J. Baker, ed.). London: Blackie Academic and Professional.

Tang, X. C., Nail, S. L., and Pikal, M. J. 2005. Freeze-drying process design by manometric temperature measurement: design of a smart freeze-dryer. *Pharmaceutical Research* 22:685–700.

Trelea, I. C., Passot, S., Fonseca, F., and Marin, M. 2007. An interactive tool for the optimization of freeze-drying cycles based on quality criteria. *Drying Technology* 25:741–751.

Vallan, A. 2007. A measurement system for lyophilization process monitoring. *Proceedings of Instrumentation and Measurement Technology Conference—IMTC 2007.* Warsaw, Poland, May 13, 2007, p. 5, DOI: 10.1109/IMTC.2007.379000.

Vallan, A., Parvis, M., and Barresi, A. A. 2005. Sistema per la misurazione in tempo reale di massa e temperatura di sostanze sottoposte a liofilizzazione. Italian Patent Application n. B02005A000320.

Velardi, S. A., Rasetto, V., Barresi, A. A. 2008. Dynamic Parameters Estimation Method: advanced manometric temperature measurement approach for freeze-drying monitoring of pharmaceutical. *Industrial Engineering Chemistry Research* 47:8445–8457.

Index

Page numbers followed by f and t indicate figures and tables, respectively.

For Product Safety Concerns and Information please contact our EU
representative GPSR@taylorandfrancis.com Taylor & Francis Verlag GmbH,
Kaufingerstraße 24, 80331 München, Germany

Printed and bound by CPI Group (UK) Ltd, Croydon, CR0 4YY
01/05/2025
01858482-0004